Flow Injection Analysis of Marine Samples

FLOW INJECTION ANALYSIS OF MARINE SAMPLES

M. C. YEBRA-BIURRUN

Nova Science Publishers, Inc.
New York

For permission to use material from this book please contact us:
Telephone 631-231-7269; Fax 631-231-8175
Web Site: http://www.novapublishers.com

LIBRARY OF CONGRESS CATALOGING-IN-PUBLICATION DATA

Yebra-Biurrun, M.C.
Flow injection analysis of marine samples / M.C. Yebra-Biurrun.
p. cm.
Includes index.
ISBN 978-1-60741-506-0 (hardcover)
1. Chemical oceanography. 2. Flow injection analysis. 3. Marine sediments. I. Title.
GC111.2.Y43 2009
551.46'6--dc22

2009012465

Published by Nova Science Publishers, Inc. ✦ New York

This book is dedicated:

To the memory of my father, Jesús, who will always be with me.
To my mother, Carmenchu, to my sister, Chusita, and to Rosita,
for their love, patience and support throughout the years.
To Sisi the little princess of our home.

This book is also dedicated:

To all those from whom I have learnt about Flow Injection Analysis.

CONTENTS

PREFACE

Marine chemistry is a matter of scientific interest from many points of view, above all those related to environmental contamination. Nevertheless, they are some problems in the analysis of marine samples (seawater, marine animals, macroalgaes, sediments, etc.) such as high salt content of the matrix, low analyte(s) concentrations, sample dissolution, etc., which results in complicated and tedious sample pretreatments. However, many of these drawbacks can be solved by the application of automatic methods based on flow injection analysis (FIA). The concept of FIA was first proposed in 1975, and from this date, many researchers have proposed analytical methodologies involving FIA to determine organic and inorganic analytes in marine samples. FIA methods allow automatic handling of sample and reagent solutions with a strict control of reaction parameters. Furthermore, a FIA manifold is robust in shipboard laboratories, with the advantage of minimal sample handling and thus low exposure to contamination. Flow injection methodologies improve off-line sample pretreatments, between others: preconcentration by using chelating resins, solid-phase extraction, and lixiviation procedures, resulting in rapid, efficient, safety, inexpensive and environmentally acceptable methodologies for the determination of several analytes in marine samples. Other advantages of FI methodologies are an important contribution to achieve a miniaturized, automatic, and green Analytical Chemistry.

This book aims to cover the most important advances in the analysis of marine samples employing flow injection methodologies and will be a tool for all chemists who perform analyses on a routine basis in the environmental field.

Chapter 1 presents an updated overview on flow injection techniques. Since FIA was first proposed and described to the new generations of FI techniques such as sequential injection analysis (SIA), bead injection (BI), and sequential injection lab-on-valve (SI-LOV), between others flow systems.

The importance of marine environment research is treated in Chapter 2. The marine environment is an important sink for many chemicals, some of which accumulate in the marine food chain. Heavy metals and bio accumulating toxic substances are introduced to the sea from land-based point and non-point sources, from atmospheric fallout and during marine transport of materials. This chapter presents an overview on the importance of monitoring chemicals (inorganic and organic) present at major or minor concentrations in seawater/estuarine water, sediments, seaweeds and marine animals used as seafood.

Chapter 3 summarizes and examines the manuscripts issued to date referred the application of flow injection methodologies to the determination of organic analytes usually monitored to assess quality of marine waters (acrolein, amines, bentazone, chemical oxygen demand, diethylene glycol, dissolved organic carbon, formaldehyde, halocarbons, iron-porphyrin-like complexes, organophosphorus pesticides, nitrophenol isomers, polycyclic aromatic hydrocarbons, red tide phytoplankton and surfactants).

Applications of flow injection methodologies to the determination of inorganic analytes (cationic and anionic species) in sea and estuarine water samples are described in Chapter 4 . A detailed review of flow methods applied to the determination of alkali metals, alkaline earth metals, silver, aluminium, arsenic, gold, boron, bismuth, cadmium, cobalt, chromium, copper, iron, mercury, indium, manganese, molybdenum, ammonium, nickel, lead, rare earth, rhodium, antimony, selenium, tin, titanium, thallium, vanadium, zinc, alkalinity, hydrogen peroxide halides (bromide, chloride, fluoride and iodide), nitrate/nitrite, phosphate, silicate, sulfate and sulfide is given in this Chapter.

The state-of-the-art of flow injection methodologies proposed for the determination of organic and inorganic analytes in sea and estuarine sediments is presented and discussed in Chapter 5. In this chapter are described FI methods applied to the determination of bio and molecular markers, dissolved organic carbon, polycyclic aromatic hydrocarbons, silver, aluminium, arsenic, bismuth, cadmium, cobalt, chromium, copper, iron, germanium, mercury, iridium, magnesium, manganese, nickel, lead, platinum, rare earths, plutonium, thorium, uranium, rhenium, antimony, selenium, tin, tellurium, zinc, carbonate, sulfide and silicate.

Chapter 6 presents a comprehensive review of flow injection methodologies proposed for the determination of organic and inorganic analytes in seaweeds samples. In this chapter are described FI methods applied to the determination of intracellular free amino acids, β-dimethylsulfoniopropionate, laminarin, arsenic, germanium, mercury, molybdenum, tin, iodide, phosphate, nitrate, nitrite and silicate.

Chapter 7 covers the review of the application of flow injection methodologies for the determination of organic and inorganic analytes in marine animals samples/seafoods. FI methods applied to the determination of amino acids (histidine, L-lysine and tyrosine), DNA/RNA, formaldehyde, histamine, hypoxanthine, polycyclic aromatic hydrocarbons, diarrheic shellfish poisoning, paralytic shellfish poisoning, succinate/glutamate, trimethylamine/ total volatile basic nitrogen, total lipid hydroperoxides, total volatile acids, uric acid, vitamin B12, silver, aluminium, arsenic, boron, calcium, cadmium, cobalt, chromium, copper, iron, gallium, mercury, indium, lithium, manganese, molibdenum, nickel, lead, antimony, selenium, tin, strontium, thallium, vanadium, zinc, nitrate/nitrite, phosphorous/phosphate and silicate are described in detail in this chapter.

Chapter 1

FLOW INJECTION: PAST, PRESENT AND FUTURE

ABSTRACT

In this chapter, an updated overview on flow injection (FI) techniques is presented. FI is a popular, simple and well-established unsegmented continuous flow-based technology. Since flow injection analysis (FIA) was first proposed and described in 1975 by Ruzicka and Hansen, it is extensively used by its versatility and simplicity for automating analytical methodologies. In recent years FIA has evolved and advanced into new generations of FI techniques such as sequential injection analysis (SIA), bead injection (BI), and sequential injection lab-on-valve (SI-LOV), between others flow systems. The advantages of FI methodologies are an important contribution to achieve a miniaturized, automatic, and green Analytical Chemistry.

INTRODUCTION

Flow injection analysis (FIA) was first described by Ruzicka and Hansen in Denmark and Stewart and coworkers in United States in the middle of 1970 [1-2]. FIA is a popular, simple, rapid, and versatile technique with is a well-established position in modern analytical chemistry, and widespread application in quantitative chemical analysis. This is confirmed by the number of related papers that have appeared in scientific publications since 1975, resulting in the publication of more than 20000 papers during the period 1975-2008 (as can be prove when Analytical and Chemical Abstracts are consulted), 22 monographs (two of them yet in press) [3-23], one CD-ROM Tutorial [24], a journal (Journal of Flow Injection Analysis) published by the Japanese Association for Flow Injection Analysis [25], two on-line databases [26-28] and hundreds of PhD theses. The main reason to believe that is because FIA has been shown to provide solutions for the automation of all aspects of analysis from sample pretreatment to data management. In addition, FIA has allowed execute procedures, which are difficult and, in many cases, not even feasible by traditional batch procedures. The advantages mentioned above already turned FIA into a powerful technique with wide applications in different fields (e.g. medicine, foods, agriculture, environmental, etc).

Flow injection analysis has gone through three generations, that is, the first generation (flow injection, FI) in 1970s, supplemented by sequential injection (SI) in the 1990s as the second generation, and the recently emerged lab-on-valve (LOV) system and the concept of bead injection (BI), involving bead-renewal approaches, as the third generation, which holds clear advantages for instrumental miniaturization and offers new avenues or chemical assays. The three generations have revolutionized the concept of sample pretreatment by facilitating on-line operation and coupling with various detection techniques [29-30].

Flow Injection Analysis (FIA)

Flow Injection Analysis (FIA) is the most widely used nonsegmented continuous technique because is the first generation of flow injection techniques. In the simplest form of FIA, well-defined volume of a liquid sample is injected into a continuous flow of a suitable liquid (carrier). The injected sample forms a zone, which moves downstream where is dispersed into the carrier and is transported to a flow through detector placed downstream, which records the analytical signal. Sample dispersion is controlled through the suitable choice of the injected sample volume, the carrier flow-rate, the length of the reaction coil, and the diameter of the tubing used.

FIA: Definitions, Principles and Features

In the first edition of their book published in 1981 titled "Flow Injection Analysis" [3], Ruzicka and Hansen defined FIA as "*a technique based on the sequential injection of a discrete liquid sample into a moving, nonsegmented continuous carrier stream. The injected sample forms a zone, which is transported toward a detector with the subsequent continuous detection of the target analyte*". Stewart, another FIA pioneer, in 1981 [31] defined FIA as "*the sequential insertion of discrete sample solutions into an unsegmented continuously flowing stream with subsequent detection of the analyte*". In 1984, Valcárcel and Luque de Castro [6], unified and completed previous definitions "*FIA is an unsegmented-flow technique that involves direct injection of the sample, and its controlled, reproducible partial dispersion. Neither physical or chemical equilibrium is ever reached and operational timing is highly reproducible*". In addition these authors indicate that this methodology has high precision and accuracy and requires inexpensive instrumentation in comparison with other automatic analytical methodologies. However, as result of the rapid development of this technique these first FIA definitions remained obsolete. Thus, in 1988, in the second edition of their book [10], Ruzicka and Hansen recall the absence of air segmentation, establish a new FIA definition "*information gathering from a concentration gradient formed from an injected well-defined zone of a fluid, dispersed into a continuous unsegmented stream of a carrier*", and institute the most distinctive FIA features: "*the injection of sample solution into a continuously flowing stream resulting in a transient output signal*". The same made Stewart in 1989 [32] because expanded his first definition: "*FIA can be viewed basically as an unsegmented liquid sample handling system. Once sequential liquid samples are placed in a liquid stream, the analyte can be moved, concentrated, diluted, reacted, purified, and delivered to any detector without intervention of an operator. Such operations frequently yield assays with greater accuracy, precision, throughput, and sometimes, better sensitivity than their manual counterparts*". Fang in 1992 [33] was based on two basic features of the technique to propose the following FIA definition: "*FIA is a non-chromatographic flow analysis technique for quantitative analysis, performed by reproducibly manipulating sample and reagent zones in a flow stream under thermodynamically non-equilibrated conditions*".

Nevertheless, these definitions still contain several controversies because some FIA systems:

a) Have gas segments: as occurs in hydride generation systems.
b) Are not continuous: preconcentration systems with off-line detection, the stopped-flow mode.
c) Do not accomplish sample injection.
d) Involve solid and gaseous samples.
e) Dos not perform the continuous detection of the target analyte: as occurs in indirect methodologies.
f) Do not transport the sample toward the detector: as occurs in those flow methodologies that include a separation technique.

Consequently, it is difficult to find a FIA definition that covers all features of this methodology. As suggest Martínez Calatayud [18], the difficulty of defining FIA work arises from the fact that the methodology is not one more analytical choice, rather, it has become an unconventional working philosophy for implementation of virtually any analytical operation that was formerly addressed with conventional means. Bo Karlberg, a pioneer in the development of FIA, suggested in his famous quote, an alternative to wrestling with the many definitions and descriptions of FIA: "*Flow injection analysis should not be explained. It ought to be demonstrated*".

FIA consists in the microfluidic manipulation of samples and reagents. Samples are injected into a carrier/reagent solution, which transports the sample zone into a detector while desired (bio)chemical reactions take place. Detector response yields a calibration curve quantifying the target analyte. Thus, FIA is based on three principles:

1) Reproducible sample injection or insertion into a flowing carrier stream.
2) Controlled dispersion of the sample zone.
3) Reproducible timing of its movement from the sample injection point to the detection system.

Being essential for the success of this technique that each sample analysis is performed in the same conditions as the previous one.

Transformation of FIA into flow injection technique signifies recognition of a tremendous versatility of this method, which being originally designed as a mere tool for automation of serial assays, now becomes a universal means for enhancement of instrumental methods of analysis.

FIA offers several advantages over their manual counterparts methods:

a) Computer compatibility.
b) Simplicity
c) Automatic handling of samples and solutions.
d) Strict control of reaction conditions.
e) Great precision, the mechanical performance of the assays reduces human errors.

f) Higher sample throughput.
g) Reduction of labor costs.
h) Smaller waste generation.
i) Smaller sample and reagent consumption.
j) Broad scope because can be used with a great variety of detectors.
k) All reactions take place in a closed system, which reduces contamination risk, analyte or/and sample losses, and improving safety for analysts by preventing direct exposure risks by handling toxic reagents and reaction products.

In addition, is necessary emphasize the low working pressures used (usually lower than 70 kPa) and the versatility of a FIA system because its modular nature allows its adaptation to a wide variety of needs in analytical chemistry. Changing one component for another is a fairly easy task, and manifolds can be assembled in a few minutes. This is especially useful in those cases where new investigations and innovations are involved.

The fast response of FIA makes the analytical information available at real time and enables high temporal resolution data in monitoring schemes with no need for discrete sample collection and storage. This is of great interest whenever off-line analysis or excessive sample handling is unacceptable due to the rapid transformation of the target species [34]. In addition, FIA can be applied to nearly all operations associated with classical analysis: filtration, precipitation, liquid-liquid extraction, ion-exchange, etc.

These advantages are an important contribution to achieve a miniaturized, automatic, and green Analytical Chemistry.

Dispersion Processes

As was established previously, controlled dispersion is one of three FIA principles, being the most important physical phenomenon in all FI systems. Dispersion takes place when the sample zone moves downstream through the manifold forming a well-defined concentration gradient. Thus, after the sample is inserted into the manifold, a continuous dilution takes place until the sample passes through the detector. Figure 1.1 shows modes of mass transport through a tube and as can be seen in this Figure, after sample injection into the carrier stream, the formed zone does not flow down the tube as a compact plug, otherwise the injected sample zone disperses according to the parabolic velocity profile characteristic for laminar flow. This parabolic concentration profile is formed because the sample molecules near the walls are retarded by friction while the molecules in the center of the tube are free to move more rapidly. As a consequence, the solution at the walls of the tube does not move at all whereas the solution in the center of the tube moves at twice the average flow rate. However, the development of this parabolic concentration profile is not desirable in FIA because it dilutes the sample and it spreads it out, causing a decrease in sensitivity. This inconvenient is controlled in FIA by employing conditions that promote radial mass transfer (by diffusion). Molecules left behind along the walls of the tube will tend to diffuse toward the more diluted center of the tube, where it will move more rapidly. Molecules at the leading edge of the parabolic concentration profile tend to diffuse toward the walls, slowing them down. The net effect is to reduce the degree of sample spreading and to cause the carrier and sample to mix. Since diffusion in liquids is a slow process, FIA is done at relatively slow rates so that diffusion has time to take place. Also, tubing diameters are small so that the distance from the walls to the center of the tube is small and sample molecules do not have to diffuse very far. Radial mass transfer is also induced by coiling the tubing between the injection device and detector to set up secondary flow patterns. The magnitude of this effect depends on the flow rate, tubing diameter, and the degree of coiling.

Accordingly, by minimizing sample dispersion, the height of the transient signal increases, the sensitivity and detection limits also increase and the peak width decreases. Therefore, the control of dispersion is the most important aspect of FI systems. Because it is a dynamic process, dispersion will not reach equilibrium or steady state before the sample reaches the detector. It is, however, reproducible at any instant in time, if factors such as flow rate, tubing internal diameter and length, type of reactor, and internal architecture of components that affect dispersion, are held constant. Therefore, the degree of dilution caused by dispersion during the transport of the sample from injector to detector can be controlled (through the manipulation of flow parameters and geometrical dimensions of flow conduits) in order to get the same reproducibility for calibrants and samples. Thus, it is possible to calibrate the system with standard samples, and determine the unknown concentrations of other samples. For that reason, the most important aspect of a FIA method is the concept of controlled dispersion of the sample zone, an entirely new concept in analytical chemistry at that time, and which allows the design of a FIA system suited to automate a given analytical procedure. This provides the basis for extracting reproducible readout under both physically and chemically non-equilibrium conditions [3,9].

The injected fluid zones in a non-segmented flow stream can be manipulated reproducibly to produce various degrees of dispersion. In order to provide a quantitative criterion to evaluate the extent of dispersion, Ruzicka and Hansen [3] introduced the concept of dispersion coefficient, D, that has been defined as the ratio of concentrations before and after the dispersion process has taken place in the element of fluid that yields the analytical readout:

$$D=C_0/C$$

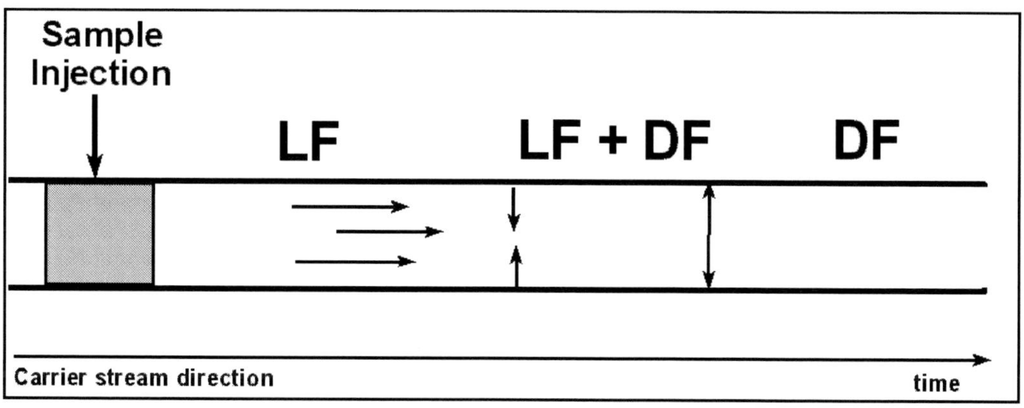

Figure 1.1. Modes of mass transport through a FIA tube. LF: laminar flow; DF: diffusion flow.

Where C_0 is the original concentration of the analyte in the solution before dispersion, and C is the concentration of that element in the dispersed fluid zone from which analytical readout is extracted. When the solution of analyte with the highest concentration is used for readout, the equation is expressed as:

$$D=C_0/C_{max}$$

Where C_{max} is the concentration of the constituent at peak maximum. D is a dimensionless value, which is equivalent to the dilution factor of the analyte within the flow system. For example, if the sample is diluted 1:1 by carrier, the dispersion coefficient is 2. FI systems are categorized in high, medium, and low dispersion systems depending on the degree of dispersion of the injected zone at the read out point. Systems with D above 10 are classified as high, those between 2 and 10 as medium, and those below 2 as a low dispersion system. The main experimental parameters influencing the dispersion of an injected fluid zone include:

- Sample volume.
- Carrier flow rate.
- Merging fluid stream(s) flow rate(s).
- Geometrical dimensions and configuration of transport conduits
- Geometrical dimensions and configuration of reactor(s).

The volume of the injected fluid zone, usually the sample, is an important factor influencing the dispersion of the FI system. As a result, the dispersion decreases with increase the sample volume. Ruzicka and Hansen [3] stated that dispersion diminishes when decreases the flow rate of the FI system. This happens because when decreasing flow rates, increase the retention time of the sample awaiting transport to the detector. In this phase the reaction between sample and reagent almost reaches the equilibrium. Hence, the peak signal will be higher in a slower flow rate. Fang [17] supposed that those conditions are only valid at extremely low flow rates, where the rate of molecular diffusion approaches that of convection. This has been experimentally demonstrated by Karlberg and Pacey [12]: with a fixed manifold, dispersion is minimally influenced by flow rate variations within a wide range (1.6-4.0 mL/min). The dispersion of the sample zone increases with the square root of the distance traveled through an open narrow tube. This rule is valid only for straight conduits. When the conduits are coiled for the sake of tidiness or knotted to improve radial mixing, the intensity of dispersion is decreased to different degrees, depending on the radius of the coil or knots. This is due to the generation of secondary flows, which limit the axial dispersion while promoting radial dispersion. The dispersion of injected zones is enhanced with increases in the inner diameter of the conduit. No generally applicable quantified relationships are available, however, owing to the complexity of the influences from other parameters.

The Role of FIA in Solving Analytical Problems

Depending on the role played by FIA as an interface between samples and instruments, there are various alternatives to insert this technique into an analytical process [35-37]:

a) Simple means of introducing samples into analytical instruments.
b) Means of automatically developing chemical reactions prior to insert the reaction products into analytical instruments.
c) A way to implement non-chromatographic separation techniques to increase selectivity and/or sensibility.
d) A tool to develop continuous separation processes involving chemical reactions.
e) A mean to carry out speciation studies and determination of several analytes in the same sample.
f) Direct analysis of heterogeneous or solid samples.
g) Coupling with high performance liquid chromatography (HPLC) instruments in precolumn or post column arrangement, depending on when takes place the FIA injection (before or after chromatographic separation).

Basic FIA Components

The word "manifold" is referred collectively to the assemblage of flow tubing, mixing coils, injection valves, etc. in a given configuration [38]. A basic FI manifold is shown in Figure 1.2, and it consists of:

- A propulsion unit that is used to propel the carrier stream through a thin tube.
- An injection device that introduces into the carrier a well-defined volume of sample solution in a very reproducible manner.
- A coil of tubing also named reaction coil, in which the sample zone disperses and reacts with the components of the carrier, forming species that are continuously monitorized by a detector.
- A detector with a readout device, which registers the FIA typical signals.

Flow Injection analysis systems can be configured in a wide variety of different modes, depending on the desired application. The most elementary FIA modes are normal (nFIA) and reversed (rFIA). The normal mode has been described above as the sample injection into a liquid stream (carrier) (Figure 1.2). In reversed FIA, occurs in the opposite way, the sample is inserted continuously into the FI manifold and acts as carrier. So, the reagent is injected into the sample stream when required for the analytical determination. The main advantages of rFIA are economy in reagents and waste disposal, the possibility of making several different determinations on the same sample solution stream, for example by the injection of several reagents, and a possible improvement of analytical sensitivity because the mixing between sample and reagent(s) is more efficient. It is necessary to carry out this FIA mode an abundance of sample, as occurs in water analysis.

Propulsion Unit

This component corresponds to the system used to deliver the carrier and reagent solutions. A peristaltic pump or piston pumps are commonly used, with a capacity of pumping between one to sixteen carrier / reagent lines. The commonest model is provided with four channels. As shows Figure 1.3, peristaltic pumps have a set of rollers on a revolving drum that squeezes flexible tubing to produce a constant, pulsing flow. Pump flexible tubes are available in different materials depending on the type of fluid to be impelled [39]. Two collars are usually glued near the end of each pump tube. The purpose of these color code collars is two-fold: to serve as points of attachment between the harnesses of the pump and to identify the inner diameter and hence the flow rate. The incompatibility of the elastic tubes of peristaltic pumps with concentrated acid or bases, and organic solvents usually forces the periodical recalibration of the FIA system or the incorporation of more expensive reagent resistant tubing. The physical adsorption of organic analytes onto the Tygon® tubes is another practical limitation commonly described.

The requirements for an ideal functioning of this unit in FIA may be summarized as follows:

- Reproducible flow-rates in order to give a predictable residence time and a constant dispersion throughout the manifold.
- The propulsion unit must stop and start instantaneously.
- The flow-path should be readily programmable to facilitate a large variety of analysis.
- Multi-channel capability for providing at least four parallel streams to ensure versatility.
- Pulse-free flow, a requisite that is not completely fulfilled by commercially available propelling systems. When pulses are unavoidable, they should be well damped by use of a suitable attenuator, if required.
- Resistance to aggressive reagents and solvents.
- Readily adjustable flow-rates, and low investment and running cost.

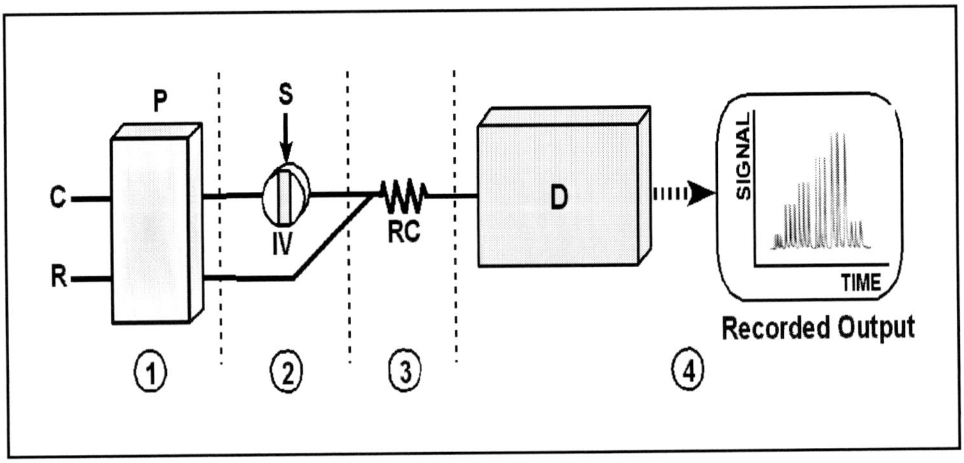

Figure 1.2. Schematic diagram showing a simple FIA manifold. (1) Propulsion unit; (2) Injection device; (3) Transport-Mixing-Reaction Unit; (4) Detection-Signal Processing Unit. C: carrier stream; D: detector; IV: injection valve; P; pump; R: reagent stream; RC: reaction coil; S: sample.

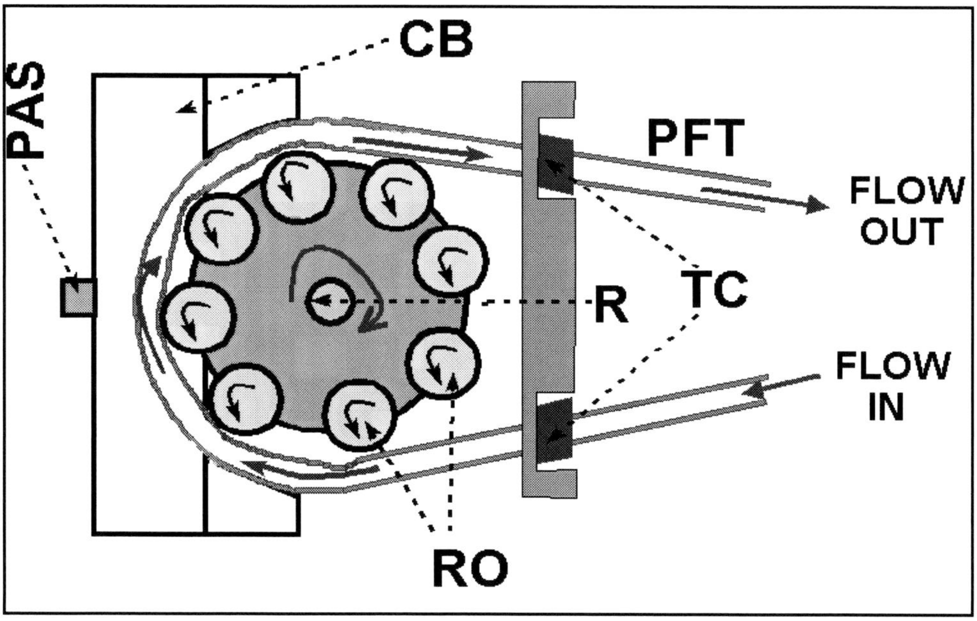

Figure 1.3. Scheme of a peristaltic pump. CB: Compression block; PAS: Pressure adjustment screw; PFT: pump flexible tubing; R: rotor; RO: rollers; TC: tube collars. The arrows indicate the sense of liquid circulation through the pump flexible tubing.

The propulsion unit can be placed at various points along the FI system. Although is usually located before the injection unit.

The flow rate of the FIA system is attained by modification of the diameter of the pump tubes or/and the roller rotation speed (revolutions per minute, rpm).

Sample Introduction Unit

Methods of sample introduction into a FI system may be divided into two main categories: volume and time based insertion.

1. Volume Based Insertion. Injection Devices

This component allows the insertion of an accurately measured sample or reagent volume into the flowing carrier or reagent stream without the need to halt the flow. The introduced volume fills a geometrically defined volumetric cavity. The volume injected (20-100 μL) is inserted as a plug into the carrier in a reproducible manner and without disturbing the flow [40-41].

When FIA was begun, the injection device used was a syringe furnished with a hypodermic needle, which, by piercing the wall of the carrier stream tube, allowed the sample introduction. Nowadays, low-pressure rotary injection valves are the most used devices for injection. The injection volume can be manipulated by changing the size of the attached injection loop.

The advantages of this insertion system are the following:

- Provide high reproducible volumes.
- A wide range of volume can be inserted.

- Facility to change the injection loop.
- Easy and rapid manual operation.
- Possibility of automatic mode change.

Nevertheless, a drawback of injection valves is that require the interruption of the flow system to be changed the injection loop.

The most ordinary, simplest and frequently used injection system is a volume-based injection unit such as a six-port rotary valve. This valve can work in two different positions: load and injection. As can be seen in Figure 1.4, in the load mode, the sample goes into the valve through port 2, fills the injection loop, which is placed between ports 1 and 4 and then, goes to waste through port 3. At the same time, the carrier stream enters to the injection valve through port 6 and goes out toward port 5. In the injection mode, the carrier enters the valve through port 6. In this mode, port 6 is connected internally with port 1, and the carrier stream sweeps the sample plug towards the detector through port 4, which is connected with port 5 (exit port). At the same time, the sample enters the valve through port 2, port 2 is connected internally with port 3 through which is sent to waste. To change between the two modes (load and injection), the injection valve must turn through an angle of 90°. Air bubbles and pressure surges must be avoided during the injection because they will modify the pattern of the flow in FIA system, affecting dispersion and precision.

Injection valves may be operated manually, or actuated pneumatically or electrically by means of a microprocessor. So, they are commercially available injection valves with variable degree of automated control and can be readily coupled to autosamples.

Proportional injectors and solenoid valves are among volume-based injection systems less commonly used in FI.

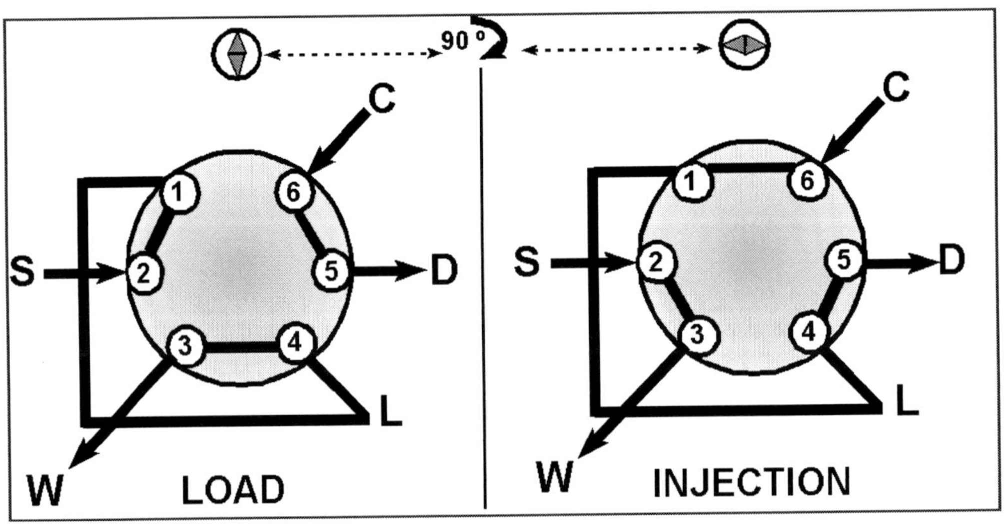

Figure 1.4. Schematic diagram of a FIA six ports injection valve and its operating mode. C: carrier; D: towards the detector; L: sample loop; S: sample; W: waste.

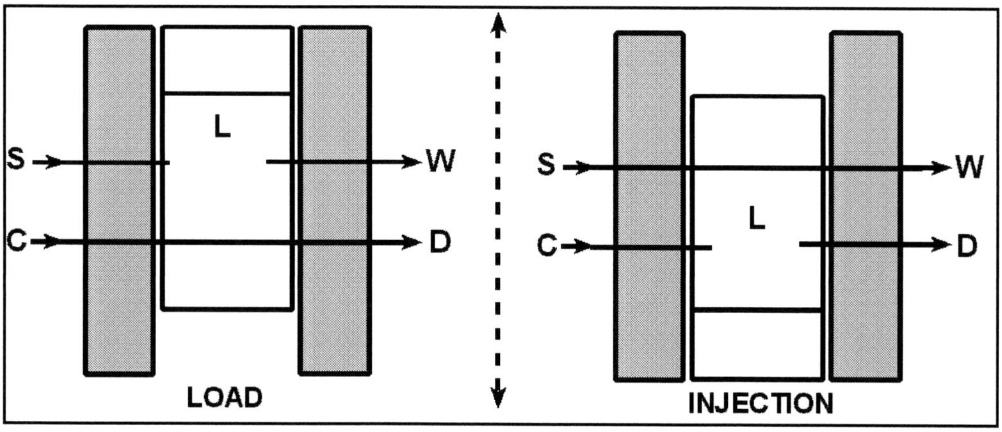

Figure 1.5. Proportional injector (commutator or slider valve) and its operating mode. C: carrier; D: towards the detector; L: sample loop; S: sample; W: waste.

Proportional injectors (commutators or slider valves) are among volume-based injection systems less commonly used in FI. They are used and developed by Brazilian investigation teams [42-51]. These injectors are based on commutation principles and introduce samples (single or multiple) into the FI system, as well as performing other hydrodynamic functions. These injectors are highly versatile and allow most FIA modes to be used. The insertion of one or several volumes of sample and/or reagents takes place simultaneously into the flow system by sliding the central block with respect to the two side blocks, which remain fixed. The central block can be in either of two positions (load and injection) (Figure 1.5) [52].

Also, other injection valves have been proposed between them, Toei developed and assembled a multifunction valve for flow injection analysis to perform many types of injection modes and to carry out a new injection mode, where many sample zones and reagent zones are injected serially or simultaneously [53]. Elsholz described a cheap injection valve based on the tubing clamp operation principle for teaching flow-injection analysis [54].

2. Time Based Insertion

Precise sample or reagent volumes can be delivered into a FI manifold by loading a sample stream for a precise time and at a precise flow rate. The injected volume can be calculated as the product of flow rate and loading time. Time-based insertions have the advantage that sample volume may be changed at will, thus providing an increase of sample throughput. However, this method is dependent on a reproducible flow rate of the sample solution, which makes frequent calibration necessary when using peristaltic pumps. In addition, the uncertainty associated to the aliquots of volumes of sample is related mainly to the precision in the control of the time of sampling [32].

Preferably, this sample insertion mode was adopted when a preconcentration technique is coupled to the FI manifold to insert large sample volumes [55-69].

Different approaches were proposed to increase reproducibility of time-based FI procedures above all involving solenoid valves [70]. Thus, in 1985, Rothwell and Woolf describe the reproducible insertion of samples into a flow stream by timed switching of a sample stream with a miniature solenoid valve and time circuit. They consider that this valve has several advantages over the ordinary rotary valve: it is cheaper and the inserted sample volume can be continuously changed as it is manipulated by temporal rather than spatial control as in rotary valves [71]. Tyson and Bysouth have used a time-based system consisting of an asymmetric three-branch network manifold for generating a family of calibration graphs of varying sensitivity. Although this system allows five calibration plots to be obtained from a single injection, attempts to obtain calibration data from a single injection into the manifold were only partially successful [72]. Burguera et al. [73] have developed a simple time-based variable-volume injector for flow injection systems by using a two-way solenoid valve and a simple timer circuit. They applied it to the atomic absorption determination of copper in water samples. The system features depend on the alternate positions of a solenoid valve. Wu et al. [74] describe a computer-controlled time-based variable-volume injector for the introduction of samples into a flow injection system. Samples with a volume in the range of 8 μL to a few milliliters are delivered precisely (usually RSD < 1.0%). Other authors also proposed the utilization of solenoid valves to manage sample and carrier delivery in FI manifolds for the determination of nitrite and nitrite [75], to extend the linear range of a FI system [76], to preconcentrate trace metals in minicolumns [77-79], for field monitoring of phosphate [80], to develop an automated FI analysis system including ultrasound filtration [81], to simultaneous multiple injection to perform titration and standard addition in monosegmented flow analysis [82], determine 3-hydroxybutyrate in animal serum and plasma [83], to insert in an on-line preconcentration FI system for multi-elemental analysis by total reflection X-ray fluorescence spectrometry [84]. The implementation of a set of solenoid valves in a FI system is the base of the multicommutation concept, which will be further discussed in a following section of this chapter.

Transport-Mixing-Reaction Unit

This unit serves to link the different parts of the FI system. One of its functions is to carry the flowing stream along the manifold and promotes the chemical reaction between sample and reagents. The reactor is the major device of this unit. These devices normally consist of straight tubes of variable length and diameter, coiled tubes (pieces of tubing coiled helically around a rigid cylinder, coils of knotted tubing or knitted flow tubing, or single bead-string reactors, may be used in order to enhance radial mixing and minimizing dispersion. However, knitted tubes require skill to make and the tubing must be flexible yet have wall thick enough to prevent collapse or narrowing at tight bends [85]. The most suitable material for making reactors is polytetrafluoroethylene (PTFE) because this material is chemically resistant and adsorbs the least solutes on its surface.

Other needed components are the connectors, which serve the purpose of joining the tubes to one another and to the other parts of the FI system. These connectors are commercially available in several kinds.

A variety of other units have been reported for specific purposes such as separation units used in non-chromatographic continuous separation systems [86]:

- Dialyzers and gas diffusion units in which the separation of analytes from a donor stream to an acceptor stream is carried out using a permeable membrane that separates the two streams.
- Solvent exaction units that contain two special components: a segmentor that creates a regular pattern of organic and aqueous segments and a separator for the organic phase.

- Packing reactors in which the packing material may be an ion-exchange or chelating resin used for analyte preconcentration or/and for removing the interfering species, immobilized enzymes used for analyte selective degradation, oxidizing or reducing agents that act directly on the analyte or generate reagents "in-situ" that react with the analyte, etc.

Detection-Signal Processing Unit

The variety of detectors used with FI makes possible and explains the widespread use of FIA in conventional analytical techniques. An important feature of FIA is that it requires no special type of detector. In order to perform the monitoring of the analyte(s) in a continuous mode, the usual location of the detection system is on-line and placed after the flow injection manifold.

The interface between a FI system and a conventional non-destructive detector is a commercial or laboratory-made flow cell of appropriate shape and dimensions (Figure 1.6). In the case of destructive detectors such as flame atomic absorption spectrometry, the flow cell is not necessary and the FI manifold is simply connected to the nebulizer of the atomic absorption spectrometer via a Teflon tube.

Spectrophotometry, nephelometry, fluorescence, chemiluminescence, atomic absorption, flame photometry, potentiometry with ion-selective/modified electrodes or field effect transistors, amperometry with sensors and biosensors and voltammetry with wire-type or rotating disk electrodes are the most important detection techniques used in FIA.

FI manifolds can also be coupled to a chromatographic system. Thus, can be coupled with high performance liquid chromatography (HPLC) systems with pre or postcolumn coupling [86-88]. The coupling mode depends on the pursued objective: a pre-column position is mainly used for implementing a continuous separation step prior to chromatographic individual separation and, in a smaller extension, for developing precolumn derivatizing reactions, meanwhile a post-column position is most often used for derivatization purposes. Precolumn FI-HPLC assemblies in which the FI manifold includes a separation unit mainly involved the use of microcolumn with the following objectives:

- Trace preconcentration, using different materials such as ion exchangers, chelating resins, etc., which provide high preconcentration factors.
- Sample clean up, by using the differences in the interaction between the components of a given sample and the sorbent.
- Sample storage, because the relatively inert character of many sorbent materials. This is of special interest when samples have to be collected in remote places.
- Protection of the analytical system, as the solid-phase microcolumn acts as a protective filter, lengthening the usable lifetime of the separation unit.
- Precolumn derivatization, by using a sorbent impregnated with the reagent, a solid redox reagent, etc.

Also other separation steps such as liquid-liquid extraction, membrane extraction, and dialysis have been coupled to HPLC through a FI manifold.

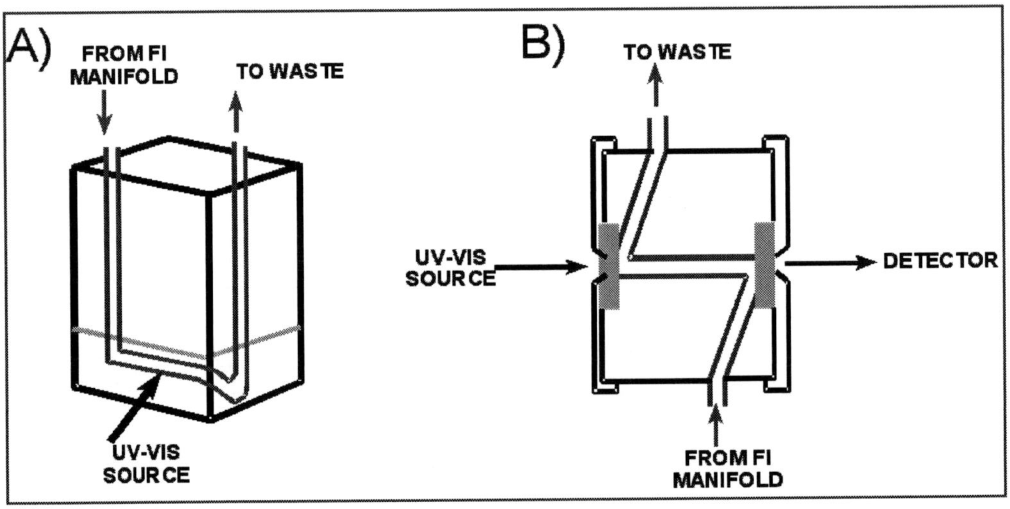

Figure1.6. Spectrophotometric flow-cells. (A) "U" Configuration and (B) "Z" Configuration.

Although it is possible to carry out a FI methodology in a manual mode, where the detector is connected to a chart recorder, most current FI systems are controlled by computers that perform data acquisition. Consequently, the signal-processing unit of the FI system is an instrument for data acquisition and control. The way most commonly used for control consists of using a computer with suitable control/data acquisition software.

FI manifolds can easily be built using commercially available components or by purchasing complete FI instruments that range from simple manual devices to automated systems for near continuous monitoring. It is important to indicate that advances in computerization, microfluidics and hardware have facilitated the development of new FI techniques. Thus, the modern FI system usually consists of a high quality multichannel peristaltic pump, an injection valve, a coiled reactor, a detector such as a photometric flow cell, and an autosampler. Additional components may include a flow through heater to increase the speed of chemical reactions, debubblers, a separation unit such as minicolumns, filters, etc.

There are also advanced and commercial systems that may be fully or partially automated. FI analyzers are now produced commercially by companies in Europe, United Kingdom, North and South America and Japan, and they are commercially offered atomic absorption spectrophotometers with a flow injection system for cold vapor and hydride generation. Automation is achieved by carrying out analyses in a flow system where a pump is used to continuously draw sample and reagent solutions into different lines or segments of plastic tubing, as well as push them forward through the system. By connecting a detector at the end of the sample's flow path, automated detection of the processed sample is ensured.

Portable FI analyzers were used to field monitoring of contaminants in water samples [89-93]. Also, micro FI systems were proposed (μFI). These systems are based on electroosmotic flow [94-96]. In the last years were developed μFI systems in which, electroosmotic flow is controlled in a microchip [97-99].

Flow Injection Signal

An analytical FI signal is a transitional signal in the shape of a peak (Figure 1.7). The basic parameters that define it are the following:

a) Peak height or peak area, related to the analyte concentration.
b) Start-up time (ts), which is defined as the time interval elapsed between injection of the sample and signal production at the detector.
c) Residence time (tr), which is the time elapsed between injection and maximum signal (time during which the analyte is inside the reactor). This time determines the sample throughput.
d) Baseline restoration time (tb), time elapsed from the appearance of the maximum signal to the reappearance of the baseline. This time determines the sensitivity.

The FIA response peak is a result of two processes, both of kinetic nature, the physical process of dispersion of the sample zone within the carrier stream, and the chemical process of the formation of chemical species. There have been many attempts to obtain a mathematical explanation for the influence of parameters such as flow rate, sample volume, reaction rate, among others on the characteristics of the peak height and residence time that take place in FI systems [100-101].

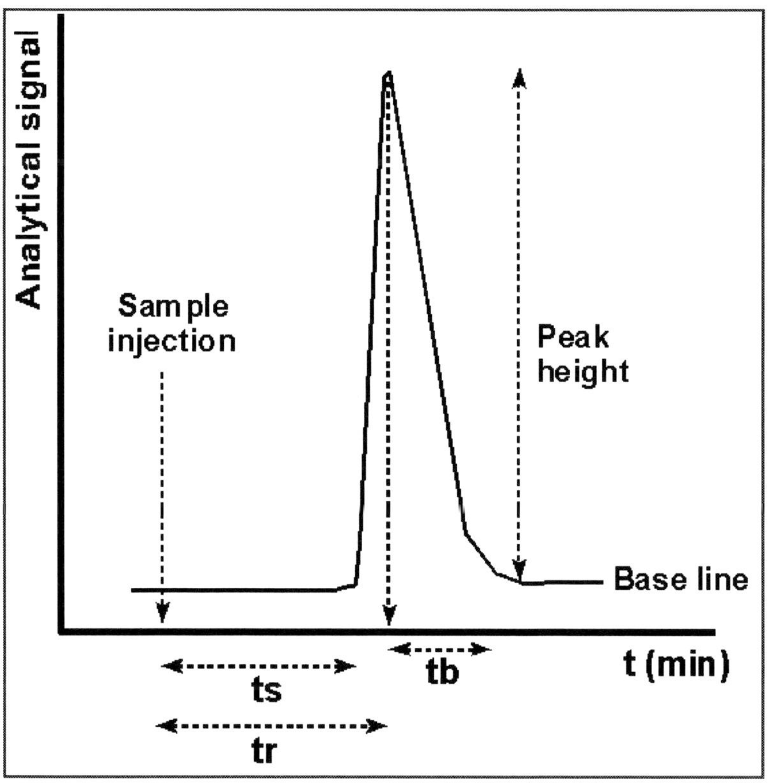

Figure 1.7. Flow injection signal.

FIA Techniques

Multiple Detection FIA Techniques

Multiple detection can be used with FIA techniques. This may be performed obtaining more data of a certain measured parameter, or by using a single detector that performs simultaneous measurements at different working parameters of the analytical instrument. Multiple determination of one or more analytes from a sample may be sequentially realized, performing a number of injections equal with the number of analytes to be determined or simultaneously, determining more analytes from the same injected sample. Thus, two or more signals are recorded in different ways:

- Using a single detector (the signals are delayed).
- Using a dynamic detector, a physical parameter is measured continuously within a certain range along the dispersed sample zone.
- Using more detectors of the same type displaced in series or in parallel.

Usually, FIA configurations with multidetection are assembled in series (Figure 1.8A) or by splitting the sample zone after injection, and diverting a portion of the sample volume sample through parallel flow lines to multiple selective detectors The sample can also be split into two streams immediately after injection, passed though reaction coils, which are of differing residence time, and recombined before passing through a single detector (Figure 1.8B) [102-106]. Multidetection flow-injection techniques are considered for the manipulation of analytical sensitivity and for broadening the determination range of an analyte with maximum accuracy [107], metal speciation [108], kinetics determinations [109-110], for simultaneous determinations of two or more components from the same injected sample [111], among others.

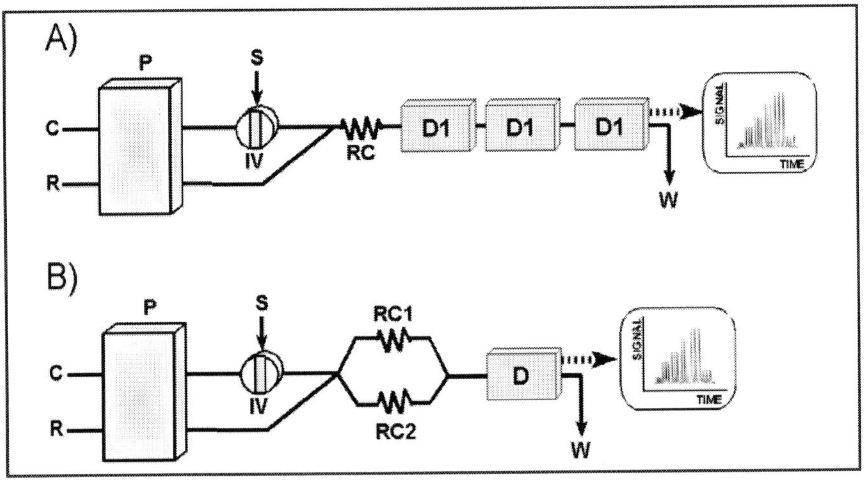

Figure 1.8. Multiple detectors in FIA. (A) Detectors in series; (B) Split system with single detector. C: carrier stream; D, D1, D2 and D3: detection systems; IV: injection valve; P; pump; R: reagent stream; RC, RC1 and RC2: reaction coils; S: sample; W: waste.

Gradient Techniques

As consequence of the controlled dispersion in space and time, FIA is distinguished from the other continuous flow techniques by the well-defined concentration gradient formed when an analyte is injected into the carrier stream. The gradient approach makes possible to perform procedures not feasible by conventional continuous flow analysis because formed concentration provides numerous sections from which an analytical signal can be recorded.. In this sense, many gradient techniques have been developed, including titration, gradient dilution and calibration, gradient scanning, stopped-flow technique, and zone sampling or split zone [112].

FIA titrations are the earliest variants of the gradient techniques. Although technical developments have advanced through various stages since the method was first described by Ruzicka et al. in 1977 [113], the basic principles remain the same. So, after injection, the sample was mixed with titrant in a gradient chamber, where chemical reactions occurred before detection. In this sense, no reaction stoichiometry was reached as in conventional titrations. The time interval elapsed between signal appearance and its return to the baseline was considered as the analytical parameter [114-115]. This FIA technique requires large dispersion, which is achieved by use of a mixing chamber. This device has a large volume as compared with the rest of the FI manifold [116].

Gradient dilution is based on selecting for the analytical readout any point other than the FIA peak maximum. This is useful, when high analyte concentration causes readout to be out of detector range [117]. In gradient calibration technique, after the preliminary determination of the dispersion coefficient at different points of the flow injection peak, the calibration can be performed by the injection of a single standard solution and measurement of the signal magnitude after a given period of time. Its main goal is to avoid the usual repetitive calibration by means of a series of diluted solutions [118].

The gradient scanning technique is an extension of gradient dilution in that it uses a dynamic detector to monitor a physical parameter by repeatedly scanning it over a preset range. This technique was originally suggested by Janata and Ruzicka [119] and provides a very convenient means of study the chemical and physical phenomena, which take place when solutions are mixed.

The FIA stopped-flow approach is based on a combination of stopped-flow measurements and gradient dilution with the aim of increasing the residence time, keeping the reactor short and decreasing the flow-rate of the carrier stream. [120-121]. Stopped-flow was performed stopping the pump for a certain period at a predetermined time after injection. This approach may be used to:

- Enhance the sensitivity of analyses based on reactions with slow kinetics.
- Discriminate analyte signal from background signal in samples having a large blank.
- If the sample plug is stopped at the detector, to obtain kinetic data about the evolution of the reaction of interest over a preset stop time by use of the gradient of the peak obtained during the stop period.
- Increase the residence time within the system to achieve a greater extent of development of the derivatizing reaction without increasing the dispersion.

Zone sampling was proposed by Reis et al. in 1981 [122]. In this FIA technique a selected portion of a dispersed zone is sampled and introduced into another carrier stream. This process provides two advantages: 1) an aliquot is taken, which is the further diluted, reminiscent of classical volumetric dilution, and 2) the injection into a second stream may allow further chemistry to occur. Variable dilution factors can be achieved with this FIA technique simply by changing the time span between the first and the second injection. Zone sampling is useful to implement the standard addition method with a FIA system by using only one standard solution, regardless of the number of the required additions and the addition levels [123]. Split zone is applied for automated dilution and is a FIA technique based on a variant to zone sampling, which relies on the reproducible cleavage of a portion of the trailing edge of the sample concentration profile. Dilutions may be achieved by cleaving a portion of the trailing edge of the sample zone. This technique involves synchronizing the stop and go sequences of three peristaltic pumps to accomplish a reproducible cleavage of a dispersed sample zone; thereby reducing the area of the resulting peaks and increasing the dispersion coefficient. The dilutions may be expressed in terms of the dispersion coefficient or the inverse mole fractions [124].

Multicommutation Flow Analysis

Multicommutation flow analysis (MCFA) is a approach characterized by the use of solenoid valves, creating a flow network, where solutions can be accessed randomly. Accordingly, the fundamental device in multicommutation is the three-way solenoid valve. The implementation of three-way solenoid valves in a flow system presents a functional structure controlled by a computer.

Introduction of sample and reagents into the analytical path can be performed by aspiration through a single pump channel placed after the detection system, and by selecting the positions of the respective valves. With a method of introduction of sample based on time, the uncertainty associated to the aliquots of volumes of sample is associated mainly to the precision in the control of the time of sampling. The error in this case is minimum if electronics are used. From a conceptual point of view, the main contributions of the multicommutation to the flow analysis is the substitution of "volumes" of insertion by "times" of insertion, which allows to develop time-based sampling methods, and, the notion of the flow assembly like a system active, versatile and easy to reshape (flow network) [125-129].

Flexibility is, with no doubt, the main advantage of multicommutation over other flow techniques. In fact, Zagatto et al. [126] considered that multicommutation can unify all concepts already proposed in flow analysis, considering the possibility of accommodating different flow modalities (FIA, SIA) in a system with just solenoid valves.

The number of applications of MCFA has experienced remarkable increase during last years. These applications are mainly focused in environmental, agronomical, pharmaceutical, biological, food and industrial samples [130-137].

All Injection Analysis

All injection analysis (AIA) is developed by Itabashi et al. [138-139] and it is based on a circulation system to minimize reagent consumption and waste. In systems based on all injection analysis, all reagent solutions are injected into a reaction coil and all solutions are circulated for a definite time. By this circulating process, the amount of consumption of the reagents is extremely eliminated, even in intermittent measurements. AIA manifolds can be used for various analytical reaction systems without rearranging the construction of the FIA assembly. The device assembly is easier to construct in the system and the amount of waste fluid as well as the consumption of the reagents are extremely lower than those of the conventional FIA system.

AIA systems were applied with environmental samples such as soil and sediment extracts [140-141].

SEQUENTIAL INJECTION ANALYSIS (SIA)

The second generation of FI techniques, termed sequential injection analysis (SIA), was established in 1990 by Ruzicka and Marshall at the University of Washington [142]. This methodology was proposed as a new feasible and mechanically simpler alternative to FIA for handling solutions. This technique for automatic sample analysis is based on the same principles as FIA (controlled partial dispersion and reproducible sample handling), and it offers different possibilities with a series of advantages and disadvantages in relation with its parent technique. The advantage of SIA over FIA is that SIA typically consumes less than one-tenth the reagent and produces far less waste, an important feature when dealing with expensive chemicals, hazardous reagents, or online/remote site applications. Disadvantages of SIA are that it tends to run slower than FIA and major difficulties in the mixture of sample and reagents.

SIA is a single-line system, completely microcomputer controlled, that can be configured to perform most operations of conventional FIA, with no or minimal physical reconfigurations of the manifold, allowing to perform determinations of different analytes. SIA has proven to be a technique that can be designed to operate in a multi-parametric way, which is of special interest when considering the design of the environmental monitors [143].

Principles of Operation and SIA Components

The sample volume in conventional FIA manifolds is inserted into a carrier stream and subsequently merged downstream with reagents. However, SIA is a fully automated discontinuous flowing technique based on the sequential aspiration of precise volumes of sample and reagent(s) in a holding coil, which are afterwards dispersed into the reaction coil by flow reversal. As a consequence of the sequential and discontinuous operation, the injection frequency as well as the consumption of reagents and sample is evidently reduced compared to FIA.

The most basic SIA system comprises an automatic bi-directional pump or syringe, a multiposition selection valve, a reactor and holding coil, a detection system, and finally a computer that controls the functions of each component (Figure 1.9). The SIA system is initially filled with a carrier stream into which a zone of sample and a zone of reagent(s) are sequentially aspirated into a holding coil. In this way, a stack of well-defined zones is obtained. By means of a flow reversal, a composite zone is formed in the holding coil, as sample and reagent zones penetrate mutually, due the parabolic profile induced by differences between flow velocities of adjacent streamlines and to combined axial and radial dispersion.

Flow reversals and flow acceleration further promote mixing. The multiposition valve is then switched to the detector position, and the flow direction is reversed, propelling the sample/reagent combined zones through the detection system, where the reaction product is monitored [144-146].

The most basic SIA manifold named above (Figure 1.9) will be described briefly following.

Pump

According to Ruzicka and Marshall [142], the pump together with the valve serves as a precision volumetric transport device and should have zero inertia and zero elasticity, requirements that preclude the use of peristaltic pumps. Thus any computer controllable piston pump capable of forward and reversed movement, would be suitable. Thus, the requirements for the pump are that it is precise, reproducible, bi-directional, and able to measure small volumes. Computer control is imperative.

The low-pressure syringe pumps for liquid driver were introduced by Gübeli et al. in 1991 [147]. These pumps provided a sinusoidal flow as a result of non-uniform motion of the piston in the cam-driven piston pump, giving rise to a flow no constant, but reproducible.

Also, some researchers have used a peristaltic pump, being it main advantage is higher analytical frequency as there is no need to aspirate wash solution, in contrast to syringe pumps, which require priming before use and have a limited reservoir capacity [148]. On the other hand, the disadvantage of the peristaltic pump arises from the need of fairly elastic tubes, which have a much shorter life. In order to circumvent this inconvenient, Cladera et al. [149] proposed an auto-burette to propel the flow in SIA.

Nevertheless, robust syringe pumps as liquid drivers rather than other flow pumps have been the most widely used to aspirate zones and propel the stack of zones through the detector because enable the manipulation of sample and reagent volumes at the low μL level with high precision.

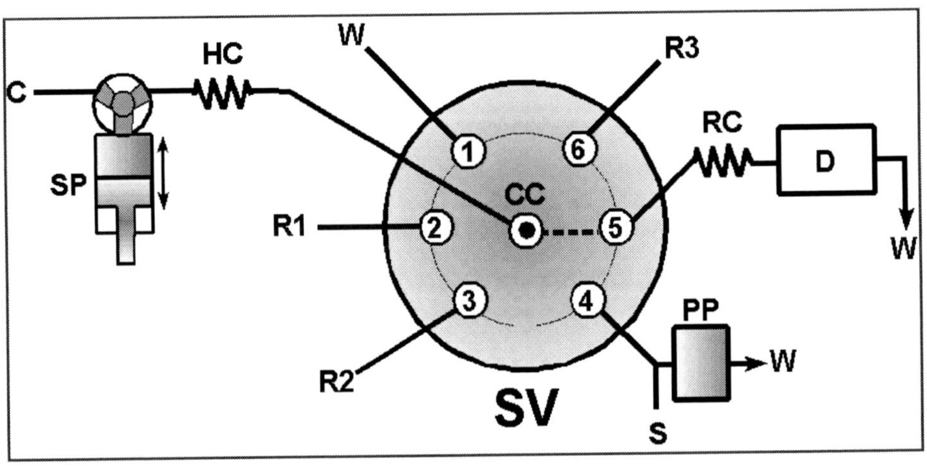

Figure 1.9. Schematic diagram of a SIA manifold. C: carrier stream; CC: communication channel; D: detector; HC: holding coil; PP; peristaltic pump; R1, R2 and R3: reagent streams; RC: reaction coil; S: sample; SP: syringe pump; SV: multiposition selection valve; W: waste.

For methods that involve separation (gas/liquid or chromatography) an additional pump will be necessary.

Selection Valve

In SIA, a multiposition selection valve, which facilitates the use of different chemistries without reconfiguration of the manifold, is the heart of the system.. Small dead volume and zero cross contamination between ports are essential features of a good selection valve. The multiposition valve can be connected to various sample and reagent containers, typically via Teflon tubing. Its central port (central communication line or communication channel) is connected to a high precision stepper motor syringe pump that is used to aspirate well defined volumes of sample and reagent solutions into the holding coil from their containers. Under precise computer control, the selector valve alternately chooses sample and reagent and stacks them into a tubular conduit.

The electrically activated selection valve must allow random access of the ports. Usually, valves are available with between 6 and 28 ports, but the 10 port multi-position valve is by far the most widely used in SIA systems.

Reactor and Holding Coil

The reaction coil is the part connecting the selection valve with the detector. This component is varied depending on the involved chemistry and the amount of dispersion required. However, the reaction coil is usually kept as short as possible to avoid excessive dilution of the formed product zone [150]. As occurs in FIA systems, the most suitable material for making reactor coils is PTFE because this material is chemically resistant and adsorbs the least solutes on its surface.

The holding coil is placed between the pump and the common port of the multi-position selection valve. This additional coil is made from PTFE tubing wound around a plastic tube wound around a plastic tube. The main function of this tubing is as a holding reservoir into which sample and/or reagent is sequentially aspirated. The holding coil volume should be large enough to prevent the stack of zones from being forwarded from the holding coil to the detector before the chemical reaction is taken place. It is essential to investigate the optimum tubing size and length to assess the best sensitivity and precision of the SIA system [151].

Detection System

UV-VIS spectrophotometry is the most frequently used detection technique in SIA systems [145], but in the last years it has been closely followed by atomic spectroscopic techniques such as atomic absorption (AA) [152-154], inductively coupled plasma (ICP) [155-157], inductively coupled plasma-mass spectroscopy (ICP-MS) [158-160], and electrochemical techniques as adsorptive cathodic stripping voltammetry (ACSV) or anodic stripping voltammetry (ASV) [161-164] and potentiometry [165]. As for FIA methodologies, the only requirement to couple a detection system to a SIA manifold is that in the case of non destructive detectors, they be equipped with a flow cell. In addition, low dead-volume and immunity to bubbles are key requirements.

Computer

One of the main drawback of SIA methodologies is the necessity of controlling the overall analytical instrumentation and physical variables (namely, volumes and flow-rates) by available computer programs. Thus, the core of a SIA manifold is the flow program. Microprocessor control with appropriate software is imperative to control the flow direction, rate and timing of the pump, the position of the multi-port valve and to collect, control and process the data. In this sense, many SIA procedures were proposed with in-house built software developed in LabVIEW [166-168], Microsoft QuickBasic [169-170], MATLAB [171-172], Microsoft Visual Basic [173], etc. Though several commercialized DOS or Windows based software packages are also available: FIALab (FIAlab Instruments, USA) [174-177], the FlowTEK package designed by Marshall and coworkers [178] (MINTEK, Randburg, South Africa) [179-181], Atlantis (Lakeshore Technology, Chicago, IL, USA) [182-183], the DARRAY package developed by the research group led by Prof. Cerdà at the University of the Balearic Islands, Spain) [149, 184-186], etc.

Other SIA Components

Solution handling peripherals include gas-liquid separators, liquid-liquid extractors, microcolumns, additional pumps, valves and external microreactors. There almost seems to be no limit to how much the basic system can be expanded by adding on detectors and components [145].

New Generations of SIA

SIA has evolved into distinct flow methodologies: bead injection (BI), sequential injection chromatography (SIC), multisyringe flow analysis (MSFA), and sequential injection laboratory on valve (lab-on-valve, LOV). This last considered as the third generation of FI techniques.

Bead Injection

Bead injection (BI) is an approach to assays based on the microfluidic manipulation of a precise volume of suspended beads that serve as a solid-phase carrier for reagents or reactive groups [187-188]. BI operates in SIA mode, but instead of reagent solutions, uses beads with a diameter of 10-150 μm as reagent carrier. The beads are trapped in the flow-through detector where the adsorption, reaction and elution of analytes are monitored in real time. The beads do not damage or block the channels of the multiposition valve.

The BI protocol comprises five steps. In the step 1, an exact volume of a bead suspension is aspirated and loaded into a flow cell as beads are trapped into a distinct geometry. In step 2, the beads are perfused with a buffered carrier stream, and the baseline for subsequent measurement (spectroscopic or electrochemical) is established. In step 3, the sample is injected and the analyte is trapped on the bead surfaces. In step 4, the analyte is treated with auxiliary reagents or eluted from the bead column. Finally, in step 5, the spent beads are automatically discarded from the flow cell at the end of the assay cycle. The major benefit introduced by BI is automatic surface renewal, a critical feature when assay surfaces become contaminated or otherwise dysfunctional with repetitive use [189-192].

Exist several bead materials, which require different methods of handling. So, suspensions of dense beads as those made of vitreous carbon are sturdy and can be directly agitated by magnetic stirring. Beads with similar densities to buffer solutions as Agarose and Dextran are fragile and should be stirred by gently rotating the suspension container. Also, it is necessary compensate the optical properties of the beads because some beads as Sepharose, Sephadex and Cytodex are fully transparent in the visible region (VIS), increasing absorbance toward the ultraviolet (UV) region. However, porous glass beads are suitable only in the VIS range, and white polymeric beads as Polysorb are only suitable for reflectance mode monitoring [193].

BI is applicable not only for chemical analysis but especially in the biological field. Thus, this flow technique can be an invaluable tool in biochemical and cellular studies. Applications describing renewable surface immunoassay, bioligand interaction, measurements of extracellular and intracellular pH, oxygen consumption and variations in intracellular calcium concentration were developed [194-198].

Sequential Injection Chromatography

The concept of Sequential Injection Chromatography (SIC) was proposed and developed by Šatínský and Solich [199] shortly after monolithic C18 columns became commercially available. It is a hybrid technique of sequential injection analysis (SIA) and liquid chromatography. Briefly, in contrast to other chromatographic techniques, SIC uses flow programming to select injected sample volumes and to produce concentration gradients for chromatographic separations. Other investigators have also used monolithic columns connected to low pressure pumps, however without flow programming, which is the essential part of SIC [200].

Usually, SIC protocol comprises five stages. 1) Sample injection, 2) Analyte adsorption, 3) Eluant injection, 4) Analyte elution, and 5) Analyte detection. Typical SIC configuration consists of an integrated SIA setup where a column is positioned in a flow line of a multi-position valve before the detector. The column of the chromatographic system must be constituted by a highly porosity material generating low back-pressure, such as monolithic or restricted access material. Operationally, the well-defined sample zone was injected in the system and it is led towards to the column for separation. Then mobile phase, acting as carrier, is employed to elute each compound from the monolithic column at relatively high flow rate. The height or area of the detected peaks is proportional to analyte concentration. Basically all detectors equipped with a flow-through cell can be used to SIC setup, covering a wide range of detection modes (UV-VIS, fluorescence, etc).

Sequential injection chromatography (SIC) is a good alternative of high performance liquid chromatography (HPLC) for fast analysis of simple samples. One of the major advantages, against conventional SIA, is that it allows simultaneous efficient separation and quantitation of more than two compounds. The SIC combines the advantages of SIA and liquid chromatography: automation, miniaturization and low sample and mobile phase consumption, high sample throughput, operational simplicity, robustness, reliability, lower instrumentation cost compared HPLC and portability of the analytical instrument.

Implementation of short monolithic chromatographic column into SIA opens new area on-line chromatographic separation of multicompound sample in low-pressure flow system, with the advantage of flow programming and possibility of sample manipulation.

The SIC applications have focused on sample injection without pre-treatment and typical applications include the following fields: pharmaceutical analysis, drug-protein interactions study, bioanalysis and food analysis [201-204].

Multisyringe Flow Analysis

Multisyringe flow injection analysis (MSFA) is described for first time by Cerdà et al. [205-206]. This flow technique combines the advantages of FIA methodologies with the versatility, robustness and reagents saving of SIA. MSFA relies on a device designated by multisyringe burette, manufactured by Crison. This device is the main element of the manifold and is a multiple channel piston pump including up to four syringes of six optional sizes for simultaneous liquid driving. The syringes are connected in block to the same stepper motor of an usual automatic burette, allowing the simultaneously movement of them. All is controlled by computer software through a serial port. A three-way solenoid valve is coupled at the head of each syringe. The valves allow the injection of reagents only at the precise moment to perform the analytical determination or the return to the stock bottle when they are not needed, reducing the consumption of sample and reagents.

Although it is not possible to modify the pumping rates for the individual lines of each syringe, the variety of syringes volumes (0.5, 1, 2.5, 5, 10 and 25 mL) leads to obtain different flows through the different manifold lines, increasing the versatility of the method. The possibility to acquire three successive peaks with only one filling of the syringe increases the sample frequency [207].

The majority of applications of MSFA are focused on environmental samples (water, soil extracts, fertilizers), but the application to process liquors from metallurgical industry, food, and biological samples was also proposed. Moreover, about half of the applications described rely on UV-VIS spectrophotometric detection. Other types of detection systems were also used, namely those based on chemiluminescence, potentiometry, fluorometry, hydride generation atomic fluorescence spectrometry (HGAFS) and optical fiber reflectance [208-215].

SEQUENTIAL INJECTION LAB-ON-VALVE (SI-LOV)

SIA in Lab-On-Valve format (SI-LOV) is considered as the third generation of flow injection. It is a technology that has won an enthusiastic acceptance in the research community for its versatility, and in the routine laboratory for its reliability. This flow technique was introduced by Ruzicka in 2000 with the aim of miniaturize flow techniques for downscaling assays to micro and submicroliter level [216], for example, in solution handling in reagent-limited assays, in handling highly hazardous chemicals, or where waste production is a critical parameter. The miniaturized LOV system potentially offers facilities to allow any kind of chemical and physical processes, including fluidic and microcarrier bead control, homogenous reaction, liquid-solid interaction and in-valve, real-time optical detection of various reaction processes with optical fibers. SI-LOV is the platform for reagent based assays, bead injection technique (BI) and sequential injection affinity chromatography [217-219].

The lab-on-valve conception proposed by Ruzicka [216] not only provides a more flexible approach for flow manifold design, but also opens a promising avenue for the miniaturization of instrument setup. This feature is especially important where rare and expensive samples/reagents are employed.

Principles of Operation and SI-LOV Components

As the name implies, a SI-LOV unit (Figure 1.10) works as an integrated laboratory, that is, to incorporate all necessary facilities for a variety of analytical schemes. Thus, it contains merging points for sample and reagent mixing or sample dilution, potential column reactors for on-line sample processing and a flow-through multipurpose cell, which is furnished with optical fibers for real-time detection of targeted reaction processes.

Fabricated as a 3-plane monolithic structure mounted atop a multiposition selection valve, in order to facilitate as many fluidic operations as possible.

A LOV platform consists of:

(1) A transparent, monolithic structure made of Perspex.
(2) A multiposition selection valve as the main component of the structure.
(3) A propulsion unit, usually a syringe pump, characteristic of SI, to circulate the required liquids through the system.

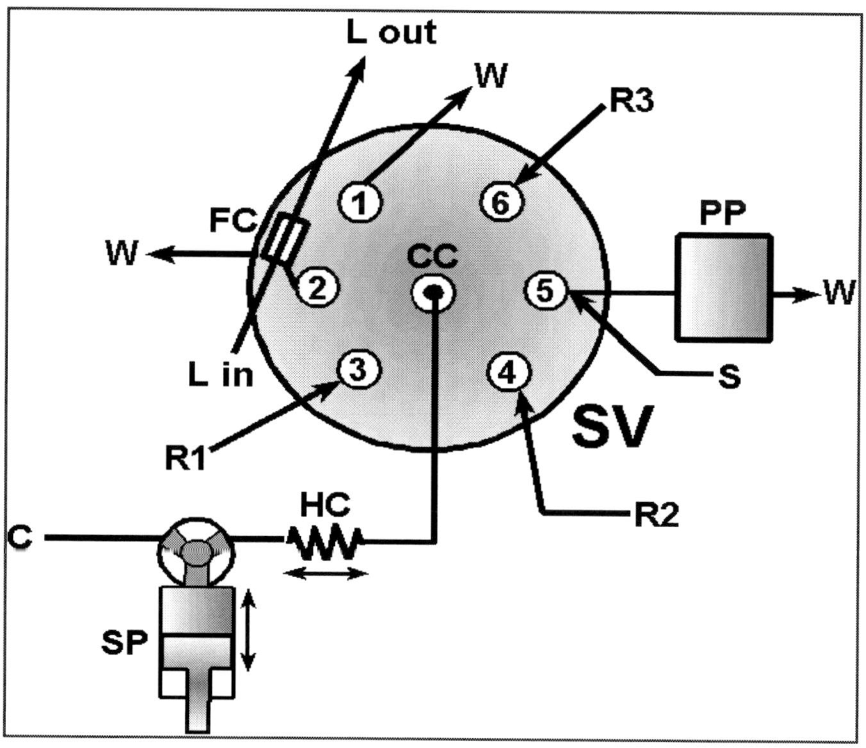

Figure 1.10. Scheme of a SI-LOV unit incorporating an integrated flow-cell. C: carrier stream; CC: communication channel; FC: flow-cell; HC: holding coil; L: light; PP; peristaltic pump; R1, R2 and R3: reagent streams; S: sample; SP: syringe pump; SV: multiposition selection valve; W: waste.

The central flow channel placed on top of a conventional multiposition selection valve (termed communication channel, CC) is designed to incorporate, integrate and facilitate all the necessary operations comprising an analytical determination to be executed in an on-line fashion. This central flow channel connects the liquid driver, i.e. a high-precision syringe pump, which offers the driving force for fluidic manipulations, with the peripheral ports of the valve (identified as 1–6 in Figure 1.10), via a holding coil (HC). The pump is employed for aspiration of sample and reagents sequentially from individual ports and placing them into the HC. Afterwards, the stacked zones are propelled forward, allowing mixing and reaction to take place, the solution finally being guided to the integrated flow-cell (port 2) for in-valve detection of the generated species because the system is compatible with UV-VIS and fluorescence spectroscopy when furnished with optical fibers [220-221].

In addition, an adequate selection of auxiliary units is essential in order to achieve the required configuration. Auxiliary units more frequently used are: one or two syringe pumps, an auxiliary six-port selection valve, one or two holding coils, an auxiliary peristaltic pump, a mixing coil, and, T connections. These units confer on LOV the high versatility, which, together with other characteristics, makes it an attractive tool in the biochemical field [222].

Resulting SI-LOV is portable with a footprint half the size of a laptop computer. LOV allows miniaturization of the flow channel; thus, a miniaturized SI-LOV unit is operated with microliter levels of sample and reagents (a typical sample and reagent volumes are between 5 and 25 μL/assay, while the consumption of carrier solution is 100 to 250 μL) and waste production is typically 0.1–0.2 mL per assay. Thereby downscaling reagent-based assays are obtained the consequent advantages of low reagent consumption and minimized production of chemical waste. Other advantage of LOV concept is that it makes function of instrument transparent to the user, and for routine assays provides a format that is easy to reproduce, so that if serial assays are optimized one instrument, the assay protocol and its software control can be transferred to other instruments, in another location with ease.

Therefore, these systems are positioned between traditional flow techniques that operate at the mL scale and the more futuristic designs of the micro total analysis systems (μTAS) concept, which is supposed to work at the nL scale.

The SI–LOV manifold is also known as a meso-fluidic system in order to distinguish it with microfluidic systems and conventional flow injection setups. It is called as "meso" from its capability to manipulate fluid between "micro" and "macro". Despite of this, most authors refer to SI–LOV manifolds to as micro-fluidic systems [222].

Versatility of LOV design is due to the way in which connector tubing and optical fibers are mounted into the LOV module. The LOV unit can be configured as a jet-ring-cell to execute a BI technique, that is, to perform sorptive extraction procedures with appropriate column reactors packed with renewable surfaces (immobilized enzymes or ion-exchangers) for analyte isolation/preconcentration prior to arrive at the detector where takes place real-time monitoring of the changes in the optical properties of the beads after the liquid-solid interaction by means of absorbance, fluorescence and/or reflectance spectroscopy [223-225].

Although most frequently used detection technique is UV-VIS spectroscopy [226], LOV has proven itself as a sublime front end to capillary electrophoresis, allowing appropriate pretreatment of the sample before introduction into the capillary [227-228]. Most recently, potentiometry [229-230], electrospray mass spectrometry (ES-MS) has been enhanced by using LOV as a sample pretreatment tool [231]. Furthermore, determination of low levels of metal ions in complex matrices has been automated using LOV as a front end to electrothermal atomic absorption spectrometry (ETAAS) [232] and inductively coupled plasma mass spectrometry (ICP-MS) [233]. This novel approach exploited BI renewable microcolumn to fluidically manipulate Sepahdex microbeads, furnished with ion-exchange functional groups for matrix removal and metal ion capture. The bead were not only automatically metered, packed and eluted and the eluate passed to the ETAAS, but a procedure for transferring the loaded beads directly into the graphite tube for pyrolization and atomization was also accomplished.

Applications range from clinical to industrial, environmental and life science analysis and include the determination of organic and inorganic ions [229-233], and DNAs, proteins, peptides and other macromolecules of biological interest, especially those in biological fluids of complex matrices [234-237]. In addition, it constitutes an ideal platform for miniaturized on-line sample-processing and getting real-time information by in situ monitoring [238-239].

CONCLUSION

It is especially remarkable that the during the last 33 years, the three generations of FI, the first generation (FIA), the second (SI) and the third (SI-LOV), have played very important roles as automatic replacements for the labor-intensive manual operation of sample pretreatments. These continuous flow techniques offer the figures of merit of minimized sample/reagent consumption (that is especially critical for samples with very low levels of analytes in complex matrices), waste production and the risk of sample contamination, fast analysis rate, and ease of hyphenation with various detection techniques. These characteristics have a profound impact on modern analytical sciences in the sample pretreatment and are an important contribution to achieve a green Analytical Chemistry.

Considering all the choices available at the present moment to implement automatic analysis, it is not possible to state which one is better. It will depend on the specific analysis aimed and the features associated, such as the sample throughput, sample availability, the mixing conditions of solutions throughout the system, and reagents cost and toxicity. All these factors must be considered when choosing a particular continuous flow methodology. Furthermore, the SI-LOV is a reliable flow technique for further downscaling of solution handling in reagent-limited assays and beads injection manipulations.

An other aspect to take into account is that for their intrinsic characteristics (rapidity, simplicity, low cost and versatility, continuous flow systems offer an excellent alternative for the development of sample screening systems.

REFERENCES

[1] Ruzicka, J.& Hansen, E.H. (1975). *Anal. Chim. Acta* 78, 145-157.

[2] Stewart, K. K., Beecher, G. R. & Hare, P. E. (1976). *Anal. Biochem.* 70, 167-173.

[3] Ruzicka, J. & Hansen, E.H. (1981). *Flow injection analysis*. New York, USA: Wiley & Sons Inc.

[4] Ruzicka, J. & Hansen, E.H.(1983). *Flow injection analysis* (Japanese). Kyoto, Japan: Kagakudonin.

[5] Ueno, K. & Kina, K. (1983) *Introduction to flow injection analysis. Experiments and applications* (Japanese). Tokyo, Japan: Kodansha Scientific.
[6] Válcarcel Cases, M. & Luque de Castro, M.D. (1984). *Flow-injection analysis* (Spanish). Córdoba, Spain: Imprenta San Pablo.
[7] Hansen, E.H. (1986). *Flow injection analysis*. Copenhagen, Denmark: Polyteknisk Forlag.
[8] Ruzicka, J. & Hansen, E.H. (1986). *Flow injection analysis* (Chinese). Beijing, China: Science Press.
[9] Valcarcel, M. & Luque de Castro, M.D. (1987). *Flow-injection analysis. Principles and applications*. Chichester, UK: Ellis Horwood Ltd.
[10] Ruzicka, J. & Hansen, E.H. (1988). *Flow injection analysis* (2nd ed.). New York, USA: Wiley & Sons Inc.
[11] Burguera J.L. (1989). *Flow injection atomic spectroscopy*. New York, USA: Marcel Dekker.
[12] Karlberg, B. & Pacey, G.E. (1989). *Flow injection analysis. A practical guide*. Amsterdam, The Netherlands: Elsevier.
[13] Ruzicka, J.& Hansen, E.H. (1991). *Flow injection analysis* (2nd ed) (Chinese). Beijing, China: University Press.
[14] Fang, Z.L. (1993). *Flow-injection separation and preconcentration*. Weinheim, Germany: VCH Verlags.
[15] Frenzel, W. (1993). *Flow injection analysis. Principles, techniques and applications*. Berlin, Germany: Technical University Berlin.
[16] Valcarcel, M. & Luque de Castro, M.D. (1994). *Flow-through (bio)chemical sensors*. Amsterdam, The Netherlands: Elsevier.
[17] Fang, Z.L. (1995). *Flow injection atomic spectrometry*. Chichester, UK: Wiley.
[18] Martínez Calatayud, J. (1997). *Flow injection analysis of pharmaceuticals: Automation in the laboratory*. London, UK: Taylor & Francis.
[19] Sanz-Medel, A. (1999). *Flow analysis with atomic spectrometric detectors*. Amsterdam, The Netherlands: Elsevier.
[20] Fang, Z. (1999). *Flow injection analysis* (Chinese). Beijing, China: Science Press.
[21]Trojanowicz, M. (2000). *Flow injection analysis. Instrumentation and applications*. Singapore: World Scientific. Ltd.
[22] Trojanowicz, M. (2008). *Advances in flow analysis*. Weinheim, Germany: Wiley-VCH.
[23] Kolev, S.D. & I.D. McKelvie (Ed.) (2008). *Advances in flow injection analysis and related* techniques: *Vol. 54. Comprehensive Analytical Chemistry Series*. Amsterdam, The Netherlands: Elsevier.
[24] Ruzicka, J. (2005). CD-ROM Tutorial, *Flow injection analysis: Methods and* applications (3rd ed.).
[25] Journal of Flow Injection Analysis. http://aitech.ac.jp/~jafia/english/jfia/index.html.
[26] Chalk, S.J. (1998). *Talanta* 45, 591-599.
[27] The Flow Analysis Database at the University of North Florida. http://www.fia.unf.edu/
[28] Flow Injection Database by E.H. Hansen at FIAlab Instruments site. http://www.fialab.com/database1_interface/fia_database/results_page.asp.
[29] Ruzicka, J. & Hansen, E.H. (2008). *Trends Anal. Chem.* 27, 390-393.
[30] Ruzicka, J. & Hansen, E.H. (2000). *Anal. Chem.* 72, 213A-217A.

[31] K.K. Stewart, K.K. (1981). *Talanta* 28, 789-797.
[32] Stewart, K.K. (1986). *Anal.Chim. Acta* 179, 59-68.
[33] Fang, Z. (1992). *Microchem. J.* 45, 137-142.
[34] Hanrahan, G., Ussher, S., Gledhill, M., Achterberg, E.P. & Worsfold, P.J. (2002). *Trends Anal. Chem.* 21, 233-239.
[35] Valcarcel, M. & Luque de Castro, M.D. (1990). *Fresenius J. Ana. Chem.* 337, 662-666.
[36] Luque de Castro, M.D. (1992). *Microchim. Acta* 109, 165-168.
[37] Kubán, V. (1992). *Crit. Rev.Anal.Chem.* 23, 15-53.
[38] McKelvie, I.D. (1999). *Analytical Testing Technology* 20, 20-24.
[39] Prados-Rosales, R.C., Luque-García, J.L. & Luque de Castro, M.D. (2002). *Anal.Chim. Acta* 461, 169-180.
[40] Möller, J. (1988). *Fresenius Z. Anal. Chem.* 329, 685-687.
[41] Prados-Rosales, R.C., Luque-García, J.L. & Luque de Castro, M.D. (2003). *Anal.Chim. Acta* 480, 181-192.
[42] Bergamin Filho, H., Reis, B.F. & Zagatto, E.A.G. (1978). *Anal. Chim.Acta* 97, 427-431.
[43] Krug, F.J., Pessenda, L.C.R., Zagatto, E.A.G., Jacintho, A.O. & Reis, B.F. (1981). *Anal. Chim. Acta* 130: 409-413.
[44] Krug, F.J., Bergamin Filho, H. & Zagatto, E.A.G. (1986). *Anal. Chim.Acta* 179, 103-118.
[45] Mattos, I.L., Zagatto, E.A.G. & Jacintho, A.O. (1988). *Anal. Chim. Acta* 214, 247-257.
[46] Silva, F.V., Nogueira, A.R.A., Souza, G.B. & Zagatto, E.A.G (1998). *Anal. Chim. Acta* 370, 39-46.
[47] Brito, V. O. & Raimundo, M. I. (1998). *Anal. Chim. Acta* 371, 317-324.
[48] Santelli, R.E., Bezerra, M., Domingos de Santana, O., Cassella, R.J. & Costa Ferreira, S.L. (2006). *Talanta* 68, 1083-1088.
[49] Costa de Figueiredo, E., Ruela de Souza, L., Schmidt de Magalhães, C., Wisniewski, C. & Orival Luccas, P. (2006). *J. Autom Methods Manag Chem.* 2006, 1-4.
[50] Figueiredo, E.C., Dedina, J., & Arruda, M.A.Z. (2007). *Talanta* 73, 621-628.
[51] Matos, G. D. & Arruda, M.A.Z. (2008). *Talanta* 76, 475-478.
[52] Ríos, A., Luque de Castro, M. D. & Valcárcel, M. (1987). *J. Autom. Chem.* 9, 30-36.
[53] Toei, J. (1987). *Anal. Lett.* 20, 1951-1976.
[54] Elsholz, O. (1998). *Anal. Chim. Acta* 366, 217-221.
[55] Fang, Z., Xu, S. & Zhang, S (1987). *Anal. Chim. Acta* 200, 35-49.
[56] Naghmush, A.M.,Trojanowicz, M. & Olbrych-Sleszynska, E.(1992). *J. Anal.At. Spectrom.* 7, 323-328.
[57] Welz, B., Yin, X., Sperling, M. (1992). *Anal. Chim. Acta* 261, 477-487.
[58] Carroll, M.K. & Tyson, J.F. (1993). *J. Chem.Educ.* 70, A210-A211, A215-A216.
[59] Burguera, M., Burguera, J. L., Rivas, C., Carrero, P., Brunetto, R. & Gallignani, M. (1995). *Anal. Chim. Acta* 308, 339-348.
[60] Cespón-Romero, R.M., Yebra-Biurrun, M.C. & Bermejo-Barrera, M.P. (1996). *Anal. Chim. Acta* 327, 37-45.
[61] Prasada Rao, T., Karthikeyan, S., Vijayalekshmy, B. & Iyer, C. S. P. (1998). *Anal. Chim. Acta* 369, 69-77.
[62] Fang, Z. (1998). *Spectrochim. Acta* 53B, 1371-1379.
[63] Hirata, S., Yoshihara, H. & Aihara, M. (1999). *Talanta* 49, 1059-1067.
[64] Elci, L., Arslan, Z. &Tyson, J.F. (2000). *Spectrochim. Acta* 55B, 1109-1116.

[65] Almeida, A.A., Jun, X. & Lima, J.L.F.C. (2000). *At.Spectrosc.* 21, 187-193.
[66] Yebra, M. C., Carro, N., Enriquez, M. F., Moreno-Cid, A. & Garcia, A.(2001). *Analyst* 126, 933-937.
[67] Yebra, M.C., Carro, N. & Moreno-Cid, A. (2002). *Spectrochim. Acta* 57B, 85-93.
[68] Anthemidis, A.N. & Koussoroplis, S.J.V. (2007). *Talanta* 71, 1728-1733.
[69] Dimitrova-Koleva, B., Benkhedda, K., Ivanova, E. & Adams, F. (2007). *Talanta* 71, 44-50.
[70] Harrow, J. J. & Janata, J. (1983). *Anal. Chem.* 55, 2461-2464.
[71] Rothwell, S. D. & Woolf, A. A. (1985). *Talanta* 32, 431-433.
[72] Tyson, J. F & Bysouth, S. R. (1988). *J. Anal. At. Spectrom.* 3, 211-215.
[73] Burguera, J. L., Burguera, M., Carrero, P., Marcano, J., Rivas, C. & Brunetto, M. R. (1995). *J. Autom. Chem.* 17, 25-29.
[74] Wu, H.L., Hayashibe, Y., Sayama, Y., Shibukawa, M. & Oguma, K. (1995). *Anal. Sci.* 11, 401-404.
[75] Martinez Calatayud, J., Garcia Mateo, J. V.& David, V.(1998). *Analyst* 123, 429-434.
[76] Gouveia, L. F., Costa Lima, J.L. F.& Morais, J.A. G. (1998). *Anal. Chim. Acta* 366, 271-279.
[77] Silva, M.M., Arruda, M. A. Z., Krug, F. J., Oliveira, P. V., Queiroz, Z.F., Gallego, M. & Valcarcel, M. (1998). *Anal. Chim. Acta* 368, 255-263.
[78] Giacomelli, M. B. O., Ganzarolli, E. M.& Curtius, A. J.(2000). *Spectrochim. Acta* 55B, 525-533.
[79] Ferreira, S. L.C., Lemos, V.A., Santelli, R.E., Ganzarolli, E.& Curtius, A.J. (2001). *Microchem. J.* 68, 41-46.
[80] Hanrahan, G., Gledhill, M., Fletcher, P. J. & Worsfold, P. J. (2001). *Anal. Chim. Acta* 440, 55-62.
[81] Wang, R. Y., Jarratt, J. A., Keay, P. J., Hawkes, J. J. & Coakley, W. T. (2000). *Talanta* 52, 129-139.
[82] Assali, M., Raimundo, I.M. & Facchin, I. (2001). *J. Autom Methods Manag Chem.* 23, 83-89.
[83] Pires, C. K., Martelli, P.B., Reis, B.F., Lima, J.L.F.C. & Saraiva, M.L.M.F.S. (2003). *J. Agric. Food Chem.* 51, 2457-2460.
[84] Menegario, A.A., Pellegrinotti, D. C., Gine, M.F., Nascimento Filho, V.F. (2003). *Spectrochim. Acta* 58B, 543-549.
[85] Waiz, S., Cedillo, B.M., Jambunathan, S., Hohnholt, S.G., Dasgupta, P.K. & Wolcott, D.K. (2001). *Anal. Chim.Acta* 428, 163-171.
[86] Valcarcel, M. & Luque de Castro, M.D. (1991). *Non-chromatographic continuous separation techniques*, Cambridge, UK: Royal Society of Chemistry.
[87] Luque de Castro, M.D. & Fernández-Romero, J.M. (1998). *J. Chromatogr.* 819A, 25-33.
[88] Luque de Castro M.D. & Tena M.T. (1995). *Talanta* 42, 151-169.
[89] Andrew, K. N., Worsfold, P.J. & Comber, M. (1995). *Anal. Chim.Acta* 314, 33-43.
[90] Alexander, P.W., Di Benedetto, L. T., Dimitrakopoulos, T., Hibbert, D.B., Ngila, J.C., Sequeira, M. & Shiels, D. (1996). *Talanta* 43, 915-925.
[91] Alexander, P. W., Dimitrakopoulos, T. & Hibbert, D. B. (1996). *Field Anal.Chem. Technol.* 1, 31-37.

[92] Sritharathikhun, P., Suekane, T., Oshima, M. & Motomizu, S. (2004). *J. Flow Injection Anal.* 21, 53-58.

[93] Coles, S., Nimmo, M. & Worsfold, P. (2000). *Lab. Rob. Autom.* 12, 183-193.

[94] Daykin, R. N. C.; Haswell, S. J. (1995). *Anal. Chim.Acta* 313, 155-159.

[95] Haswell, S. J. (1997). *Analyst* 122, 1R-10R.

[96] Ma, L., Oshima, M., Motomizu, S. & Hattori, T. (1998). *Bunseki Kagaku* 47(6), 375-380.

[97] Petsul, P. H., Greenway, G. M. & Haswell, S. J. (2001). *Anal. Chim.Acta* 428, 155-161.

[98] Cai, Z., Chen, H., Chen, B. & Huang, C.(2006). *Talanta*, 68, 895-901.

[99] Nishihama, S., Imabayashi, H., Matoba, T., Toya, C., Watanabe, K. & Yoshizuka, K. (2008). *Talanta* 74, 1350-1354.

[100] Wu, N. S., Chen, Y. & Zhao, H. Z. (1993). *Anal. Lett.* 26, 2679-2686.

[101] Tryzell, R. & Karlberg, B.(1998). *Anal. Chim. Acta* 364, 97-106.

[102] Luque de Castro, M.D. & Valcarcel, M. (1986). *Trends Anal. Chem.* 5, 71-74.

[103] Valcarcel, M., Luque de Castro, M. D., Lazaro, F. & Rios, A. (1989). *Anal. Chim.Acta* 216, 275-288.

[104] Luque de Castro, M. D. & Tena, M. T. (1995). *Talanta* 42, 151-169.

[105] Rocha, F.R.P., Martelli, P.B.& Reis, B.F. (2001). *Anal. Chim.Acta* 438, 11-19.

[106] Moreira, L.N., Ramos, L.A., Tarley, C.R.T. & Luccas, P. O. (2007). *J. Flow Injection Anal.* 24, 13-16.

[107] Rios, A., Lazaro, F., Luque de Castro, M. D.& Valcarcel, M. (1987). *Anal. Chim.Acta* 199, 15-27.

[108] Ruz, J., Rios, A., Luque de Castro, M. D. & Valcarcel, M. (1986). *Talanta* 33, 199-202.

[109] Fernandez-Romero, Juan M.; Luque de Castro, M. D.; Valcarcel, M. *Anal. Chim.Acta* (1989), 219(2), 191-9.

[110] Fernandez-Romero, J. M. & Luque de Castro, M. D. (1991). *J. Pharm. Biomed. Anal.* 9, 679-684.

[111] Alpizar, J., Crespi, A., Cladera, A., Forteza, R. & Cerda, V. (1997). *Electroanalysis* 9, 247-250.

[112] Hansen, E.H. (1988). *Fresenius' J. Anal. Chem.* 329, 656-659.

[113] J. Ruzicka, J., Hansen, E. H. & Mosbaek, H. (1977). *Anal. Chim.Acta* 92, 235-249.

[114] Martelli, P.B., Reis, B.F., Korn, M. & Costa Lima, J.L.F. (1999). *Anal. Chim.Acta* 387, 165-173.

[115] Kahlert, H., Pörksen, J.R., Behnert, J. & Scholz, F. (2005). *Anal. Bioanal. Chem.* 382, 1981-1986.

[116] Toci, J. (1988). *Fresenius' J. Anal. Chem.* 330, 484-488.

[117] Hansen, E. H. & Ruzicka, J. (1976). *Anal. Chim.Acta* 87, 353-363.

[118] Sperling, M., Fang, Z. & Welz, B. (1991). *Anal. Chem.* 63, 151-159.

[119] Janata, J. & Ruzicka, J. (1982). *Anal. Chim.Acta* 139, 105-115.

[120] Olsen, S., Ruzicka, J. & Hansen, E.H. (1982). *Anal. Chim.Acta* 136, 101-112.

[121] Olsen, S., Ruzicka, J. & Hansen, E.H. (1982). *Anal. Chim.Acta* 136, 101-112.

[122] Reis, B. F., Jacintho, A. O., Mortatti, J., Krug, F. J., Zagatto, E. A. G., Bergamin, H. & Pessenda, L.C.R. (1981). *Anal. Chim.Acta* 123, 221-228.

[123] Silva, E. C., Araújo, M.C.U., Honorato, R.S., Costa Lima, J. L. F., Zagatto, E.A.G. & Brienza, S.M.B. (1996). *Anal. Chim.Acta* 319, 153-158.

[124] Clark, G.D., Ruzicka, J. & Christian, G.D. (1989). *Anal. Chem.* 61, 1773-7778.

[125] Reis, B.F., Giné, M.F., Zagatto, E.A.G., Lima, J.L.F.C. & R.A. Lapa. (1994). *Anal. Chim.Acta* 293, 129-138.

[126] Zagatto, E.A.G., Reis, B.F., Oliveira C.C., Sartini, R.P. & Arruda, M.A.Z. (1999). *Anal. Chim.Acta* 400, 249-256.

[127] Catalá Icardo, M., Martínez Calatayud, J. & García Mateo, J.V. (2002). *Trends Anal. Chem.* 21, 366-378.

[128] Rocha, F.R.P., Reis, B. F., Zagatto, E. A. G., Lima, J.L.F.C., Lapa, R.A.S. & Santos, J.L.M. (2002). *Anal. Chim.Acta* 468, 119-131.

[129] Feres, M.A., Fortes P.R., Zagatto E.A., Santos. J.L. & Lima J.L. (2008). *Anal. Chim.Acta* 618, 1-17.

[130] Martelli, P. B., Reis, B. F., Kronka, E.A.M., Bergamin, F.H., Korn, M., Zagatto E.A.G., Lima, J.L.F.C. & Araujo, A.N. (1995). *Anal. Chim.Acta* 308, 397-405.

[131] Araújo, A.N., Costa Lima, J.L.F., Reis, B.F. & Zagatto E.A.G. (1995). *Anal. Chim.Acta* 310, 447-452.

[132] Oliveira, C.C., Sartini, R. P., Reis, B.F. & Zagatto E.A.G. (1996). *Anal. Chim.Acta* 332, 173-178.

[133] Kronka, E.A.M., Reis, B.F., Korn, M.& Bergamin, H. (1996). *Anal. Chim.Acta* 334, 287-293.

[134] Vieira, J.A., Reis, B.F., Kronka, E.A.M., Paim, A.P.S. & Giné. M.F. (1998). *Anal. Chim.Acta* 366, 251-255.

[135] Sa, F., López Malo, D. & Martínez Calatayud, J. (2007). *Anal. Lett.* 40, 2872-2885.

[136] Ródenas-Torralba, E., Morales-Rubio, A., Lavorante, A.F., Freire dos Reis, B. & de la Guardia M. (2007). *Talanta* 73, 742-747.

[137] Infante, C.M., Masini, J.C. & Rocha, F.R. (2008). *Anal. Bioanal. Chem.* 391, 2931-2936.

[138] Itabashi, H., Kawamotoand, H. & Kawashima, T. (2001). *Anal. Sci.* 17, 229-231

[139] Itabashi, H. (2003). *Current Advances in Materials and Processes* 16,1608.

[140] Itabashi, H. & Morimoto, H. (2004). *Tokyo konfarensu Koen Yoshishu*, 2004, 305.

[141] Amornthammarong, N., Anujaravat, P., Sereenonchai, K., Chaisuwan, P., Sastranurak, P., Wilairat P. & Nacapricha, D. (2005). *Talanta* 68, 480-487.

[142] Ruzicka, J.& Marshall, G. D. (1990). *Anal. Chim. Acta* 237, 329-343.

[143] Hansen, E.H. & Wang, J. (2005). *Anal.Lett.* 37, 345-359.

[144] Ruzicka, J. (1991). *Anal. Sci.* 7, 635-637.

[145] Lenehan, C.E., Barnett, N.W. & Lewis, S.W. (2002). *Analyst* 127, 997-1020.

[146] Economou, A. (2005). *Trends Anal.Chem.* 24, 416-425.

[147] Gübeli, T., Christian, G.D. & Ruzicka, J. (1991). *Anal. Chem.* 63, 2407-2413.

[148] Ivaska, A. & Ruzicka, J. (1993). *Analyst* 118, 885-889.

[149] Cladera, A., Tomàs, C., Gómez, E., Estela, J. M. & Cerdà, V. (1995). *Anal. Chim. Acta* 302, 297-308.

[150] Tesfaldet, Z.O., van Staden, J.F. & Stefan, R.I. (2004). *Talanta* 64, 1189-1195.

[151] Cladera, A., Gomez, E., Estela, J. M. & Cerdà, V. (1996). *Talanta* 43, 1667-1674.

[152] Marqués, M.J., Morales-Rubio, A., Salvador, A. & de la Guardia, M. (2001). *Talanta* 53, 1229-1239

[153] Zhanga,Y. & Adeloju, S.B. (2008). *Talanta* 74, 951-957.

[154] Peña, F., Lavilla, I. & Bendicho, C. (2002). *Anal. Chim. Acta* 456, 283-292.

[155] Ek, P., Huldén, S.G. & Ivaska, A. (1995). *J. Anal. At. Spectrom.* 10, 121-126.

[156] Katarina, R.K., Lenghor, N. & Motomizu, S. (2007). *Anal. Sci.* 23, 343-350.

[157] Sabarudin, A., Lenghor, N., Oshima, M., Hakim, L., Takayanagi, T.Y,. Gao, H. & Motomizu, S. (2007). *Talanta* 72, 1609-1617.

[158] Jiménez, M.S., Velarte, R. & Castillo, J. R. (2002). *Spectrochim. Acta* 57B, 391-402.

[159] Akinlua, A., Torto, N. & Ajayi, T.R. (2008). *Fuel* 87, 1469-1477.

[160] Lopes, C.M.P.V., Almeida, A.A., Saraiva, M.L.M.F.S. & Lima, J.L.F.C. (2007). *Anal. Chim. Acta* 600, 226-232.

[161] Daniel, A., Baker, A.R. & van den Berg, C.M.G. (1997). *Fresenius' J. Anal. Chem.* 358, 703-710.

[162] Kefala, G. & Economou, A. (2006). *Anal. Chim. Acta* 576, 283-289.

[163] Vieira dos Santos, A.C. & Masini, J.C. (2006). *Anal. Bioanal. Chem.* 385, 1538-1544.

[164] Chuanuwatanakul, S., Dungchai, W., Chailapakul, O & Motomizu, S. (2008). *Anal. Sci.* 24, 589-594.

[165] Zárate, N., Araújo, A.N., Montenegro, M.C.B.S.M. & Pérez-Olmos, R.(2003). *Am. J. Enol. Vitic.* 54, 46-49.

[166] Economou, A., Tzanavaras, P.D. & Themelis, D.G. (2005). *J. Chem. Educ.*5, 1820-1822.

[167] Economou, A. & Voulgaropoulos, A. (2003). *J. Autom Methods Manag Chem.* 25, 133-140.

[168] Lenghor, N., Jakmunee, J., Prazen, B.J., Synovec, R.E., Christian, G.D. & Grudpan, K. (2006). *Anal. Sci.* 22, 147-160.

[169] Morais, I.P.A., Renata, M., Souto, S. & Rangel, A.O.S.S. (2005). *J. AOAC Int.* 88, 639-644.

[170] Reis Lima, M.J., Fernández, S.M.V. & Rangel, A.O.S.S. (2006). J. Food Sci. 67, 3280-3283.

[171] Pasamontes, A. & Callao, M.P. (2004). *Anal. Chim. Acta* 515, 159-165.

[172] Calvo, D., Durán, A. & del Valle, M. (2007). *Anal. Chim. Acta* 600, 97-104.

[173] Makchit, J., Upalee, S., Thongpoon, C., Liawruangrath, B. & Liawruangrath, S. (2006). *Anal. Sci.* 22, 591-597.

[174] Abate, G., dos Santos, L.B.O., Colombo, S.M. & Masini, J.C. (2006). *J. Braz. Chem. Soc.* 17, 491-496.

[175] Idris, A.M., Assubaie, F.N. & Sultan, S.M. (2007). *J. Autom Methods Manag Chem.* 2007, 1-7

[176] Hirakawa, K., Katayama, M., Soh, N., Nakano, K. & Imato, T. (2006). *Anal. Sci.* 22, 81-86.

[177] Baxter, P. J., Christian, G.D. & Ruzicka, J. (1994). *Analyst* 119, 1807-1812.

[178] Marshall, G.D. & van Staden, J.F. (1992). *Anal. Instrum.* 20, 79-100.

[179] Beyene, N.W., van Staden, J.F. & Stefan, R.I. (2004). *Il Farmaco* 59, 1005-1010.

[180] van Staden, J.F. & Botha, A. (1999). *Talanta* 49, 1099-1108.

[181] Beyene, N.W., van Staden, J.F. & Stefan, R.I.(2004). *Anal. Chim. Acta* 521, 223-229.

[182] Pollema, C.H. & Ruzicka, J. (1993). *Analyst* 118, 1235-1240.

[183] Luo, Y., Al-Othman, R., Ruzicka, J. & Christian, G.D. (1996). *Analyst* 121, 601-606.

[184] Muñoz, A., Mas Torres, F.J., Estela, M. & Cerdà,V. (1997). *Anal. Chim. Acta* 350, 21-29.

[185] Mas, F., Cladera, A., Estela, J.M. & Cerdà, V. (1998). *Analyst,* 123, 1541–1546.

[186] Mateos, J.J., Gómez, E., Garcías, F., Casas, M. & Cerdà, V. (2000). *Applied Radiation and Isotopes* 53, 139-144.

[187] Pollema, C.H., Ruzicka, J., Christian, G.D. & Lemmark, A. (1992). *Anal. Chem.* 64, 1356- 1361.

[188] Ruzicka, J.& Ivaska, A. (1997). *Anal. Chem.* 69, 5024-5030.

[189] Ruzicka, J. & Scampavia, L. (1999). *Anal. Chem.* 71, 257A-263A.

[190] Rama, M.J., Medina, A.R. & Díaz A.M. (2003). *Anal Bioanal Chem.* 376, 527-533.

[191] Hartwell, S.K., Grudpan, K. & Christian, G.D. (2004). *Trends Anal. Chem.* 23, 619-623.

[192] Ruedas-Rama, M.J., Ruíz-Medina, A. & Molina-Díaz, A. (2005). *Anal. Sci.* 21, 1079-1084.

[193] Solé, S., Merkoçi, A. & Alegret, S. (2001). *Trends Anal. Chem.* 20, 102-110.

[194] Lähdesmäki, I., Beeson, C., Christian, G.D. & Ruzicka, J. (2000). *Talanta* 51, 497-506.

[195] Ruedas-Rama, M.J., Ruiz-Medina, A. & Molina Diaz, A. (2004). *J. Pharm. Biomed. Anal.* 35, 1027-1034.

[196] Ruedas-Rama, M.J., Ruiz-Medina, A. & Molina Diaz, A. (2004). *Talanta* 62, 879-886.

[197] Gutzman, Y., Carroll, A.D. & Ruzicka, J. (2006). *Analyst* 131, 809-815.

[198] Ruzicka, J., Carroll, A.D. & Lähdesmäki, I.(2006). *Analyst* 131, 799-808.

[199] Šatínský, D., Solich, P., Chocholouš, P. & Karlíček, R. (2003). *Anal. Chim. Acta* 499, 205-214.

[200] Adcock, J.L., Francis, P.S., Agg, K.M., Marshall, G.D. & Barnett, N.W. (2007). *Anal. Chim. Acta* 600, 136–141.

[201] Satínský, D., Huclová, J., Solich, P. & Karlícek, R. (2003). *J. Chromatogr. A* 1015, 239-244.

[202] Satínský, D., Neto, I., Solich, P., Sklenákova, H., Conceição, M., Montenegro, B.S. & Araújo, A.N. (2004). *J Sep Sci.* 27, 529-536.

[203] Šatínský, D., Huclová, J., Ferreira, R.L.C., Montenegro, M.C.B.S.M. & Solich, P. (2006). *J. Pharm. Biomed. Anal.* 40, 287-293.

[204] Chocholouš, P., Solich, P. & Šatínský, D. (2007). *Anal. Chim. Acta* 600, 129-135.

[205] Cerdà, V., Estela, J.M., Forteza, R., Cladera, A., Becerra, E., Altimira, P. & Sitjar P. (1999). *Talanta* 50, 695-705.

[206] Albertús, F., Horstkotte, B., Cladera, A. & Cerdá, V.(1999). *Analyst* 124, 1373-1381

[207] Miró, M., Cerdà, V. &. Estela, I.M. (2002). *Trends Anal. Chem.* 21, 199-210.

[208] Miró, M., Estela, J.M. & Cerdà, V. (2005). *Anal. Chim. Acta* 541, 57-68.

[209] Albertús, F., Cladera, A. & Cerdà, V. (2000). *Analyst* 125, 2364-2371.

[210] Albertús, F., Cladera, A., Becerra, E. & Cerdà, V. (2001). *Analyst* 126, 903-910.

[211] Horstkotte, B., Elsholz, O. & Cerdá, V. (2005). *J. Flow Injection Anal.* 22, 99-109.

[212] Segundo, M.A. & Magalhães, L.M. (2006). *Anal. Sci.* 22, 3-8.

[213] Magalhães, L.M., Segundo, M.A., Reis, S., Lima, J.L.F.C., Estela, J.M. & Cerdà, V. (2007). *Anal. Chem.* 79, 3933-3939.

[214] Cerdà, V., Forteza, R. & Estela, J.M. (2007). *Anal. Chim. Acta* 600, 35-45.

[215] Fajardo, Y., Ferrer, L., Gómez, E., Garcías, F., Casas, M. & Cerdà, V. (2008). *Anal. Chem.*, 80, 195-202.

[216] Ruzicka, J. (2000). *Analyst* 125, 1053-1060.

[217] Scampavia, L.D. & Ruzicka, J. (2001). *Anal. Sci.* 17, i429-i430.

[218] Wang, J. & Hansen, E.H. (2003). *Trends Anal.Chem.* 22, 225-231.

[219] Wang, J. & Fang, Z.(2004). *Fenxi Huaxue* 32, 1401-1406.
[220] Wang, J. & Hansen, E. H. (2005). *Trends Anal. Chem*. 24, 1-8.
[221] Chen, X..W. & Wang J.H. (2007). *Anal. Chim. Acta* 602, 173-180.
[222] Luque de Castro, M.D., Ruiz-Jiménez, J. & Pérez-Serradilla, J.A. (2008). *Trends Anal. Chem*. 27, 118-126.
[223] Long, X., Miró, M., Hansen, E.H., Estela, J.M. & Cerdà, V. (2006). *Anal. Chem*. 78, 8290-8298.
[224] Ruzicka, J. (2000). *Analyst* 125, 1053–1060.
[225] Quintana, J.B., Miró, M., Estela, J.M. & Cerdà, V. (2006). *Anal. Chem*. 78, 2832-2840.
[226] Hansen, E.H. & Miró, M. (2008). *Appl. Spectrosc. Rev*. 43, 335-357.
[227] Wu, C.H., Scampavia, L., Ruzicka, J. (2002). *Analyst* 127, 898-905.
[228] Wu, C.H., Scampavia, L., Ruzicka, J. (2003). *Analyst* 128, 1123-1130.
[229] Jakmunee, J., Patimapornlert, L., Suteerapataranon, S., Lenghor, N. & Grudpan, K. (2005). *Talanta* 65, 789-793.
[230] Amorim, C.G., Araujo, A.N. & Montenegro, M.C.B.S.M. (2007). *Talanta* 72, 1255-1260.
[231] Ogata, Y., Scampavia, L., Carter, T.L., Fan, E. & Tureček, F. (2004). *Anal.Biochem*. 331, 161-168.
[232] Wang, Y., Chen, M.L.& Wang, J.H. (2007). *Appl. Spectrosc. Rev*. 42, 103-118.
[233] Wang, J. & Hansen, E.H. (2003). *Trends Anal. Chem*. 22, 836-846.
[234] Carroll, A.D., Scampavia, L., Luo, D., Lernmark, A. & Ruzicka, J. (2003). *Analyst* 128, 1157-1162.
[235] Chen, X., Wang, J. & Fang, Z. (2005). *Talanta* 67, 227-232.
[236] Chen, X., Wang, W. & Wang, J. (2005). *Analyst* 130, 1240-1244.
[337] Edwards, K.A. & Baeumner, A.J. (2006). *Anal. Chem*. 78, 1958-1966.
[238] Lähdesmäki, I., Park, Y.K., Carroll, A.D., Decuir, M. & Ruzicka, J. (2007). *Analyst* 132, 811-817.
[239] Chen, X.W., Xu, Z.R., Qu, B.Y., Wu, Y.F., Zhou, J., Zhang, H.D., Fang, J. & Wang, J.H. (2007). *Anal Bioanal Chem*. 388, 157-163.

Chapter 2

THE MARINE ENVIRONMENT: SAMPLES AND ANALYTES

ABSTRACT

The marine environment is of particular interest because is an important sink for many chemicals, some of which accumulate in the marine food chain. Heavy metals and bio accumulating toxic substances are introduced to the sea from land-based point and non-point sources, from atmospheric fallout and during marine transport of materials. Pollution of the marine environment is a major concern to countries having coastal and marine areas to the overall maintenance and control of the coastal ecosystem. This chapter presents an overview on the importance of monitoring chemicals (inorganic and organic) present at major or minor concentrations in samples from the marine environment (seawater/estuarine water, sediments, seaweeds and marine animals used as seafood).

INTRODUCTION

The marine environment has high intrinsic value as a global common good. Oceans and seas provide 99% of the available living space on the planet, cover 71% of the Earth's surface and contain 90% of the biosphere, and consequently a large share of global biological diversity on the planet. Marine ecosystems conceal a rich and as yet largely unknown biological and mineral potential. The marine environment is characterized by saline water. Ocean saline concentrations average 3.5 parts per thousand whereas estuary saline concentrations range between 5 and 25 parts per thousand creating a wide range of ecosystem environments [1].

The oceans have been estimated to produce more than 35% of the primary production of the planet and they play a critical role in energy and nutrient cycling, supplying minerals and other natural resources, energy, and habitat for sustaining living resources. In addition, provide a medium for recreation, learning and enlightenment and play a key role in climate and weather processes. Near shore ecosystems are supported by the ocean and the interrelationship between oceanic and land systems. This interrelationship can affect the profit and growth potential of many economic sectors, including natural resource harvesting, commercial and recreational fishing, manufacturing, tourism, and waste assimilation.

However, the marine environment is facing a number of threats including loss or degradation of its biodiversity and changes in its structure, loss of habitats, contamination by dangerous substances and nutrients and impacts of climate change [2-3].

Marine pollution is defined as the introduction of substances or energy into the marine environment, which leads to detrimental changes. Pollution can, therefore, include among other chemical substances (both inorganic and organic), heat and noise. This pollution can reach the sea via direct discharges, indirect discharges and/or accidental releases [4-5].

Direct discharges include releases from vessels, from ships, discharges of municipal and industrial wastewater via pipelines, and dumping of waste materials, such as dredged material, into ocean waters. Plastic and other marine litter, such litter causes direct harm to certain animals [6].

Indirect discharges of untreated sewage and agricultural fertilizers (nitrates and phosphates) from rivers and diffuse sources or the percolation of ground water, or deposited as dust or dissolved in rain from the atmosphere are the source of one to two-thirds of pollutants contributing to the degradation of coastal and marine waters. These and include sediments, nutrients, pathogens, and toxic compounds. The difficulty in controlling these pollutants is the diverse array of sources: runoff and seepage from agricultural and urban areas, and air deposition onto land and into water, and the multiple methods of transport. It has been estimated that indirect loadings account for more than half of the suspended solids, nutrients, fecal coliform, and metals entering coastal waters annually. Pollutants from agricultural and pasture lands include sediments, fertilizers, pesticides, herbicides, and animal wastes which contain bacteria and nutrients. Excessive nutrients can stimulate the growth of algae and other plants and organisms, which in turn deplete the levels of dissolved oxygen and harm aquatic life. Furthermore, too many nutrients can also trigger toxic algal blooms. Bacteria and pesticides from agricultural and pasture lands can kill aquatic life and contaminate seafood. Increases in concentrations of carbon dioxide in the atmosphere are inducing changing in the climate and so altering marine ecosystems. But they also threaten the viability of many marine organisms such as corals, mollusks, echinoderms and many others, by reducing their ability to lay down calcium carbonate to form their skeletons. An increase of noise levels, originated particularly by low frequency sounds, travels well through water so it is not only used by whales and some fishes for communication, but also by scientists, navies and fishermen. Heat can be produced by coastal power plants that use seawater for cooling and discharge the warmed water at the coast. This locally disturbs the ecological balance of the marine communities, especially if it is already a low oxygen environment (gases are less soluble in warm water). On global scales climate warming can have some counterintuitive results [7-8].

Accidental releases happen because large quantities of petroleum and other toxic substances are transported in the coastal and marine environment. Oil pollution is a highly emotive subject, especially following a major tanker accident occurs [9-10].

Chemical substances that are directly toxic to all or some organisms include heavy metals such as mercury, lead, zinc, cadmium and copper [11-14], PAHs (polycyclic aromatic hydrocarbons, that come from oil and industrial manufacturing processes) [15-16], PCBs (polychlorinated biphenyls that come from the manufacture of plastics) [17-19], POPs (persistent organic pollutants that are often manufactured and used as pesticides) [20-21] and endocrine disrupters that interfere with hormonal systems. The most notorious of the endocrine disrupters are TBTs (tributyl tin) [22-23]. The presence in high amount of pollutants in the marine environment represents a potential danger for human health and for environment due to their toxicity. Population can be contaminated with pollutants by ingestion of contaminated or polluted food and water. The gravity of toxic effect depends on nature, concentration, body resistance and presence of other contaminants. For this reason, accurate monitoring of their concentration plays an important role in analytical chemistry.

Flow injection (FI) techniques are currently regarded as attractive tools for the routine control of chemical parameters in samples of marine environmental concern because allow continuously record one or more quality parameters of marine samples. Glass apparatus such as beaker, flask, and pipettes usually required for chemical analysis was omitted because most analytical operations were done automatically into the FI manifold [24-28]. This is due to their advantages when compared with other analytical methods. This is particularly as a consequence of their straightforward configurations, easy operation, full automation, acceleration of wet chemical assays including on-line sample treatment, and considerable decrease in sample and reagent consumption and waste generation, with the consequent minimization of costs for disposal. The fast response inherent in FI techniques renders the analytical information available in real time with minimum error, thereby making these approaches suitable for implementation in monitoring schemes. In fact, when the user needs to analyze hundred or thousands of samples per day, or when dealing with smaller scale testing facilities, FI techniques emerge as a more economically viable choice for its low instrument and maintenance costs. Moreover, FI offers other benefits, such as an automated operation over an extended period of time, and a detailed analyte profile. In addition, FI systems are able to measure and/or preconcentrate compounds in-situ. This is an important environmental field application for real time water analysis of unstable substances. Real time analysis also eliminates the need for addition of "preservatives" to mitigate analyte concentration changes during shipment and laboratory storage, and avoids collect a discrete sample which must be transported to the laboratory to be analyzed. Consequently, the development of automated in-situ measuring systems has become a major challenge in the field of oceanographic instrumentation.

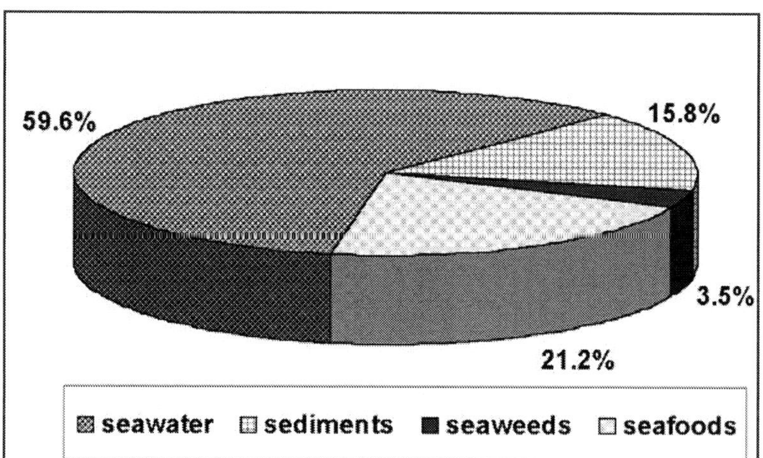

Figure 2.1. Distribution of FI determinations according to the marine sample analyzed.

Elemental analysis of toxic elements in major, minor, or even trace amounts is essential for solving several environmental pollution problems. In addition, it is important highlight relationships between numerous geochemical, biological and anthropogenic activities and ionic composition of marine matrices. For instance, although the concentration levels of major seawater components, such as chloride and sulfate or sodium and magnesium, remain relatively constant, a diversity of natural factors can be enumerated that cause these anionic and cationic constituents to be non-conservative. Those are effects of estuaries, evaporation in isolated basins, admixture with brines or interstitial waters, precipitation and dissolution, submarine volcanism, exchange between the atmosphere and sea, influence of anoxic basins, freezing, etc. Interconnections between biota or ammonium development in intertidal sediments, phosphorus cycle in sea surface sediments and the nutrient loads through benthic release, on the one hand, and aquaculture in a coastal sea or a river estuary, on the other, have also received a great deal of attention. Other related topics of particular relevance include the relationship between a phytoplankton growth rate and a red tide occurrence in a heavily eutrophic bay, eutrophication and concomitant issues in the sea and environmental assessment based on the measurement of nutrients before future industrialization. Last but not least, formation, disappearance and genotoxicity of disinfection byproducts following chlorination or ozonation and involving an array of (highly) toxic species of bromine, i.e., bromate, hypobromite, brominated trihalomethanes and others, are in the first instance related to the concentration of bromide in seawater and adjacent aqueous compartments.

Assays typically performed by FI methodologies include the analysis of the following marine samples: seawater (including estuarine water), sediments, marine animals used as seafood (fish and shellfishes) and seaweeds. The importance of the determination of analytes in these samples involve aspects of food and environmental analysis, which are treated within of this chapter. Thus, several inorganic and organic analytes, pollutants and nutrients are quantified in marine samples by using FI techniques. As can be seen in Figure 2.1, seawater samples are those more analyzed with a 59.6% of the FI determinations (information obtained by searching Chemical Abstracts between 1975 and September 2008 using Scifinder Scholar and the keywords "corresponding marine sample" and flow injection). By the other hand, it is significant that marine samples as marine seabirds, marine mammals or marine air are not still analyzed by FI methodologies.

SEAWATER AND ESTUARINE WATER

Seawater is water from a sea or ocean. It is a complex mixture of 96.5 percent water, 2.5 percent salts, and smaller amounts of other substances, including dissolved inorganic and organic materials, particulates, and a few atmospheric gases. The density of surface seawater ranges from about 1.020 to 1.029 g/mL, depending on the temperature and salinity. Seawater in equilibrium with atmospheric CO_2 is slightly alkaline, with a pH medium of around 8.0-8.4 and a salinity of about 3.5% [29-30]. Salinity and pH of estuarine water depend on the river that feeds the estuary and the ocean from which it derives its salinity. The degree of mineralization of inland water can be highly variable from place to place and depends upon the solubility of minerals, length of time and conditions of contact of water with minerals, and concentrations of substances through evaporation. Seawaters are more highly mineralized than inland waters, and estuarine waters are intermediate in mineralization [31].

Any ion, complex or salt that is smaller than 0.45 μm size can be considered as dissolved, and by the other hand, the particles in seawater with size greater than 0.45 μm can be considered as suspended particulates.

The chemical constituents of the dissolved fraction of seawater are classified into major, minor, and trace elements. Thus, the three most abundant ions of seawater are chloride (Cl^-), sodium (Na^+) and sulfate (SO_2^{4-}). Of the many minor dissolved chemical constituents, inorganic phosphorus and inorganic nitrogen are among the most notable, since they are important for the growth of organisms that inhabit the oceans and seas [32]. Seawater also contains various dissolved atmospheric gases, chiefly nitrogen, oxygen, argon, and carbon dioxide. Some other components of seawater are dissolved organic substances, such as carbohydrates and amino acids, and organic-rich particulates [33]. The composition of seawater is shown in Table 1. In open oceans, the total dissolved concentration of these constituents may vary as a function of seawater salinity.

Changes in precipitation, pH, water temperature, wind, dissolved CO_2, and salinity can affect water quality in marine waters. The quality of coastal water is of great importance to the overall maintenance and function of the coastal ecosystem. Water quality impacts both directly and indirectly on the diversity and abundance of marine communities as well as recreational use of the coast [34].

Usually, artificial seawater was used to optimize new analytical methodologies. Mostly of these artificial samples are prepared with a salinity of about 3.5%. Thus, between other, the following procedures are proposed:

- Artificial seawater obtained by dissolving 23.926 g NaCl, 4.008 g Na_2SO_4, 0.667 g KCl, 0.196 g $NaHCO_3$, 0.098 KBr, 0.026 g H_3BO_3, 0.003 NaF, 10.831 g $Mg_2Cl{\cdot}6H_2O$, 1.5188 g $Ca_2Cl{\cdot}2H_2O$ and 0. 0240 g $Sr_2Cl{\cdot}6H_2O$ in 1 L of ultrapure water [38].
- Artificial seawater (salinity 3.42%) obtained by dissolving 32 g NaCl, 14 g $MgSO_4{\cdot}7H_2O$ and 0.15 g $NaHCO_3$ in 1 L of ultrapure water [39].
- Artificial seawater (salinity 3.51%, with a final pH of 8.04) obtained by dissolving 234.71 g NaCl, 14.60 g Na_2SO_4, 1.94 g $NaHCO_3$, 106.54 g MgC12, 14.60 g CaC12 and 7.42 g KCl in 10.0 L of ultrapure water [40].
- Artificial seawater (salinity 3.5%) obtained by dissolving 41.5 g NaCl and 15 g $MgSO_4{\cdot}7H_2O$ in 1.5 L of ultrapure water [41].

In addition, is commercially available a synthetic seawater from Sigma-Aldrich [42], which also was used to perform optimization and to verify the correct applicability of a new analytical procedures developed [43].

In this same context, several certified reference materials (CRMs) with a seawater matrix have been used to check accuracy of a new analytical procedure. They are an essential part of the quality assurance necessary for the reliable analytical measurement of nutrients and pollutants in seawater. Thus, the CRMs more used and actually available are the following:

Table 1. Elemental composition of sea water

	Mean concentration (μg/L)		
Element	A[a]	B[b]	C[c]
Ag	0.04	-	-
Al	2	2	0.54
As	3.7	4	1.7
Au	0.004	0.004	-
B	4440	5000	4500
Ba	13	2	14
Bi	0.02	0.02	-
Br	67300	67000	67000
C	28000	142000 (as bicarbonate)	27600
Ca	412000	411000	412000
Cd	0.11	0.05	0.08
Cl	19.35×10^6	19.35×10^6	19.354×10^6
Co	0.02	0.05	-
Cr	0.3	0.3	-
Cu	0.25	0.5	0.25
F	1300	1300	1300
Fe	2	2	55
H	110×10^6	-	-
Hg	0.03	0.03	1
I	60	60	50
In	0.00011	-	-
K	399000	399000	399000
Mg	1.29×10^6	1.29×10^6	1.29×10^6
Mn	0.2	0.2	14
Mo	10	10	-
N	640 (as $NH_3 + NO_3^-$) 15500 (as N_2)	-	420000
Na	10.77×10^6	10.76×10^6	10.77×10^6
Ni	0.56	0.5	0.5
O	883×10^6	-	-
P	60	-	70
Pb	0.03	0.03	2
Ra	8.9×10^{-8}	1×10^{-7}	-
S	905000	2710000 (as sulfate)	904000
Sb	0.24	-	-
Se	0.2	0.2	-
Si	2200	-	2800
Sn	0.004	0.01	-
Ti	1	1	-
Tl	0.019	0.01	-
U	3.2	3.3	3.3
V	2.5	2	-
Zn	4.9	2	0.4

[a] Bowen, 1979 [35].
[b] Drever, 1982 [36].
[c] Summerhayes et al., 1996 [37].
- No information available.

LGC6016 (Estuarine water for trace metals) from LGC Standards, UK; NASS-5 (seawater reference material for trace metals), CASS-4 (nearshore seawater reference material for trace metals), SLEW-3 (estuarine water reference material for trace metals) and MOOS-1 (seawater certified reference material for nutrients) from the National Research Council of Canada [44-56]; IAEA-381 (Irish sea water for radioactive isotopes) from the International Atomic Energy Agency (IAEA), Austria; BCR-403 (seawater for trace elements), BCR-505 (estuarine water for trace elements), BCR-579 (coastal seawater for mercury) from the Institute for Reference Materials and Measurements, Belgium [57-61].

The need for continuous chemical determinations for monitoring oceanic waters is now widely recognized. The concentration levels of major elements (sodium, chloride, sulfate, magnesium) in seawater are almost constant all over the sea, while trace elements offer show some local distributions. These concentration variations of trace elements often reflect some natural physicochemical and/or biochemical changes in the sea (effects of estuaries, evaporation in isolated basins, admixture with brines or interstitial waters, precipitation and dissolution, submarine volcanism, exchange between the atmosphere and sea, influence of anoxic basins, freezing, etc.), even though their concentrations are extremely low. There is a growing worldwide demand for information on the concentration levels of seawater inorganic ions responsible for or related to such issues as maintaining a healthy environment, environmental conservation and a better understanding of sea chemistries. Therefore, the accurate determination of analytes in seawater is very important to elucidate their roles in the aquatic system. On the other hand, environmental protection is necessary to protect the marine waters against contamination so as to allow adequate conditions for aquatic life and the various uses derived from it. Thus, The European Community has established several Directives on the quality of continental and marine waters. For marine ecosystems these European Directives limit the levels of temperature, dissolved oxygen, pH, suspended solids, biochemical oxygen demand (BOD), total phosphorous, nitrites, phenolic compounds, petroleum hydrocarbons, non-ionized ammonia, total ammonium, total residual chlorine, total zinc and dissolved copper in seawater to support fish life [62], and temperature, pH, suspended solids, coloration, salinity, dissolved oxygen, petroleum hydrocarbons, organohalogenated substances, metals (Ag, As, Cd, Cr, Cu, Hg, Ni, Pb, Zn), faecal coliforms, substances affecting the taste of the shellfish and saxitoxin (produced by dinoflagelates) in shellfish waters [63]. Also, the United States Environmental Protection Agency (EPA) developed water quality standards for marine waters [64-65]. So, pH, temperature, dissolved oxygen, toxic substances, color producing substances, odor producing substances, or other deleterious substances attributable to savage, industrial wastes or other wastes, bacteria, radioactivity and turbidity have limited theirs levels for swimming and other whole body water-contact sports, shellfish harvesting, fish and wildlife, etc.

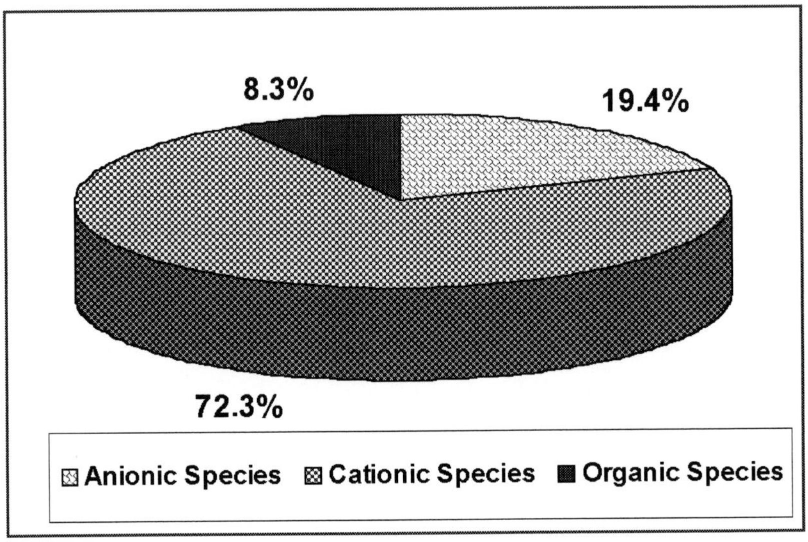

Figure 2.2. Distribution of FI determinations according to the analyte analyzed in seawater and estuarine water samples.

Taking into account all the analytes that have been determined by FI methodologies in seawater and estuarine water samples, the distribution of these FI determinations is shown in Figure 2.2. This Figure shows that the great majority of the determinations were proposed for cationic species (72.3%), while organic species are the analytes less determined by FI methodologies in these samples (8.3%) (information obtained by searching Chemical Abstracts between 1975 and September 2008 using Scifinder Scholar and the keywords "corresponding species", seawater or estuarine water and flow injection). In view of these results, it is clear that FI analytical determinations in saline waters were developed practically only for inorganic analytes.

Cationic Species Determined by FI Methodologies

Over 300 papers have been published involving the determination of cationic species in seawater and estuarine water by FI methodologies. Thus, the cations that have been determined by using FI methodologies have been the following: alkali metals (Na^+), alkaline earth metals (Ba^{2+}, Ca^{2+} and Mg^{2+}), Ag(I), Al(III), As(III)/As(V), Au(I)/Au(III), B(III), Bi(III), Cd(II), Co(II), Cr(III)/Cr(VI), Cu(II), Fe(II)/Fe(III), Hg(I)/Hg(II), In(III), Mn(II), Mo(VI), NH_4^+, Ni(II), Pb(II), rare earths (Eu, Ho, Lu, Pu, Tb, Tm and U), Rh(III), Sb(III)/Sb(V), Se(IV)/Se(VI), Sn(II), Ti(III), Tl(I)/Tl(III), V(IV)/V(V) and Zn(II). Trace elements as As, Se, Cr, etc. can be found as anions in seawater, and Sn and Hg can occur as organotin compounds and organomercurials, respectively, but they have been included here because most of the analytical procedures exhibit speciation. As can be observed in Figure 2.3, the cationic species for those which have been proposed more new FI methodologies have been Pb(II), Cu(II) and Cd(II) with a percentage of publication of 11.7, 11.1 and 10.0%, respectively (information obtained by searching Chemical Abstracts between 1975 and

September 2008 using Scifinder Scholar and the keywords "corresponding cationic species", seawater or estuarine water and flow injection).

Details and characteristics about the cationic species determined by FI methodologies in saline waters are discussed below.

Alkali Metals

The only alkali metal determined by a FI methodology in saline waters was Na(I). Analysis of Na(I) is a common procedure within the frame work of determining the chemical composition of seawater [66-67].

Alkaline Earth Metals

The alkaline earth metals determined by a FI methodology in saline waters were Ba^{2+}, Ca^{2+} and Mg^{2+}.

Barium is of oceanographic interest since it is a nonconservative stable trace element. In spite of the relatively short 11000 year oceanic residence time for Ba, ocean biology largely determines its distributions in the ocean interior. The distribution of barium in seawater is of great significance to the study of the marine environment. The concentration of barium in the oceans has been proposed as a potential indicator of marine drilling contamination, as an indicator of biologically productive oceanic areas, and as a chemical analogue of radium, which may be useful for the study of deep water circulation processes. The barium concentration in seawater is regulated by the sulfate concentration, based on the solubility of $BaSO_4$. Dissolved concentrations in the major oceans range between 5.6-28 μg/L and profiles show lowest concentrations near the surface and enrichment at depth [68-69]. The determination of calcium and magnesium concentrations in seawater allows calculate total water hardness [70-71]. The constancy in chemical composition of seawater is the result of longer residence time of the major ions in the ocean than 10^7 years and the mixing of seawater. However, among the major ions in seawater, the residence time of calcium is the smallest (8.0×10^5 years) and calcium is the most biophile element. Therefore, it is expected that the vertical profile of calcium distribution in the ocean bears some resemblance to those of nutrients such as nitrate, sulfate and silicate [72]. Calcium in seawater is involved in calcification processes. Chemically, mineralization of the shell can be considered as the formation of calcium carbonate and hydrogen ions from calcium and bicarbonate ions in the cuticle as follows: $Ca^{+2} + HCO_3^- \leftrightarrow CaCO_3\downarrow + H^+$. Although this reaction is reversible, the high pH value of the fluid of the exoskeleton (pH=8.2) favors the deposition of calcium carbonate [73].

Silver

Silver is a metal of commercial importance that has been recognized as a toxic element owing mainly to its abundance in marine environments. It is valued for its resistance to corrosion and for its use in alloys, medicine and jewelry. Because of its marked antibacterial properties, its compounds and alloys have been widely used to disinfect the water used for drinking and recreational purpose, in dental and pharmaceutical preparation, and in implanted prosthesis [74]. It is also used in electronic devices, photographic material, mirrors, and cloud seeding. Monitoring of dissolved silver concentrations in estuarine and coastal waters is therefore of great importance for water quality management. Anthropogenic sources of silver to these waters include sewage discharges, run off from operating and disused mines, and release from contaminated sediments [75]. Dissolved silver is taken up by marine organisms, and in oceanic environments it shows a recycled behavior, parallel to silicate. Oceanic dissolved-silver concentrations are considerably lower than coastal concentrations, increasing from 0.3 pmol/L in surface waters to 22.8 pmol/L in deep oceanic waters. Open ocean depth profiles for dissolved silver show recycling, with depleted surface-water concentrations and enrichment with depth. The geochemistry of silver determines its low dissolved concentrations and toxic potential. Silver sorption onto particles is strong, resulting in enhanced removal in estuarine systems. Furthermore, the toxicity of silver depends on not only its total concentration, but also its speciation. Dissolved-silver speciation is expected to change markedly in estuaries, owing to changes in chloride concentration. Formation of the neutral chlorocomplex, $AgCl^0$, may increase bioaccumulation of silver [76-78]. The low polarity of this complex will increase the diffusion of silver across biological membranes, so dissolved-silver measurements in near-shore waters and estuaries with intensive human activities should be undertaken in order to assess the effects of this element on the functioning of marine ecosystems.

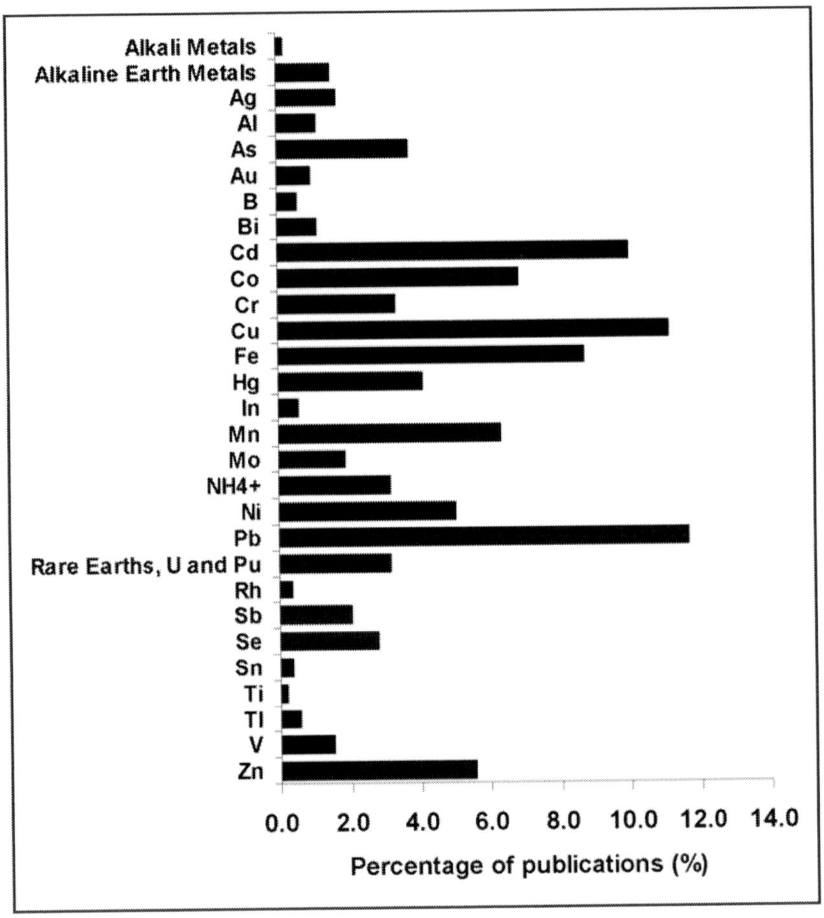

Figure 2.3. Distribution of FI determinations of cationic species in seawater and estuarine water.

Aluminium

Aluminun is an element not an essential element in humans, and it has been linked with a number of disorders in man including Alzheimer's disease, Parkinson's dementia and osteomalacia. Aluminium occurs ubiquitously in the environment, in the form of salts and oxides [79]. Aluminium exhibits a complex chemistry in natural waters. Solution pH and the presence of fluoride, silicon, iron and natural chelating agents in the water may all influence the speciation of aluminium and consequently its bioavailability. Trivalent Al^{3+} and the hydrolyzed ions $Al(OH)^{2+}$ and $Al(OH)_2{}^{+}$ represent the species of principal concern, with organically bound Al and polymeric forms considered far less toxic. The determination of aluminum is an important task to study geochemical cycles in the ocean. Aluminum concentrations are extremely low in open seawater (< 0.1 nmol/Kg), although aluminum is the most abundant metallic element in the Earth's crust (8.13%). Aluminum shows unique vertical profiles that are probably controlled by dissolution from atmospherically derived dust particles and bottom sediments balanced against scavenging by particulate matter. Vertical profiles obtained from the world's oceans have shown interocean variability, which is the largest found thus far. Aluminum could be dissolved abundantly by water–rock interaction in these systems. Many natural organic compounds, e.g. humic and fulvic acids can release aluminium from soils and sediments [80].

Arsenic

The occurrence of arsenic in natural waters has received much attention during recent years, because of its potential toxicity on human health. Arsenic is widely present in the marine environment both as a consequence of natural sources mainly related to the process of rocks erosion, and anthropogenic sources including smelters, use of arsenical pesticides, fertilizers and wood preservatives agents. In seawater, dissolved arsenic is present at low levels (around 1 μg/L), being distributed among arsenite (As(III)), arsenate (As(V)), monomethylarsenate (MMA), dimethylarsenate (DMA) and unknown organic forms. The toxicity of different arsenic species varies in order As (III) > As (V) > MMA > DMA. For this cause, speciation of arsenic in environmental samples is gaining increasing importance in the last years [81]. Many studies provide evidence that As(V) is the most abundant species in oxygenated seawater, As(III) rarely accounts for more than 20% of total arsenic, whereas the concentration of organic arsenic in surface seawater is lower than 0.07 nM. However arsenic speciation in seawaters depends on many factors as climatic variations, phytoplankton and bacteria population levels of nitrogen and degree of pollution [82]. Particularly, it was noticed that in sewage contaminated coastal waters the 83% of total dissolved inorganic arsenic was in the reduced form. Marine phytoplankton readily assimilates As(V) and incorporates some of it into the cell. Most of the As(V) is reduced, methylated and released to solution. It was calculated that 15-20% of the total dissolved As in seawater is reduced by phytoplankton during blooms on the continental shelf [83].

Gold

Concentrations of Au in Atlantic and Pacific seawater samples are on the order of 50 fM. Mediterranean deep waters contain higher concentrations of Au (100–150 fM). This is attributed to proximal aeolian dust and/or riverine sources [84]. Gold is predicted to be present in oxic seawater as Au(I) species as the neutral species [$Au(OH)(H_2O)$], but if thermodynamic equilibrium is not attained, it may also be present in other soluble forms including complexes of Au(III) and reduced elemental colloids. Under anoxic conditions, in the presence of sulfide, $[Au(HS)_2]^-$ and the metallic form are predicted to be the major species. Progress in understanding the behavior of Au in the weathering cycle and in the oceans will require further mapping of its distributions in natural waters combined with an assessment of its redox behavior [85].

Boron

The toxic effect of boron to humans is not confirmed yet. However, some harmful effects have been noted. They depend on dose applied and the time and frequency of exposure. Health of an investigated being and some other environmental factors should be considered also. Taking into account the results on the insignificant relationship between tumor and boron exposure US-EPA classifies boric acid to the group E of carcinogenic species [86]. In nature, boron appears mostly as boric acid and borax, $Na_2B_4O_7.10H_2O$. In aquatic system it exists primarily as undissociated boric acid and as borate ions. Boron was described as a significant constituent of seawater, with an average concentration of 4.5 mg/L. Dissolved boron in the ocean contributes to the alkalinity balance and to the buffering properties of seawater. Although borate ($B(OH)_4^-$) represents only about 5% of the total alkalinity at typical seawater pH of 8.2, boron compounds have to be taken into account in quantitative calculations of carbonate system parameters. The boric acid-borate equilibrium also affects physical properties of seawater. For example, sound absorption in the ocean in the 1 kHz frequency range involves chemical equilibrium with relaxation rates, which have been identified as the boric acid-borate relaxation rates. Recently, the stable isotopes of boron have received attention as they provide a means of reconstructing the pH of the paleocean [87].

Bismuth

Bismuth, which naturally occurs as bismuthinite (Bi_2S_3), bismite (Bi_2O_3) or bismutite [$(BiO)_2CO_3$] is one of the micro constituents of the earth's crust. Its amount is comparable to that of antimony and cadmium. It has been widely utilized in medicines medicine due to its antacid action and mildly astringent action in gastrointestinal disorders and cosmetics and it is also used as an additive to aluminum alloys in order to improve their mechanical properties, in semiconductors. Although the absorption of Bi(III) in the human organism is generally low, several cases of nephrotoxic, neurotoxic, and kidney damage symptoms attributable to the use of Bi(III)-containing pharmaceutical formulations have been reported. Bismuth is present in seawater at concentration of about 10^{-10} M (ca. 20 ng/L). Bi is present in seawater dominantly as $Bi(OH)_3^0$ [88-89].

Cadmium

Cadmium is known to be a hazardous environmental pollutant with toxic effects for the living organisms in the marine environment. Adverse effects of cadmium are produced not only because of its high toxicity even at trace concentrations, but also due to bioaccumulation processes along the food chain. As water plays an important role in nature as carrier, among others, the anthropogenic inputs of cadmium and industrial pollution can be easily widespread over the marine aquatic medium. Therefore, cadmium can stay accessible to living organisms, including microorganisms and microalgae which are at the first steps of the food chain and involved in important biological processes and so, information about Cd concentration in the marine environment is very important in evaluation of its pollution level. Cadmium concentrations increase from 0.01 μg/L at the surface to 0.1 μg/L below 1 Km [90]. A strong association of the vertical distribution of cadmium concentration with that of inorganic nutrients, mainly phosphate and nitrate, has been recognized for two decades, suggesting that the oceanic biogeochemistry of this metal is controlled by the organic matter cycling processes [91]. Computer modeling results for cadmium speciation in seawater show that cadmium forms relatively strong chloro complexes and was calculated that in seawater no cadmium would be adsorbed on inorganic colloids. However, colloidal organic matter as humus or plankton can adsorb cadmium and alter the speciation dramatically [92].

Cobalt

Cobalt is an essential element in biological compounds like vitamin B12 and some metalloproteins. However, fewer studies have looked at cobalt–enzyme interactions in the marine environment. Cobalt is known to substitute for zinc in the enzyme carbonic anhydrase when marine phytoplankton is cultured under zinc-limiting conditions. Cobalt is an interesting element in seawater from the geochemical viewpoint because the hydrothermal activity can be recognized chemically by monitoring anomalous concentrations of cobalt in deep seawater. Generally, the oceanic cobalt concentration is extremely low, in open-ocean waters, cobalt has a relatively unique profile. Concentrations are generally low in surface waters 10-40 pM, which increase to a maximum in the upper thermocline 30-100 pM, and then decrease to values in deep waters of 10–30 pM. Because the low concentration of this metal in seawater, points to the possible role of cobalt as a biolimiting nutrient. The discharge of various cobalt radionuclides from nuclear installations to coastal waters and their accumulation by marine organisms has also increased the interest in the fate of this element [93]. Calculations for seawater suggest that 56.25% of total dissolved cobalt(II) occurs as the free ion, Co^{2+}, 39.53% as $CoCl^{+}$, 2.76% as $CoCl_2^{0}$, 1.02% as $CoSO_4^{0}$, and less than 1.0% as $CoNH_3^{2+}$, $CoCO_3^{0}$, $CoHCO_3^{+}$, $CoCl_3^{-}$, CoF^{+} and $CoBr^{+}$. Cobalt concentrations in open-ocean surface waters were low with cobalt speciation being dominated by its complexation to natural organic ligands with concentrations that tend to be higher than that of cobalt [94-96]. The available data on cobalt distribution in seawater show surface minima, a maximum within the upper thermocline as a result of atmospheric input, and depletion at depth due to its removal from seawater, probably in association with MnO_2.

Chromium

Chromium in environmental water and seawater occurs in two thermodynamically stable oxidation states, Cr(III) and Cr(VI). The degree of toxicity depends on the chemical form. Chromium (VI) is known to be toxic and carcinogenic and Cr(III) may be considered as essential for mammals and included into the glucose, lipid and protein metabolism, rendering Cr speciation analysis important. Chromium has entered the environment via many industrial applications, including galvanization, steel, paint and pigment production, and leather industries. Thus, the resulting anthropogenic contamination of chromium is observed in the coastal sediments and seawater [97].

Thermodynamic calculations predict that in seawater at pH 8.1, chromium should exist almost exclusively as Cr(VI). At this pH hexavalent chromium should be mainly as the chromate ion CrO_4^{2-}, with minor amounts as $HCrO_4^{-}$, H_2CrO_4, and $Cr_2O_7^{2-}$, while Cr(III) should be present to 85% as the aquahydroxy species $Cr(OH)_2^+$ $(H_2O)_4$, 13.5% as CrO_2^-, and $Cr(OH)^{2+}$ $(H_2O)_5$ as a minor species. If equilibrium exists between Cr(III) and (VI), it is possible to calculate from stability constant data their relative proportions in natural water. Elderfield [98] gives a theoretical expression for seawater: log(Cr(VI)/Cr(III)) = 6pH + 3pE - 65. Taking values for seawater of pH = 8.1 and pE (redox potential) = 12.1, this equation indicates that hexavalent chromium is the thermodynamically favored form. Consequently, it would be expected that all chromium in natural waters should be present as chromate. However, experimental measurements do not support this. This can be motivated by the following:

(a) species other than Cr^{3+} and CrO_4^{2-} play a more important part in determining speciation than assumed;
(b) the rate of exchange between forms is so slow that the observed ratio of Cr(VI) to Cr(III) does not reflect equilibrium conditions;
(c) the analytical data are of suspect accuracy;
(d) biological systems are responsible for maintaining concentrations of Cr(III).

All four explanations are likely to be true to some extent. Complexation by natural organic matter is sure to play a part in the speciation of Cr(III). This in turn might account for a residual fraction of Cr(III), which does not tend to be converted to Cr(VI). Complexation might also retard oxidation or create a locally reducing environment, which would stabilize Cr(III) [99]. Natural levels for total chromium in open seawater appear to lie in the range 0.1-0.5 μg/L, with the Cr(III) content being 0.001-0.36 μg/L. The relative proportions of each valency state appear to be controlled by in situ redox processes [92].

Copper

Copper is a heavy metal employed extensively in chemical industries and domestic activities. At low levels of concentration, it is considered to be an essential micronutrient for the normal metabolism of some living organisms. However, at higher concentrations, becomes toxic because it can bind to the cell membrane and alters the transport process through the cell wall. Increases in copper concentration in estuarine and coastal areas have resulted from industrial and domestic waste discharge, disposal of mining washings, refineries, and the use of copper as a base compound for antifouling paints [100]. Dissolved Cu(II) may exist in seawater as the free hydrated Cu^{2+} ion, complexed with inorganic ligands as $Cu(CO_3)^0$, or chelated with organic ligands as CuL. Thus, it was proven that more than 80% (generally 99%) of dissolved copper in surface waters is organically complexed and a high percentage combined with inorganic colloidal particles. Of particular importance in estuarine and coastal seawater is the extent to which copper, at elevated and potentially toxic concentrations, exists as relatively inert, hydrophilic chelates formed with natural or anthropogenic Cu-binding organic ligands. Studies indicate that the overwhelmingly dominant form of dissolved copper in many coastal regimes is as such Cu-organic chelates. Examples include chelates formed with anthropogenic ligands such as EDTA, or with strong biogenic Cu(II) chelators produced by marine cyanobacteria in response to copper stress. These hydrophilic Cu-chelates, such as $Cu(EDTA)^{2-}$ or CuL, are chemical species thought not to be directly available for uptake by microorganisms and, thus, are considered to be biologically inert forms of copper. In coastal waters there is most likely a continuum of ligands that appear to be relatively hydrophilic and cover a range of molecular weights with a large fraction being less than 1000 Da nominal molecular weight. It was demonstrated that Cu-binding ligands can be released by cyanobacteria into seawater and, when the sequestering capacity of these ligands is exceeded, Cu can exert acute toxic effects on cyanobacteria above a free copper ion level of 10^{-11} M. In coastal areas, high concentrations of Cu-binding ligands generally reduce metal toxicity [101-102].

Iron

Despite its enhanced abundance in the earth's crust (5.6%), iron is present in very low iron concentrations in the oceans (0.05–2 nM). This metallic element plays an important role in plant metabolism where it is essential for photosynthetic and respiratory electron transport, nitrate reduction, chlorophyll synthesis and detoxification of reactive oxygen species [103]. The marine biogeochemistry of iron has significant implications for regulating carbon dioxide flux at the air/sea interface and thus climate change. Due to a combination of the low solubility of iron in oxic seawater and the comparatively high iron demand by phytoplankton, iron concentrations in oceanic surface waters can be extremely low (< 1 nM). As a consequence, it is easy to contaminate samples during both the sampling and the analytical processes. Coastal waters typically contain higher dissolved iron concentrations than open ocean waters because of the proximity to terrestrial, fluvial and continental shelf sources. In seawater, iron exists in different physico-chemical forms and physical fractionation is traditionally performed using membrane filtration techniques (0.2–0.45 μm, cellulose acetate or polycarbonate) to differentiate between dissolved and particulate iron fractions. The proportion of iron in the dissolved phase in the ocean varies strongly between regions. In coastal waters, iron predominantly occurs in the particulate form (e.g. 0.05–10 μM in the North Sea) with lower dissolved concentrations between 5 and 400 nM. In contrast, in the oligotrophic region of the central North Pacific, dissolved iron concentrations (0.02–0.4 nM) exceed the particulate iron fraction in the majority of the water column. Further size fractionation has been performed to quantify colloidal iron [104]. Furthermore, iron occurs in seawater in two oxidation states: iron(II) and iron(III). These oxidation states are involved in the formation of soluble inorganic and organic complexes, colloids and particulate phases. The iron(III) oxidation state predominates in oxygenated waters and is highly insoluble through the formation of oxyhydroxides. Iron(II) is thermodynamically unstable in oxygenated seawater and is rapidly oxidized to iron(III). Potential sources of iron(II) are photoreduction of iron(III) in surface waters, atmospheric deposition and diffusion from sediments. Only a small fraction of dissolved iron(III) occurs in a free hydrated (Fe^{3+}) or inorganically complexed form, and 80–99% is strongly complexed by organic ligands, possibly produced by iron limited phytoplankton or bacteria. This organic complexation prevents iron(III) from forming insoluble oxyhydroxides, thereby maintaining enhanced dissolved iron concentrations in seawater. Thermodynamic speciation calculations indicate that a major fraction (76%) of the total iron(II) exists in a free hydrated form (Fe^{2+}), with the remaining fraction as $FeCO_3^0$. However, other studies have indicated the possibility of complexation of Fe(II) by organic ligands, thereby maintaining Fe(II) concentrations in seawater by decreasing the rate of oxidation to Fe(III). The concentration of free hydrated Fe^{3+} is thought to be too low to satisfy the iron demand for primary productivity. The majority of the dissolved iron(III) (<99%) in seawater resides in organic complexes. It was demonstrated that iron(III) complexed by organic ligands (siderophores and porphyrins) is available for uptake by different types of phytoplankton. Furthermore, solubilization and uptake of colloidal and particulate iron has been suggested to take place by protozoan and zooplankton grazers [105].

Mercury

Although mercury is one of the most toxic metal ions, methylated forms of mercury are even more toxic. Therefore, the study of the mercury cycle is important because of the toxicity of methylmercury, its accumulation in biota, and its biomagnification in the aquatic food chain. The link between industrial emissions and mercury levels in the oceans is less clear. On the one hand, significant quantities of reactive inorganic mercury are deposited in the oceans and methylmercury is found in different types and sizes of marine fish, sometimes at very high concentrations [106]. Although the levels of methylmercury in the surface layer of the open oceans are low it is present in ocean waters at picomolar concentrations, the results suggest that methylmercury is formed in the oceans. Inorganic mercury (Hg(II)) is a relatively reactive species in the environment. It occurs primarily as inorganic and organic complexes and the concentration of the free metal ion (Hg^{2+}) is exceedingly small, especially in seawater systems. Thus, it is the reactivity of the inorganic and organic complexes that defines the degree to which Hg is converted between its various oxidation states, and forms. Mostly, inorganic Hg is present as Hg(II), but reduction to elemental Hg (Hg^0), which exists in solution as a dissolved gas, appears to readily occur in most natural waters. Complexes in the intermediate oxidation state (Hg(I)) are not stable, nor found, in unpolluted surface waters. In addition to the redox transitions, Hg(II) can be taken up by microorganisms (plankton and bacteria), some of which methylate the Hg(II) complexes, forming methylmercury, in which the oxidation state of Hg is still Hg(II). Thus, the coastal zone plays an integral role in the global mercury cycle not only as a sink for terrestrially-derived Hg but also as a potential source of methylated Hg to the ocean. Current estimates suggest that about 50–80% of the total Hg input via rivers, and from direct deposition to estuaries, is trapped within the estuarine zone. Terrestrial input contributes < 10% of the total Hg input to the ocean. However, it is likely a larger fraction of the methylmercury because the relative concentrations of this in riverine and coastal surface waters (1–5% of the total Hg) are greater than in ocean waters and open ocean rain (1% or less) [107-108].

Indium

In the last years, the demand for indium is increasing mainly due to its use in the field of electronics industry which has been caused by being utilized as a material of liquid crystal display, light emitting diode, solar cell and lead-free solder, etc. Other industrial uses of indium include plating, soldering, manufacturing semiconductors, etc. Indium is toxic and causes the sickness and vomiting in the case of oral intake. Thus, its determination of in sea water is important since it is accumulated in marine creatures and is finally taken into human beings. Although its crustal abundance is small (~ 0.1 ppm), indium is widely distributed in nature but never occurs uncombined. It tends to be enriched in zinc blende making this mineral the main source of indium at present [109]. However, the short ocean residence time also leads to very low concentration of indium in seawater.

Manganese

Manganese is an essential micronutrient for all organisms, but is also toxic at high concentrations affecting the central nervous system with symptoms resembling those of Parkinson's disease. Under typical ocean conditions, manganese in seawater can be found in three different oxidation states, as soluble Mn(II), Mn(III) and Mn(IV). Manganese is known to be a highly reactive trace element and plays an important role in the geochemical cycles of trace metals in the ocean. In oxic environments, dissolved Mn(II) is easily oxidized and precipitated as manganese oxide under the alkaline ocean conditions, which is a powerful scavenger for trace elements. The manganese oxide in the sediments is reduced to Mn(II) and is regenerated in the water column under mild reducing conditions [110]. The concentrations of dissolved Mn in seawater have been documented to range from as low as 0.08nM to as high as 10nM, depending on whether the measurements were made in coastal or open ocean. A large amount of manganese is also injected into deep water by hydrothermal vents in the active oceanic crust. An O_2 evolving enzyme, containing manganese, is important for oxidation of water to O_2 during photosynthesis. Unlike iron and zinc, the vertical profiles of manganese in the open ocean generally show the scavenging type, indicating that manganese is not depleted for phytoplankton growth [111].

Molybdenum

Molybdenum is one of the biologically essential microelements for all classes of organisms, because of its important role in enzymatic processes, and stimulating the synthesis of proteins and nucleic acids. Apart from its biochemical functions, molybdenum is widely used in the steel industry in alloys, pigments, lubricant, and chemical catalysis, which can increase the level of molybdenum released to environment. However, excess molybdenum may induce physiological effects in human beings and particularly animals. High doses of soluble molybdates cause anorexia, ataxia and anemia in animals. High concentrations of ammonium dimolybdate were also found to be toxic to fish [112]. Molybdenum and iron are two trace elements both required for nitrogen fixation in seawater. The predominant form of molybdenum occurrence in seawater is oxyanion MoO_4^{3-} (at the typical seawater pH) and the over-all average concentration of dissolved molybdenum in seawater is indicated about 9-11 μg/L.

Ammonium

Ammonia (NH_3) is an important nitrogen species in the natural environment. As a dominant gaseous base in the air, it plays a very important role on the acid-base chemistry of the atmosphere and greatly influences the atmospheric sulfur cycle in the remote marine boundary. Being a gaseous compound, ammonia exchanges at the air-sea interface although its flux is not well quantified in a variety of environments. Ammonia can easily dissolve in water and become ammonium ion (NH_4^+). In the ocean, ammonium is the dominant form, with ammonia as a minor component. Ammonium is also one of the most commonly used nutrients by marine phytoplankton. Compared to nitrate, phytoplankton generally prefers ammonium because additional energy is required for them to reduce nitrate to ammonium [113]. However, high concentrations of ammonia are toxic to marine organisms such as fish, shrimp, abalone, and sea urchin, especially larvae or juveniles of these species.

Nickel

Nickel is an essential metal to plants and some animals, being a component of the enzyme unease and of five other important enzymes. It is moderately toxic element but can cause allergic reactions and certain nickel compounds are carcinogenic. This metal is released into the aquatic environment from dissolution of rocks and soil, atmospheric fallout and biological cycles. Also, the high consumption of nickel products in industrial activities and sewage waste-waters inevitably leads to environmental pollution. In seawater nickel is dominant as free ion Ni^{2+} and its concentration ranged between 0.2-0.7 μg/L [114]. Values appeared to increase significantly with depth for the first 200 m, thereafter reaching a constant value.

Lead

Lead is toxic towards aquatic life because is a cumulative poison, and even trace levels may cause long-term health risks. Lead produces structural alterations in chromosomes and binds strongly to mitochondrial membranes. Some sources including antiknock reagent used in gasoline, smelting operation, and combustion of coal have contributed greatly to lead pollution [115]. Anthropogenic lead is considered to be one of the most severe contaminants for the environments. Since anthropogenic Pb is easily transported via atmosphere, sea surface water is polluted by lead mainly through precipitation and deposition of aerosols. The lead concentrations, which can be found in non-polluted seawater, are very low (0.5-30 ng/L). In seawater, $PbCO_3$ (83%) and lead chloro complexes, $PbCl_2$ (11%) are computed to be the predominant inorganic complexes and it is reported that a 40-80% of the lead is associated with inorganic colloids. There is evidence that tetraalkyl lead compounds can be formed in aquatic sediments. These organolead compounds are considerably more toxic than inorganic lead [116-117].

Rare Earths, Uranium and Plutonium

Rare earth elements (REEs) have been widely used as micro-additives in functional materials, catalysts, medicines, diagnosis reagents of magnetic resonance imaging and cosmetics as well as fertilizers, resulting in a potential risk to the environment. Continuous exposure to low concentrations of REEs could cause adverse health effects because of their bioaccumulation along the food chain [118]. REEs (Eu, Tb, Ho, Tm, Lu, etc.) form an extremely coherent group in terms of chemical behavior and they have been investigated intensively as marine-geochemical tracers of chemical processes for characterizing water masses and ocean circulation and fingerprints of source materials. This can be explained by their consistent behavior and small but systematic variation of particulate affinities among elements. In the oceans, the REEs show "nutrient-like" profiles in that their concentrations increase with depth. In addition, due to suitable residence time and sensitive affability to scavenging, REEs have been demonstrated to respond quickly to variations of vertical particulate flux [119]. Therefore, information concerning REEs' concentration and their relative distribution in seawater is very important for environment science and marine geochemistry.

Uranium is a naturally occurring radioactive element, which is important for nuclear technology. Seawater is a major source of uranium. The total estimated quantity of uranium in seawater is around four and a half billion tones. Thus, the oceans have the potential to become the most eco-friendly and long sustainable resource for uranium. The chemical forms of uranium in an artificial seawater are determined by use of the ion-association model. The data on the stoichiometric association constants of the ion pairs are used in calculation to evaluate the distribution of the major ionic species. Also, the more accurate values of the thermodynamic stability constants are used for the uranyl complexes. The calculation indicates that a dominant species of uranium is the tri-carbonate complex $UO_2(CO_3)_3^{4-}$ in artificial seawater at 25°C [120].

Plutonium is a highly toxic element, which is released into the marine environment as a result of atmospheric nuclear weapons testing and discharges from the nuclear fuel facilities. The chemical form of plutonium in the marine environment is very complicated because plutonium shows four oxidation states (III, IV, V, and VI) and complexation with inorganic and organic ligands in oxic waters. The chemical speciation of plutonium is still a significant topic in marine radioecology, although its achievement is very difficult because $^{239,240}Pu$ concentrations in seawater are extremely low (less than 10^{-17} mol/L). in a computer model was predicted that Pu (IV) would be hydrolyzed and adsorbed by sediments and particulate matter. It is considered that plutonium in surface waters is removed by particle scavenging processes [121].

Rhodium

Rhodium is one of the least abundant of the Pt-group elements in the Earth's crust, It is present at about 0.001 μg/mL in the crust of the earth. Rhodium metal is known for its stability in corrosive environments, physical beauty and unique physical and chemical properties. Rhodium occurs naturally with other platinum metals only at very low concentration. The main sources of rhodium pollution are catalytic converters containing rhodium. Since then, approximately 73% of the world production of rhodium is consumed in the production of autocatalysts [122].

Antimony

Dissolved antimony is present in seawater at very low concentration levels, around 0.2 μg/L. Depending on biological activity, dissolved antimony can be distributed between different chemical forms: mainly inorganic Sb(III) and Sb(V) forms and eventually methylated antimony forms representing less than 10% of total antimony in the open ocean. The toxicity and biological behavior of antimony depends on the oxidation state, the presence of binding partners and potential ligands, and the solubility of the compounds. Sb(III) shows a high affinity for red blood cells and sulfhydryl groups of cell, while erythrocytes are almost impermeable to Sb(V) [123]. Therefore, the determination of total antimony is not sufficient to evaluate its toxic effects in seawater, indeed, inorganic Sb(III) is 10 times more toxic than inorganic Sb(V) or Sb organic compounds.

Selenium

Selenium is a ubiquitous trace element that exists in multiple chemical forms in seawater in different oxidation states and organic metalloid compounds. The metalloid selenium in natural concentration is essential for the growth of various photosynthetic species and for the vital functions of marine organisms, but it would be toxic to fish and birds in abnormal high concentrations. The enzyme, glutathion peroxidase, requires it as a cofactor to catalyse reactions [124]. Total selenium is found in seawater at low concentrations (within 30–200 ng/L; but, they can reach 400 ng/L in the estuaries submitted to high anthropogenic pressure. Anthropogenic selenium inputs to the marine environment come mainly from refining, fossil fuels combustion (coal, oil), mining and agriculture because of its use as pesticide and antifungal agent Though being useful, selenium is also known for its toxicity, which depends not only on its total concentration, but also on its speciation. In seawater, dissolved selenium is, indeed, found as inorganic compound at the following oxidation degrees Se(IV) ($HSeO_3^-$, SeO_3^{2-}) and Se(VI) (SeO_4^{2-}). The toxicity of selenium is affected by its chemical form, so selenite (Se(IV)) is more biologically available to phytoplankton growth than selenate (Se(VI)), the latter can be more toxic. Dissolved selenium in seawater exists as Se(VI), Se(IV), and dissolved organic selenide (Se(II), primarily seleno amino acids in peptides), although from the view point of thermodynamics, only selenate should be stable in oxic seawaters [125].

Tin

Tin has been extensively used in industry for the last five decades, and as a result of this, it has entered in marine ecosystems. Nevertheless, it is also an essential trace element in plants, animals and humans. One of the biggest utilizers of tin is certainly the food industry, where significant amounts of tin and its compounds are used for canned food packaging. It is also used as a thermal and light stabilizer for poly(vinyl chloride), from which tin and its compounds can be leached and thus enter the food chain as inorganic tin or organotin species. In addition, tin compounds have also been used as components in several fungicides, molluskicides, ovicides, rodent repellents, in preparations for food preservation, in marine anti-fouling paints, for anti-corrosive protection, thinning and as electrochemical catalysts. It has been proven that the group of trisubstituted tin compounds of the series R_3SnX, e.g. tributyl-, triphenyl and tricyclohexyltin are the most toxic, while triethyltin acetate has been recognized as the most toxic organotin compound to mammals. Tributyltin (TBT) compounds are present in marine water due to the antifouling paints used in ships; dibutyltin (DBT) and monobutyltin (MBT) compounds come from their use as catalysts, as PVC stabilizers, and also from the degradation of TBT compounds [126-127].

Titanium

Titanium is a widespread element in nature and it is used for many purposes. The main problem with titanium ion is to maintain it in the dissolved state. The solution of titanium (IV) in the pH range below 2 is very stable. Above pH 2, Ti(IV) exists in the solution only as a complex with various organic ligands (oxalate, tartrate, citrate, etc.) Titanium(IV) occurs in seawater at very low quantities (2 ± 300 pM) [128].

Thallium

The study of thallium is important due to its high pollutant load. Thallium is widely but sparingly distributed over the earth, mainly in rock formations and soils containing potassium feldspars and micas, with an abundance in the order of 1 mg/Kg. Thallium is also found in fossil fuels and in the sea at a concentration of around 0.01 μg/L. The most important sources of thallium for exposure at the population level are air emissions from coal burning power plants. Although pesticides containing thallium are no longer sold in certain countries, they can still be found stored in many homes. As a result of such storage, accidental contamination of food has occurred. Thallium has also been used in the past for therapeutic purposes, as for example to produce hair fall in the treatment of ringworm. Absorption of thallium compounds is rapid following ingestion, inhalation, or skin contact. Thallium is widely distributed in human tissues, with the highest concentration in the kidney. It is excreted in both animals and humans by the kidney and the intestines, and to a small degree via the hair. Thallium and its compounds have been identified as potential toxicants to living organisms [129]. Thallium exists in aquatic systems in two oxidation states, Tl(I) and Tl(III). Tl(I) is more stable and highly toxic than Tl(III), with small doses of 8 mg/Kg and 40 ng/mL being lethal to humans and aquatic species, respectively. It is an analogue of potassium, and therefore easily enters into cells and inhibits all the metabolism of potassium. TI(III) is unstable, but is stabilized by forming complexes with dissolved organic matter. Some thermodynamic models predict that seawater Tl should be exclusively or predominately univalent, but others indicate that Tl exists primarily in the trivalent state. The dominant Tl(I) species in seawater are expected to be Tl^+ and TlCl, whereas Tl(III) should occur either as sparingly soluble $Tl(OH)_3$ or in the form of chloro or hydroxy complexes. Equilibrium calculations showed that in seawater at pH 8.1, trivalent thallium should predominate [130-131].

Vanadium

Vanadium is emitted into the environment from refineries and the iron, steel, and chemical industries. Among the latter are the phosphate industries, a major source of vanadium pollution. Unlike organic pollutants, vanadium is not biodegradable and it may build up in certain ecosystems to the level, which may be toxic to living organisms. Owing to its toxic and essential nature in biological systems (e.g., to marine algae and green plants) there has been considerable interest in the determination of vanadium in marine environmental samples. Vanadium has many existing forms, among which V(IV) and V(V) are the dominant oxidation forms. It has been approved that V(V) is largely responsible for the restraining effect to adenosine triphosphatase in the biological bodies, while V(IV) has little effect [132].

Zinc

Interest in Zn concentrations in the ocean stems from its dual role as a required nanonutrient and as a potential toxicant due to its widespread industrial and marine usage. The major inputs of Zn to surface seawater include atmospheric deposition (both natural and anthropogenic in origin), fluvial runoff, and up welled waters. Zinc is an essential micronutrient used in numerous enzyme systems involved with essential metabolic processes including carbon and phosphorus acquisition. Low zinc concentrations can limit the growth of marine phytoplankton. Zn speciation in the surface ocean is dominated by complexation with natural organic ligands, with about 95% of the total Zn occurring as organic complexes [133]. Thus, the Zn^{2+} concentration in the open ocean can be extremely low at concentrations of only around 10^{-12} mol/L, which is slightly higher than that thought to limit oceanic phytoplankton growth. Computer models of seawater suggest that inorganic zinc is divided between Zn^{2+} (27%), chloro complexes (47%) and $ZnCO_3$ (17%). The bioavailability or toxicity of Zn is believed to be controlled by the free Zn^{2+} concentration as opposed to the concentration of total Zn or Zn–synthetic ligand complexes [134].

Anionic Species Determined by FI Methodologies

Over 80 papers have been published involving the determination of anionic species in seawater and estuarine water by FI methodologies. Thus, the parameters that have been determined by using FI methodologies have been the following: alkalinity, hydrogen peroxide halides (bromide, chloride, fluoride and iodide), nitrate/nitrite, phosphate, silicate, sulfate and sulfide. As can be observed in Figure 2.4, the anionic species for those which have been proposed more new FI methodologies have been phosphate, nitrate/nitrite and halides with a percentage of publication of 25.2, 20.6 and 19.6%, respectively (information obtained by searching Chemical Abstracts between 1975 and September 2008 using Scifinder Scholar and the keywords "corresponding anionic species", seawater or estuarine water and flow injection).

Details and characteristics about the anionic species determined by FI methodologies in saline waters are discussed below.

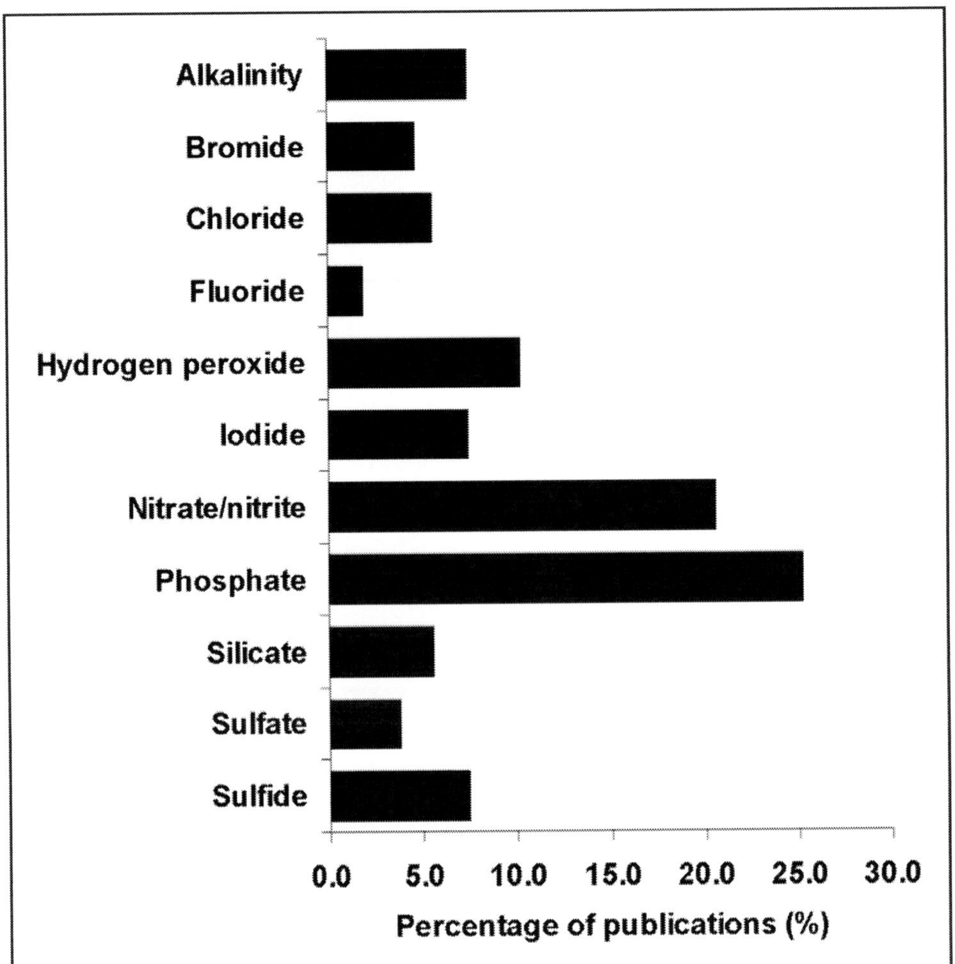

Figure 2.4. Distribution of FI determinations of anionic species in seawater and estuarine water.

Alkalinity

Alkalinity is an important parameter related to carbon dioxide chemistry. pH is the most commonly measured chemical parameter in natural waters and its knowledge is necessary for the understanding of speciation of trace elements in natural waters. In practice, the pH of seawater primarily depends on the concentration of CO_2 species. Any environment that is exceedingly acidic or alkaline is detrimental to marine life, but fortunately oceanic waters have excellent acid-base buffering power due to interactions with carbonate minerals and atmospheric carbon dioxide (CO_2-HCO_3^--CO_3^{2-}). In oceanic waters, pH is generally maintained at 7.5-8.5, although it is believed that the continuous release into the atmosphere of CO_2 from the burning of fossil fuels results in a reduction in ocean pH [135].

Halides

Bromide ion (Br^-) is one of the trace constituents of seawater, and its average concentration in seawater is approximately 60–70 mg/L. Intrusion of seawater can significantly elevate the levels of bromide ion in the surface and ground waters near the sea. Bromide concentrations in water depend primarily on the geochemistry of the materials in which the water has come into contact, with most episodes of freshwater contamination leading to increased salinity probably occurring with a concomitant increase in bromide, e.g., runoff following winter salting of roads. Anthropogenic sources of bromide alone do exist, e.g., bromine containing pesticides and fuel additives, but their contribution to overall bromide levels would generally be expected to be insignificant. However, in localized areas, intensive application of brominated pesticides, e.g., the soil fumigant methyl bromide, can produce a major contribution to bromide levels [136]. Some oxidizing agent in such waters may oxidize bromide ion to liberating reactive gas species bromine (Br_2) and hypobromous acid (HOBr), which can affect to ozone chemistry. Moreover, HOBr reacts with the naturally occurring organic matter in competition with the hypochlorous acid. Also, bromide can combine with many kinds of organic pollutants, which may present in natural waters, to form toxic compounds of bromo-derivatives, which can cause serious harm to human health and environment [137].

Chloride ion (Cl^-) is a major constituent of seawater and its most abundant ion. It is considered a conservative element in the marine environment because its supply and loss rates are similar. Chloride forms complexes with several heavy metals (including Ca and Mg), but concentration of complexing metals, even the major cations, are too low to significantly affect the overall activity of chloride in seawater. Therefore, it is assumed that almost all Cl^- in seawater is present as free form [33]. Sodium chloride in seawater makes the seawater a good conductor of electricity and more corrosive than fresh water.

Fluoride (F^-) is a conservative major element of seawater with a concentration of 1.3 mg/kg at 3.5% salinity and it is uniformly distributed throughout the world's oceans. When coral aragonite ($CaCO_3$) forms in seawater, it incorporates fluoride from the surrounding seawater. Marine aragonites have been found to contain as high as 1000 ppm F, an amount that could not be explained by simple adsorption mechanisms. Fluoride forms complexes with a number of cations and is a component of many low solubility minerals including fluorspar, rock phosphate, cryolite, apatite, mica, and hornblende. The study of fluoride concentrations is of particular interest because of its inclusion in the list of elements believed to be essential for animal life and also because of its toxic effects at higher concentrations. Sources of human exposure to fluoride include air, dental products, and foods and beverages. In non-industrial areas, concentrations in air are generally low. However, industrial production of phosphate fertilizers, dust from high fluoride minerals, volcanic activity and burning of high fluoride coal in both industrial and domestic settings make air a significant exposure route in some regions of the world [138].

Iodine is one of the most abundant biophilic, redox sensitive minor elements in the oceans, which is involve in a range of inorganic and biologically mediated oxidation–reduction reactions. It is an essential micronutrient in seawater and sea products. It exists in seawater primarily in its inorganic forms, as iodide, I^-, and iodate, IO_3^-, along with a small fraction of organic iodine compounds, and plays a special role in biological processes, such as geochemically or biologically active processes and hazardous contaminant, etc. Iodide in the oceans is produced by biologically mediated reduction of iodate, also favorable under reducing conditions. While the concentration of total dissolved iodine present in oxic seawaters stays around at 0.45 μM, those of I^- and IO_3^- are spatially more variable. The concentrations of iodine and iodide are usually very low in seawater except coast, estuary and bay water, where the pollution due to human activities is unavoidable. Iodide, which is thermodynamically unstable in oxygenated water, is usually a minor species in seawater compared to iodate [139]. Elevated concentrations of I^-, at about 0.1 to 0.2 μM, are found primarily in the surface water. In the deep sea, the concentration of iodide drops to around its detection limit of about 0.005 μM. Correspondingly, the concentration of IO_3^- is frequently at a minimum, at about 0.3 to 0.4 μM, in the surface waters and it increases with depth to become the only readily detectable form of dissolved iodine in the deep water. The distribution of iodide in seawater, being varied with depth and geographical location, provides the important clues of marine environment. In addition, determination of iodide and iodate in environmental samples attracts more attention because iodine may play a role in taste and odor problems [140].

Hydrogen Peroxide

Hydrogen peroxide (H_2O_2) is both a strong oxidant and reductant that can react with a variety of chemical species in natural waters. Hydrogen peroxide is produced in seawater by photochemical reactions, which may involve dissolved organic matter, though precipitation also can contribute significantly to the input of hydrogen peroxide to surface waters. Photodissociation of H_2O_2 produces hydroxyl radical (•OH), which is one of the most important oxidizing species in natural water. The distribution of hydrogen peroxide in the ocean is useful for understanding surface-water mixing processes, redox chemistry, and oxidative stress on biota. H_2O_2 has been studied by numerous investigators because of its high concentration relative to other reactive oxygen species and because of its potential chemical and biological reactivity. Thus, H_2O_2 can play an important role in controlling the speciation of a number of ecologically and geochemically important redox-sensitive trace elements, such as iron, copper, chromium and arsenic. The concentration of H_2O_2 ranges from less than 5 nM in deep seawater to 300 nM in surface seawater [141].

Nitrates/Nitrites

Nitrogen is an essential element for phytoplankton growth in aquatic environments. Nitrogen enters estuaries from several natural and anthropogenic sources. The elemental gas N_2 is the most abundant form of nitrogen in estuarine waters, but for biogeochemical processes, the most important forms are the dissolved inorganic species, i.e. nitrate, nitrite and ammonia. Some types of bacteria and blue-green algae, however, can fix nitrogen gas, converting it to organic nitrogen and making it available to other species [142].

Nitrate is the predominant form of bioavailable inorganic nitrogen in oceanic waters (1-500 μM) and therefore a major nutrient to support primary productivity in the surface ocean. Nitrate is a growth-limiting micronutrient in the euphotic surface layers of sea water, where light can penetrate, and its concentration in coastal waters is regulated by advective transport of nitrate to the surface layers and assimilation by organisms, e.g. phytoplankton. This gives rise to seasonal trends in nitrate concentrations in coastal waters. Winter cooling within the water column causes vertical mixing, thereby allowing the upwelling of nutrients to enrich the surface layers, whereas in late spring and summer, nitrate concentrations are depleted by increased biological activity. Because both physical and biogeochemical processes alter the spatial and temporal distribution of nitrate concentrations in the ocean, the determination of nitrate concentrations can provide essential information for understanding such processes and nitrogen cycles in the open ocean ecosystem. Nitrate concentration in seawater derived from rock weathering, atmospheric deposition, diffuse run-off from agricultural land and point source discharges. High concentrations of nitrate can originate from nitrification processes or directly from sewage or fertilizers, and may lead to water eutrophication, especially in areas with poor water renovation [143]. Consequently, perturbations on the nitrogen cycle can occur, with the subsequent alteration of marine productivity.

Nitrite is an intermediate species in the nitrogen cycle and it is toxic for fishes. It is formed during the biodegradation of nitrate and ammoniacal nitrogen or nitrogenous organic matter. Nitrite is an important indicator of faecal pollution of natural water. Usually, nitrite occurs in seawater at concentrations between 0.1–50 μM. The toxicity of nitrite is primarily due to its interaction with blood pigment to produce methemoglobinalmia. The reaction between nitrite and secondary or tertiary cumene leads to the formation of N-nitroso compounds, some of which are known to be carcinogenic, tetratogenic and mutagenic nitrite [144].

Phosphate

Phosphorus is an essential and often limiting nutrient for plant, animal, algal and bacterial growth, that is, for living organisms in both terrestrial and aquatic environments. The total phosphorus content of natural waters comprises both particulate and dissolved forms, the latter being operationally defined as the fraction, which passes through a 0.45 μm membrane. This total dissolved phosphorus fraction can be further subdivided into dissolved inorganic phosphorus and dissolved organic phosphorus. The dissolved inorganic phosphorous fraction comprises orthophosphate and condensed phosphates with orthophosphate being the most frequently determined form and also referred to as dissolved reactive phosphorus. The role of phosphorus in marine ecosystems is well recognized, as primary production could be controlled by the availability of phosphorus in nutrient-limited oceans [143]. Orthophosphate (predominantly HPO_4^{2-}) is considered the most important phosphorus species in seawater that is immediately biologically available. Phosphate availability might control oceanic carbon production in oligotrophic marine seawaters. Phosphate is one of the most important electrolytes and an essential component of all living organisms, organic phosphate forming a part of ATP, nucleic acids and of the macromolecular cell membrane structure of most organisms. Phosphate plays a crucial role also in the environment, being responsible for the phenomenon of eutrophication, water pollution caused by excessive plant nutrients. This process can be greatly accelerated by human activities that increase nutrient loading rates to water, and especially by phosphorus inputs. Furthermore, orthophosphate, plays a key role in photosynthesis. Because of the biological uptake of phosphate, the concentrations of phosphate in the surface water are usually down to nanomolar level. Phosphate is one of the factors limiting phytoplankton growth and enrichment of phosphate in the aquatic environment results in accumulation of phytoplankton, which leads to a red tide in the coastal ocean [145].

Silicate

The silicic acid ($Si(OH)_4$), also called silicate, represents the main chemical species of dissolved silica in sea water, and is considered as a core parameter of oceanography. The concentration of dissolved silica is a valuable tracer of water masses motions or mixing processes. On the biogeochemical point of view, silicate concentration drives the growth of a large part of the siliceous phytoplankton and organisms in aquatic environments in the open ocean or coastal environments. Concentrations of available silica are often low (<1 μM) in surface waters of high productivity. Although most of the silica is recycled near the surface, a fraction reaches the sediment surface where it may dissolve. As a result, deep waters may have dissolved silicon concentrations as high as 200 μM. Upwelling returns the dissolved silicon to the surface and the cycle is repeated [146].

Sulfates

About half the sulfate in seawater is present as the free SO_4^{2-} ion, 20% is present as $MgSO_4$ ion pair, 20% as $NaSO_4^-$ ion pair, and the remainder as ion pairs with potassium and calcium. The sulfate in seawater may come from weathered minerals or through the decay of ocean organisms. The average concentration of dissolved sulfate in seawater is 28 mM at a salinity of 3.5%. The dissolved sulfate in seawater converts the ferrous ion in the basalts to ferric ion and the sulfate is converted to dissolved H_2S [147]. The high concentrations of sulfate in seawater allow that it acts as an electron acceptor for sulfate-reducing bacteria, which leads to the production of sulfide.

Sulfide

Sulfide is one of the most important parameters to monitor in water matrices due to its high toxicity for aquatic organisms. Sulfide (as bisulfide, HS^-) is known to form stable complexes with metal ions. Besides, hydrogen sulfide controls the bioavailability of heavy metals in anoxic environments. In sediment-water environments, the oxygen availability (redox conditions) plays a relevant role in the biogeochemical cycle of nutrients and heavy metals. As depth increases, redox potential decreases, sulfate reduction takes place and metal sulfide salts are formed. Metal sulfides have an extremely low solubility, and concentrations of free sulfide in sediment pore waters are generally low. Sulfide enters the aquatic environment when organisms die or as byproducts from bacterial processes. The predominant species of sulfide in pH 8 seawater is bisulfide, which is converted to H_2S at pH values below 7 ($pK_1 = 6.98$). Biogenic sulfide produced by bacterial sulfate reduction is one of the most important indicators in anoxic marine environments, particularly in confining bays, polluted estuaries and aquaculture areas. The increase of its concentration usually indicates worsened conditions [148].

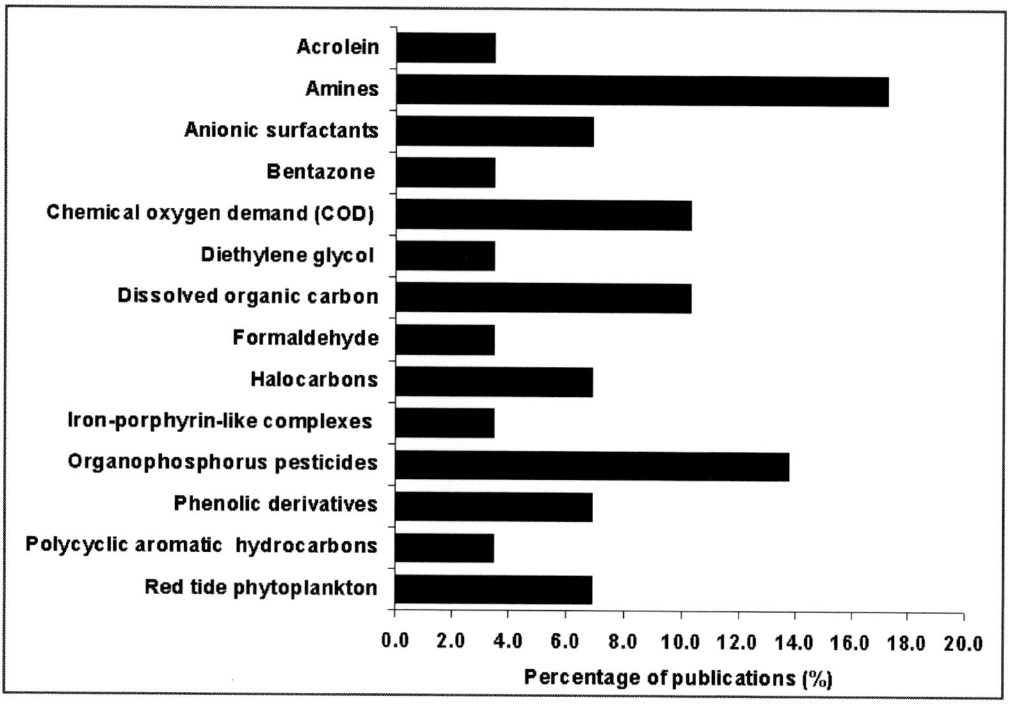

Figure 2.5. Distribution of FI determinations of organic compounds in seawater and estuarine water.

Organic Species Determined by FI Methodologies

Over 35 papers have been published involving the determination of organic species in seawater and estuarine water by FI methodologies. Thus, the parameters that have been determined by using FI methodologies have been the following: acrolein, amines, anionic surfactants, bentazone, chemical oxygen demand (COD), diethylene glycol, dissolved organic carbon (DOC), formaldehyde, halocarbons, iron-porphyrin-like complexes, malathion, methyl-parathion, nitrophenol isomers, polycyclic aromatic hydrocarbons and red tide phytoplankton.. As can be observed in Figure 2.5, the organic species for those which have been proposed more new FI methodologies have been amines, organophosphorus pesticides, COD and DOC with a percentage of publication of 17.2, 13.8, 10.3 and 10.3%, respectively (information obtained by searching Chemical Abstracts between 1975 and September 2008 using Scifinder Scholar and the keywords "corresponding organic compound", seawater or estuarine water and flow injection).

Details and characteristics about the organic species determined by FI methodologies in saline waters are discussed below.

Acrolein

Acrolein (2-propen-1-one; acrylaldehyde; acrylic aldehyde; allyl aldehyde; aqualin; magnacide B; magnacide H; NSC 8819; prop-2-en-1-al; propenal, H_2C=CH-CH=O) is considered a prioritary pollutant according to the US Environmental Protection Agency [149]. Acrolein can be found in the gas phase of cigarette smoke and environmental tobacco smoke, in exhaust emission from heavy-duty diesel vehicles and diesel engines and in different aqueous media as result of increasing use of chemicals in modem agriculture as herbicides containing acroleine utilized in agricultural canals as an injectable aquatic herbicide to control aquatic weeds. Acrolein is highly toxic to both plants and animals, producing a variety of adverse effects [150].

Amines

Nitrogen, a bio-essential element in the marine environment that is found in a variety of organic forms in oxic seawater, mono-, di-, and tri-methylamine derivatives (CH_3NH_2, $(CH_3)_2NH$ and $(CH_3)_3N$ abbreviated MMA, DMA and TMA, respectively). Methylamines are polar, volatile, water soluble species, which undergo extensive hydrogen bonding to form basic solutions (pK 9.25-10.77). Due to their low molecular weight, ability to participate in phase transfer processes and importance in marine nitrogen fertility, detoxification and osmoregulation are widely distributed and dynamic within the marine environment. Due to their volatility, they may play a significant role in the regulation of atmospheric acidity, are capable of gaseous evasion across the air-sea, interface, thus introducing alkali and reduced nitrogen into the troposphere. Methylamines are of additional importance due to the conversion of secondary amines (for example DMA) to their N-chloro-derivatives in chlorine disinfected wastewaters and the implication of secondary and tertiary amines in the synthesis of carcinogenic nitrosamines in aqueous media, air, soils and foodstuffs [151]. Naturally occurring dissolved free primary amines include several classes of compounds such as aliphatic amines, hexosamines and amino acids, resulting from metabolic processes of degradation, hydrolysis and excretion at various levels of the food chain. Although no data are available on the specific composition of the whole group of compounds, dissolved free amino acids (DFAA) have been widely studied because can take part in a variety of abiotic reactions. While their main source in seawater is primary production, they account for only a minor fraction of dissolved organic matter (DOM) because of their high turnover rate which makes them an important source of carbon and nitrogen for heterotrophic bacteria, and to a lesser extent for phytoplankton [152].

Anionic Surfactants

Surfactants are a group of chemicals that comprise both polar and nonpolar regions, and are classified according to the nature of the hydrophile as anionic, cationic, nonionic, or amphoteric. Such amphiphilic properties allow surfactants to dissolve in both oil and water, (ad)sorb at interfaces, and solubilize hydrophobic compounds in micelles and at or within sorbed layers. Surfactants are, therefore, critical in a number of technologies, including detergency, emulsification, dispersion, coating, petroleum recovery, and adhesion. Surfactants are found in a plethora of formulations, but their major use by far is in household products. In the aquatic environment, the behavior, fate, and effects of surfactants are largely governed by their rate of degradation, tendency to form aggregates (or micelles) and propensity to interact with natural particles. Regarding ionic surfactants, particle-water interactions may be extremely complex as, for a given chemical, a number of different sorption mechanisms are potentially involved, including ion exchange, ion pairing, polarization of Π electrons, and hydrophobic bonding. Estuaries and coastal waters receive large quantities of surfactants through sewage and industrial effluents. Anionic surfactants in solution have negatively charged. This is the most widely used type of surfactant for laundering, dishwashing liquids and shampoos because of its excellent cleaning properties and high. Anionic surfactants are particularly effective at oily soil cleaning and oil/clay soil suspension. Still, they can react in the wash water with the positively charged water hardness ions (calcium and magnesium), which can lead to partial deactivation [153]. The most commonly used anionic surfactants are alkylbenzene sulfonates (LAS), alkyl ethoxysulfates (AES) and their non ethoxylated homologs, the alkyl sulfates (AS).

Bentazone

Acidic herbicides have been identified in natural waters as a consequence of its wide usage. Among these herbicides, bentazone (BTZ, 3-isopropyl-1H-2,1,3-benzothiadiazin-4-(3H)-one-2,2-dioxide, Figure 2.6) is commonly used in rice culture, but also other crops and fruit tree cultivation, and is, as a consequence, frequently found in estuarine waters. Next to rainfall wash-off, farms connected to sewage drains are another important source of pollution, e.g. as a result of cleaning spraying equipment. This herbicide easily enters the aquatic environment due to its high polarity ($\log K_{OW} = -0.46$ at pH 7) [154]. Bentazone is poorly biodegradable, but direct photolysis can minimize pollutant concentration in sunlight-accessible environmental compartments like surface water or top soil.

Chemical Oxygen Demand

Chemical oxygen demand is an important parameter for assessing the concentration of organic contaminants in water resources because it can reflect the pollution degree of water. Chemical oxygen demand (COD) is defined as the amount of oxygen equivalents consumed in oxidizing the organic compounds of samples by strong oxidizing agents such as dichromate or permanganate. Because the degradation of organic compounds requires oxygen, their concentrations can be estimated by the amount of required oxygen. The chemical oxygen demand (COD), biological oxygen demand (BOD) and total organic carbon (TOC) are three main indexes used to assess the level of organic pollution in aqueous systems. Although TOC analysis is valid for organic compounds and the BOD reflects the biodegradable part of the pollutants, the COD represents the pollution load of most wastewater discharges [155]. As a pollution monitoring parameter, COD has the advantage of speed and simplicity over BOD, and requires less equipment compared to TOC determination.

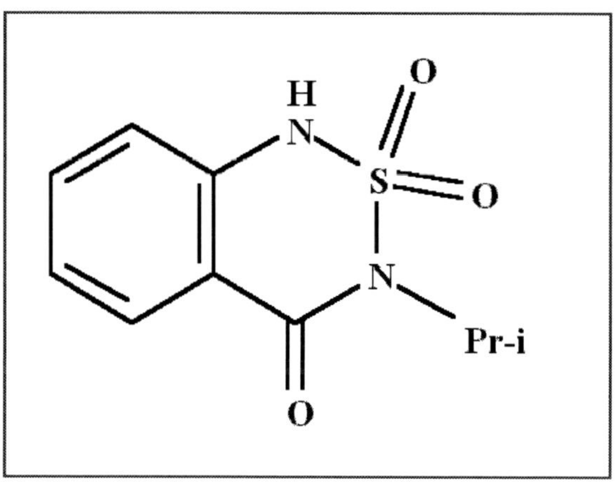

Figure 2.6. Bentazone.

Diethylene Glycol

Diethylene glycol (HO-CH_2-CH_2-O-CH_2-CH_2-OH) is a widely used industrial chemical with a potential for human exposure, which can be found in produced formation water discharges. In fact, this chemical, together with ethylene glycol and triethylene glycol, is commonly used in the dehydration processes of natural gas. Due to their high hygroscopic properties, glycols can absorb water from the natural gas stream, a characteristic that makes them the most widely used dehydration agents in the gas industry. The ''water-laden'' glycols are usually regenerated (stripped of water) in a stripping column. Because of the efficiency of the regeneration processes occurring in the glycol-based dehydration devices is less than 100%, diethylene glycol is introduced in the marine environment by the offshore extraction plants, through the discharge of produced formation water. Benthic and pelagic organisms might be affected by diethylene glycol exposure and due to its chemical properties, diethylene glycol can act as a co-solvent for other environmental pollutants of the produced formation water discharges [156].

Dissolved Organic Carbon

Dissolved organic carbon (DOC) forms the vast majority of the organic matter in seawater and on a global scale its contribution to the carbon pool is of the same order of magnitude as that of atmospheric CO_2. Thus, DOC in the marine environment is a key component of the global carbon cycle. It is the result of production, consumption and transformation processes, due to biological activity, in the different layers of the water column. However, its depth distribution is determined not only by the localized biological activity, but also by the water mass circulation in the basin. In surface waters, new DOC originates mainly by in situ biological production. Release of exudates by phytoplankton, excretion by zooplankton, dissolution of fecal pellets, cell lysis from viral infection, and egestion of microzooplankton grazing on phytoplankton are potential sources of DOC. Usually, this freshly produced DOC is rapidly consumed by bacterioplankton and is either integrated in the microbial loop, through protist predation, or respired to CO_2. In the intermediate and deep layers the DOC production and consumption are due mainly to the activity of the bacteria and protists on sinking particles and sediments [157].

Formaldehyde

Formaldehyde (HCHO) is a key reactive intermediate in the methane oxidation chain. This compound can be directly emitted to the atmosphere by anthropogenic (automobile exhaust gases and industrial emissions) and natural sources, but it can also be formed in the atmosphere as an intermediate product of the photo-oxidation of methane and other hydrocarbons emitted by natural and anthropogenic sources. It plays an important role (as a free radical source) in the reactions occurring in the troposphere. HCHO inhibits S(IV) oxidation to H_2SO_4 and it is a precursor of HCOOH. Moreover it influences the oxidizing capacity of the troposphere through interactions of H_2O_2 with •OH and •HO_2 radicals. Low-molecular-weight (LMW) carbonyl compounds are of special interest in the oceanic organic compounds cycles. Photochemical oxidation processes, such as photodegradation of humic substances and other UV-absorbing organic compounds, constitute a major source of LMW carbonyl compounds in subsurface water. Surface film present at the air-sea interface (marine microlayer, 1 to 1000 μm thick) is the crossing area between the bulk water (liquid phase) and atmosphere (gas phase), as well as an environmental compartment with own properties and structural composition [158]. Thus, exchange with the atmosphere may be a source for highly soluble compounds, such as formaldehyde.

Halocarbons

Low molecular weight volatile, halogenated organic compounds (VHOC) consist of a vast number of chlorinated, brominated, iodinated and fluorinated organic compounds. They originate from biogenic as well as from industrial sources. Traditionally, brominated and iodinated volatile halocarbons are thought to have mainly a natural origin, but chlorinated compounds such as trichloroethylene, tetrachloroethylene and chloroform, also have natural sources. Marine algae are well known sources of VHOC, although the purpose of this production is not known. It has long been suggested that the VHOC play a role in the algal defense against epiphytes or grazing by herbivores, since these compounds are toxic. However, recent studies reveal that the formation of volatile halocarbons could result from scavenging of strong oxidants, like hypochlorite and hydrogen peroxide, in the algae cells [159]. VHOC produced in the oceans are emitted to the atmosphere.

Iron-porphyrin-like Complexes

Iron in seawater limits phytoplanktonic growth in the ocean. The solubility and bioavailability of iron is largely controlled by its chemical speciation that is still poorly understood. Almost all dissolved iron, supposed to be the major source of iron for microorganisms, is bound by organic ligands that are still largely unidentified. Two distinct classes of binding ligands were discovered according to their conditional affinity constant for iron. The stronger ligands are suspected to be siderophores. These molecules are excreted by bacteria and marine organisms to acquire iron. Porphyrins are molecules biosynthesized in almost all living organisms. They are characterized by a tetrapyrrolic ring that allows the complexation of different metal cations. This is the case for iron, leading to Fe–porphyrin (Fe–Py) complexes that are found in various metalloproteins involved in essential biochemical functions, such as photosynthesis, respiration, and nitrogen fixation. Catalase, peroxidase, cytochrome, hemoglobin, myoglobin are some examples of proteins containing Fe–Py complexes playing a key role in enzymatic reactions involved in the electron transfer or the oxygen transport. Although these molecules might be released in aquatic environments following cell degradation or passive excretion, they have not yet been identified as dissolved species in natural waters [160].

Organophosphorus Pesticides

These compounds (e.g. malathion, methyl parathion, dimethoate and chlorpyrifos) show low environmental persistence but have a high acute toxicity because have the potential to cause serious health threats to humans and animals. Thus, their effects often lead to ecological and toxicological problems. They are known as typical enzyme inhibitors, and have been widely used for decades in agriculture, medicine, industry and even as chemical warfare agents in military practice. Exposure to organophosphates has been recognized as capable of inducing heightened sensitivity to chemicals. Malathion, a slightly toxic compound in EPA toxicity class III, is a General Use Pesticide (GUP) and is one of the earliest organophosphate insecticides developed (introduced in 1950) [161]. The acute toxicity of an exposure to these chemicals stems from the fact that they target a number of important proteins, including a group of hydrolytic enzymes, such as acetylcholinesterase, which is particularly critical for central and peripheral nervous-system functions. Organophosphates can also impair function of the immune action of white blood cells, increasing recurrent infection rates of the upper and lower respiratory system, and have other damaging immune effects [162]. Nevertheless, malathion is a widely used organophosphorus insecticide because of its relatively low toxicity to mammals and high selectivity for insects compared with other organophosphorus insecticides. There are many earlier findings that clearly warned of the genotoxic potential of technical-grade malathion in a wide range of organisms including fish species.

Phenolic Derivatives

Phenol and substituted phenols currently regarded as priority pollutants due to their toxicity and persistence. They enter the trophic chain as residues of a wealth of industrial processes, including the production of plastics, paper, pesticides, drugs and antioxidants. Although the discharge of nitrophenol-bearing wastewater is strictly prohibited, a considerable amount of them is yet still released into marine environment, creating serious environmental damages harmful to human beings and marine organisms. Furthermore, phenolic compounds are also formed during the natural decomposition of humic substances, tannins and lignins, and photolytic or metabolic degradation of herbicides and insecticides. These compounds show toxicity values from moderate to higher, the toxicity level depends on the number, position and kind of substituent. The US Environmental Protection Agency (EPA) and the European Union (EU) have included phenolic compounds as 4-nitrophenol, 2-nitrophenol, phenol, 2,4,6-trichlorophenol and 4-chlorophenol in the lists of priority contaminant because they are considered dangerous pollutants. Some waterways can be contaminated for those phenols and hazard effects may occur to the people, also to aquatic organisms, fish and other life forms [163].

Polycyclic Aromatic Hydrocarbons

Polycyclic aromatic hydrocarbons (PAHs) are a series of organic contaminants that have become ubiquitous in the environment. The US Environmental Protection Agency (EPA) lists 16 PAHs as priority environmental pollutants. Emitters of PAHs include diverse sources such as combustion of fossil fuel, wood and coal burning, and metal smelting. Nevertheless, one of the major sources of PAHs to the marine environment is a petroleum spill. Once PAHs appear in the marine environment, they can distribute in the water column and accumulate in sediment and biota. PAH concentrations in seawater vary widely, depending upon such factors as proximity of the water body to the source, source type, and season. PAHs are slightly soluble in water and after binding to particulate matter, they tend to accumulate in the bottom sediments. Levels of PAH in sediments vary, depending on the proximity of the sites to areas of human activity and on the biodegradation of these chemicals, a process reliant upon abiotic and biotic factors that are dependent on site characteristics. In general, sediments contain a PAH concentration that is a factor of 1000 or more than the overlying water column. PAH concentrations in biota depend upon their proximity to the sources of pollution, their bioavailability, and species ability to biotransform PAHs, and in mussels ranged from 33 to 150 ng/g wet weight [164]. A wide range of marine life is sensitive to PAHs compounds, and the toxicity increases generally with the molecular weight of the compound. PAHs can be metabolized in aquatic fauna to active and potent carcinogenic forms. Fish can biotransform

PAHs to hydrophilic metabolites in the liver, using an active oxidative enzymatic system. Therefore, parent PAHs are not often found in fish, and thus the determination of parent PAHs may not indicate the exposure of PAHs in higher trophic-level organisms. Analysis of hydroxylated metabolites of PAHs in fish bile is a more appropriate approach to understand the exposure to PAHs. Parent PAHs themselves are not active carcinogens; however, their metabolites, after being generated in the detoxification process in living cells, may be carcinogenic. In fact, metabolites of PAHs found in benthic fish are strongly associated with hepatic lesions and liver neoplasm. Consequently, metabolites of PAHs can be analyzed in bile and can serve as biomarkers for recent exposure to PAHs. Marine mammals are exposed to PAHs through the ingestion of contaminated prey species, particularly benthic invertebrates [165].

Red tide Phytoplankton

Red tide is a phenomenon caused by blooms of microalgae or phytoplankton where seawater becomes red. Both toxic and nontoxic red tides have occurred throughout recorded history, but in recent years, there has been a global increase in the number of these events due to coastal pollution and, probably, to other unclear factors, for example, long-distance transport of species across oceans. Some kinds of red tides cause mass fish death and thus bring serious economic problems to the countries where people take fish as one of the important protein supplies. Heterosigma carterae (formerly *Heterosigma akashiwo*, class *Raphidophycea*n) is known to be the main plankton, in addition to Chattonella marina and Chattonella antiqua, causing the red tide observed in Japan. They cause serious damage to fish farms in the coastal seawaters. It has been reported that Chatonella antiqua produces active oxygen species such as superoxide anion (O_2^-), peroxide (H_2O_2) and hydroxyl radicals (•OH), which may be responsible for toxicity of this flagellate [166].

MARINE AND ESTUARINE SEDIMENTS

Marine and estuarine sediments are any collection of particles of insoluble material, primarily rock and soil particles, which are loosely deposited on the sea floor and closely packed and consolidated under increasing lithostatic pressure. These ranging from sand to fine muds all consist of loose particles and interstitial spaces variously filled with water, air, detritus and organisms. These have traditionally been referred to differently in terms of their size. The microfauna, which includes bacteria and protists, are extremely important in interstitial sulfur chemistry and the oxygenation of the sediment. It is well known that sediments constitute an important compartment in the biogeochemical cycle of metals in soft substrates of estuaries.

Marine sediments are formed by several processes. Detrital fragments of rocks and minerals can be carried to the sea from distant, upland sources. Alternately, they can be formed in place by biological or chemical processes operating either at the site or very close. Marine sediments can be grouped into several categories, based on the source of material, the place of deposition, the chemical or mineral composition and their particle size [167].

Source of material:

- Lithogenous: those derived from rock, including particles of minerals and rocks resulting from weathering processes on land and then transported to the ocean by streams, wind or glaciers.
- Biogenous: those derived from organisms or produced by organisms which would include shells, coral fragments, coralline alga fragments as well as skeletal remains of one-celled organisms living in surface waters. Most biogenous sediments are calcareous or siliceous.
- Hydrogenous: those derived from seawater by chemical processes, for example carbonate mud and ooids as well as some phosphate and manganese nodules.
- Cosmogenous: those derived from outer space.

Place of deposition:

- Terrigenous: those deposited near land and derived primarily from land sources and therefore mostly lithogenous sediment (gravel, sand, and mud).
- Neritic: those deposited in shallow coastal waters, mainly on the continental shelves. these may include some finer lithogenous material and may include biogenous material from organisms living on or in the sea floor or in shallow coastal water.
- Pelagic: those sediments deposited in the open ocean which include those materials which have settled to the deep-ocean floor (biogenous and lithogenous) as well as those formed on the ocean floor precipitated from sea water (hydrogenous). Examples of pelagic deposits are:
- Biogenous oozes: sediment containing more than 30% biogenous material by weight.
  a) Diatomaceous ooze: accumulation of the skeletal parts of single-celled plant-like protozoa known as diatoms, most commonly associated with cold waters (sioiceous).
  b) Radiolarian ooze: remains of small animal-like protozoa that are most abundant in equatorial regions of the ocean (siliceous).
  c) Foraminiferal ooze: skeletal remains of marine animal-like protozoa (calcareous).
  d) Coccolithic ooze: fragments (platelets) of extremely small plant-like protozoa (calcareous).
- Red or brown clay: very fine lithogenous sediment that is mostly transported by wind and covers much of the sea floor below depths of 4500 m. Clay, or mud as it is commonly referred to as, is the predominant sediment in areas where there is reduced biogenous material covering nearly 40% of the deep ocean floor.

Chemical or mineral composition:

- Calcareous: composed of calcium carbonate (lime) ($CaCO3$).
- Siliceous: composed of silica (organic opal) ($SiO2$).

Particle size:

- This classification is useful for sediments consisting of minerals and rock fragments:
- Gravel or granule: particles between 2-4 mm.
- Sand: particles between 0.25-2 mm.
- Silt: particles between 0.0039-0.0625 mm.

- Clay: grain to small to see except with a high magnification microscope (0.0002-0.0039 mm).

Sediment texture plays a controlling role in the concentrations of metals and other substances spatial distribution. It is the physico-chemical characteristics of the sediments such as the fine particle fraction, which affect, to a large extent, the retention of metals in the sediment. Macrobenthic organisms are in permanent contact with the sediment and bioaccumulation takes place in tissues of organisms. This is determined by the biological availability, the metabolism and the excretion rate of contaminant or metabolites and this emphasizes the complexity and importance of biota in the process [168].

Sediments serve as sources and sinks for nutrients and toxic chemicals. Agriculture, fishing and cattle raising therefore strongly depend upon control of the pollution of sediments. Additionally, sediments act as the memory of an ecosystem, providing a good picture of its evolution over time, recording the date of environmental impact episodes and accumulating the chemicals employed or produced by human activities, thus modifying its quality and fate. Industrial, agriculture or urban activities always involve a drastic modification of their composition and can involving health risks. Thus, studies on trace elements in sediments can provide useful information in detection pollution in the aquatic environment. Human activities have lead to accumulation of toxic metals in marine sediments. However, heavy metals are not necessarily fixed permanently to the sediment but may be recycled via biological and chemical agents within sediment and water column. The presence of metals, especially heavy metals, and their species is one of the most important parameters to be determined in sediments for their characterization from the environmental point of view. Organic compounds can be modified through different types of reaction: hydrolysis, oxidation, reduction, addition, substitution etc. The half life of some of these compounds and their inherent toxicity implies the need for drastic control of their presence in natural ecosystems in order to avoid side effects.

Usually, several certified reference materials (CRMs) with a marine or estuarine sediment matrix have been used to check accuracy of a new analytical procedure. Thus, the CRMs more used and actually available are the following:

IAEA-135 (marine sediment for radioactive isotopes), IAEA-315 (marine sediment for radioactive isotopes), IAEA-368 (marine sediment for radioactive isotopes), IAEA-383 (marine sediment for organochlorine compounds and PAHs and indicative values for further organochlorine compounds, petroleum hydrocarbons and sterols), IAEA-408 (marine sediment for a range of organochlorine compounds and PAHs and indicative values for further organochlorine compounds, petroleum hydrocarbons and sterols) and IAEA-433 (marine sediment for trace metals) from International Atomic Energy Agency (IAEA), Austria [169-171]; LGC6137 (Estuarine sediment for extractable metals) and LGC6114 (RM) (harbour sediment for polychlorinated biphenyls, PCBs), LGC6156 (Harbour sediment for extractable metals) from LGC Standards, UK [172]; GBW07314 (offshore marine sediment for trace elements) and GBW07313 (marine sediment for trace elements) from the Chinese National Research Center for Certified Reference Materials (Beijing, China) [173]; HIPA-1 (marine sediment for butyltins), HISS-1 (marine sediment for trace elements and other constituents), MESS-3 (marine sediment for trace elements and other constituents), PACS-2 (marine sediment (sand) for trace elements and other constituents including butyltins), PACS-2 (harbour sediment for trace elements and organotin compounds), SOPH-1 (marine sediment for butyltins), CS-1 (marine sediment for PCBs), HS-1 (marine sediment for PCBs), HS-2 (marine sediment for PCBs) and HS-3B, 4B, 5, 6 (harbour sediment for PAHs) and SES-1 (estuarine spiked sediment for PAHs) from the NRC Institute for Marine Biosciences, National Research Council of Canada [174-175]; CRM580 (estuarine sediment for mercury and methylmercury) from the Institute for Reference Materials and Measurements, Belgium; SRM4357 (ocean sediment for radioactive isotopes), CRM462 (coastal sediment for organotin compounds), SRM1646a (estuarine sediment for metals) and MURST-ISS-A1 (Antarctic sediment for trace elements) from the National Institute of Standards and Technology (NIST), USA [176-177].

Taking into account all the analytes that have been determined by FI methodologies in marine and estuarine sediments, the distribution of these FI determinations is shown in Figure 2.7. This Figure shows that the great majority of the determinations were proposed for cationic species (87.6%) (information obtained by searching Chemical Abstracts between 1975 and September 2008 using Scifinder Scholar and the keywords "corresponding species", marine and estuarine sediments and flow injection). In view of these results, it is clear that FI analytical determinations in marine and estuarine sediments were developed practically only for inorganic analytes.

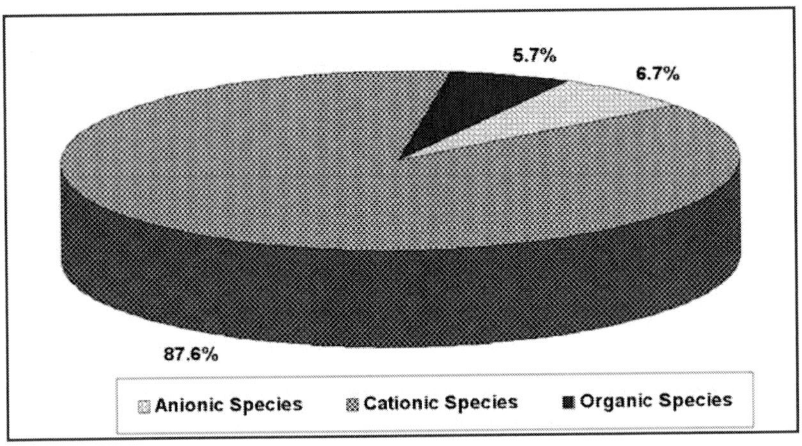

Figure 2.7. Distribution of FI determinations according to the analyte analyzed in marine and estuarine sediments.

Cationic Species Determined by FI Methodologies

Over 100 papers have been published involving the determination of cationic species in marine and estuarine sediments by FI methodologies. Thus, the elements that have been determined by using FI methodologies have been the following: Ag, Al, As, Bi, Cd, Co, Cr, Cu, Fe, Ge, Hg, Ir, La, Mg, Mn, Ni, Pb, Pt, rare earths (U, Th and Pu), Re, Sb, Se, Sn, Te and Zn. As can be observed in Figure 2.8, the cationic species for those which have been proposed more new FI methodologies have been Hg, As and Pb with a percentage of publication of 14.9, 10.1 and 9.5%, respectively (information obtained by searching Chemical Abstracts between 1975 and September 2008 using Scifinder Scholar and the keywords "corresponding species", marine and estuarine sediments and flow injection).

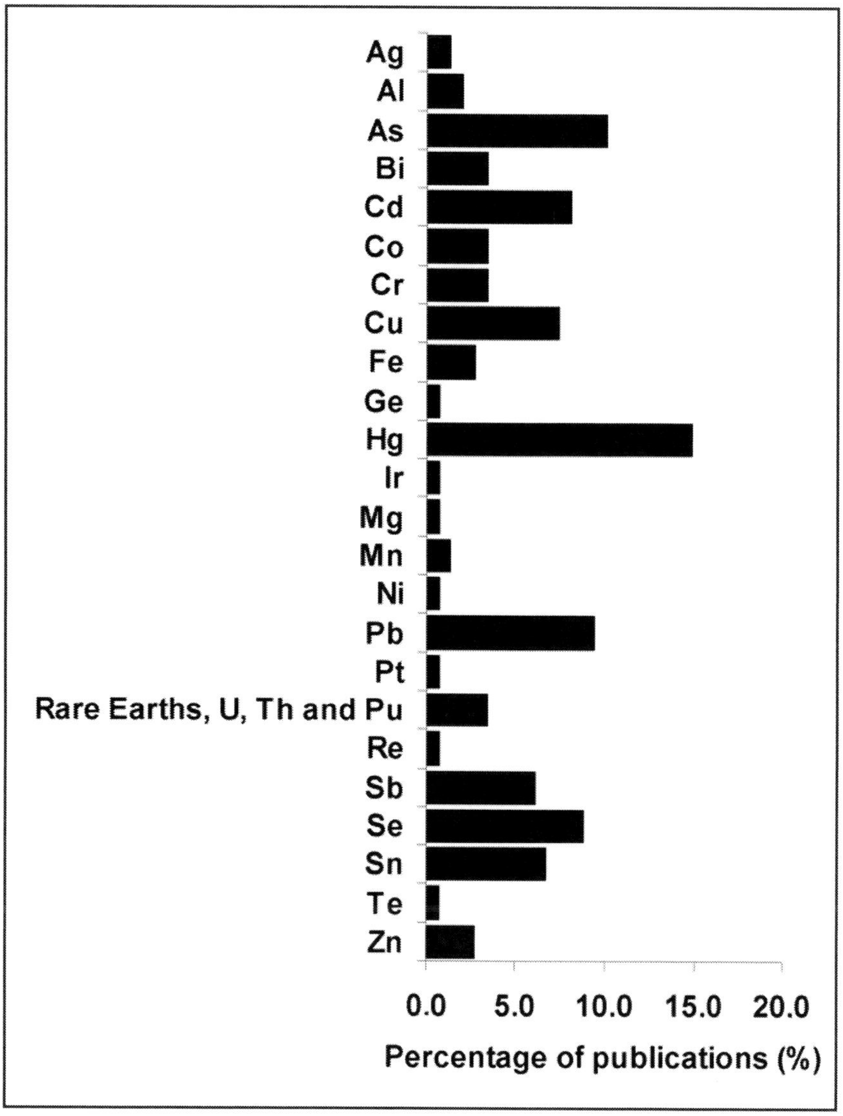

Figure 2.8. Distribution of FI determinations of cationic species in marine and estuarine sediments.

Marine sediments provide a matrix, which act as a sink for elements and compounds (including trace metals) naturally released by erosion in continental areas. The chemical composition of the sediments is directly influenced by processes and equilibrium that occur in the water column. The sequence and accumulation of sedimentary metals in core samples, for example, permits study of fluctuations in the depositional history of sediments in these environments. Once deposited in sediment, many contaminants enter a dynamic state of flux between sediment particles, interstitial waters, and the overlying water column. In shallow, tidal environments, diffusive fluxes between these reservoirs are accelerated by a variety of physical processes (e.g., tidal exchange, storms, dredging), and by the bioturbating effects of infaunal macrobenthic organisms. Being exposed to and/or consuming contaminated particles and interstitial waters, these organisms are also valuable biomonitors. Contamination in estuarine sediments by metals has been a problem since these environments often receive heavy metal wastes generated naturally through the weathering of rocks and through a variety of human activities. In estuarine environments, metals discharged from sources such as industrial and sewage effluents may accumulate in bottom sediment as the suspended particles on which they are adsorbed settle out. On the other hand, coagulation and flocculation and co-precipitation can also cause removal of dissolved heavy metals from the water column to sediment due to changes in pH and salinity of waters during estuarine mixing. Consequently, heavy metal concentrations in sediments near large population centers are often significantly higher than those in the sediments that have little history of contamination. When elevated, metals can exert toxic effects, including those that are biologically essential. Understanding of ecological and environmental consequences of metals in estuarine sediments is a complex problem, but one that must be thoroughly investigated to make good decisions for protecting exposed aquatic habitats. Thus, the ability to determine these is important for geochemical prospecting, environmental monitoring, and basic geochemical studies.

Details and characteristics about the cationic species determined by FI methodologies in marine and estuarine sediments are discussed below (only the elements that have not been commented in the previous section: marine and estuarine water).

Germanium

Germanium is a trace element in the Earth's crust, averaging about 1 ppm in whole rocks and minerals. Because of nearly identical ionic radii and electron configurations for germanium (Ge) and silicon (Si), the crustal geochemistry of Ge is dominated by a tendency to replace Si in the lattice sites of minerals. These two elements exist in seawater as similar hydroxyacids, i.e., $Ge(OH)_4$ and $Si(OH)_4$ and the uptake and regeneration profile of Ge is similar to that of Si, thus providing an interesting tracer for biogenic silica cycling in the ocean. It is distributed in the crust in the form of silicate, sulfide, sulfate and so on [178].

Iridium, Rhenium and Platinum

Rhenium, iridium, and platinum occur in trace concentrations in most earth-surface materials. Interest in the geochemistry of these elements has arisen from their use as sedimentary indicators of past meteorite impacts (Pt and Ir) or of anoxic environments (Re) as well as their economic importance. The decay of ^{137}Re (half-life 45 billion years) also offers a new method by which to date ancient sediments [179].

Platinum group elements (Rh, Pd and Pt) are used in automobile catalytic converters. Pd and Pt are employed in a converter to oxidize carbon monoxide and hydrocarbons to water and carbon dioxide, while rhodium is used to reduce nitrogen oxides to nitrogen. Pt concentrations ranged in sediments between 0.25 and 54 μg/Kg [180].

Tellurium

Tellurium is a rare element in the Earth's crust but its use in many technological processes results in local enrichment and release. The emission of inorganic tellurium compounds in the environment may create serious problems owing to the toxicity of this element. In addition and after microbial action, it can be transformed into volatile organometalloid compounds with the consequent modification of its transport pattern and toxicological behavior. Tellurium and its compounds are applied in the areas of solid-state studies, thermoelectrics, electronics and microalloying in ferrous and non-ferrous metallurgy [181].

Anionic Species Determined by FI Methodologies

Over 10 papers have been published involving the determination of anionic species in marine and estuarine sediments by FI methodologies. Thus, the parameters that have been determined by using FI methodologies have been the following: carbonate, phosphate, silicate and sulfide. As can be observed in Figure 2.9, the anionic species for which have been proposed more new FI methodologies has been phosphate with a percentage of publication of 42.9%, (information obtained by searching Chemical Abstracts between 1975 and September 2008 using Scifinder Scholar and the keywords "corresponding anionic species", marine or estuarine sediments and flow injection).

Organic Species Determined by FI Methodologies

Over 10 papers have been published involving the determination of organic species in marine and estuarine sediments by FI methodologies. Thus, the parameters that have been determined by using FI methodologies have been the following: bio and molecular markers, dissolved organic carbon (DOC) (in sediment pore waters) and polycyclic aromatic hydrocarbons. As can be observed in Figure 2.10, the organic species for which have been proposed more new FI methodologies has been bio and molecular markers with a percentage of publication of 50.0% (information obtained by searching Chemical Abstracts between 1975 and September 2008 using Scifinder Scholar and the keywords "corresponding organic compound", marine or estuarine sediments and flow injection).

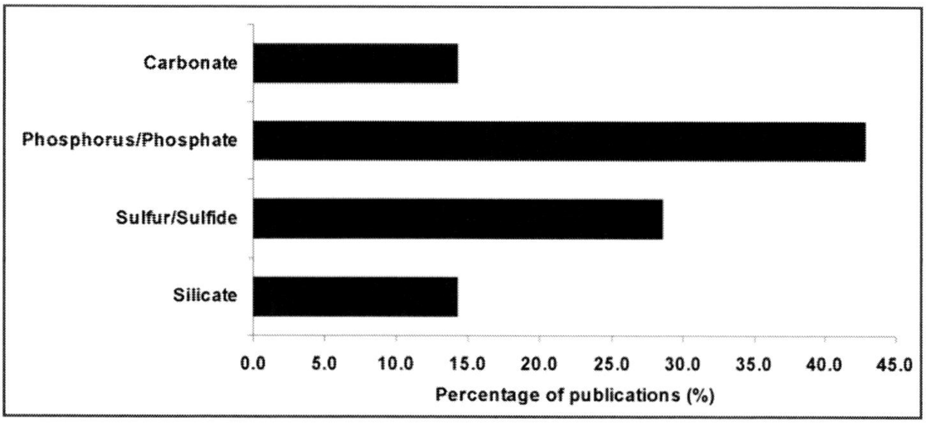

Figure 2.9. Distribution of FI determinations of anionic species in marine and estuarine sediments.

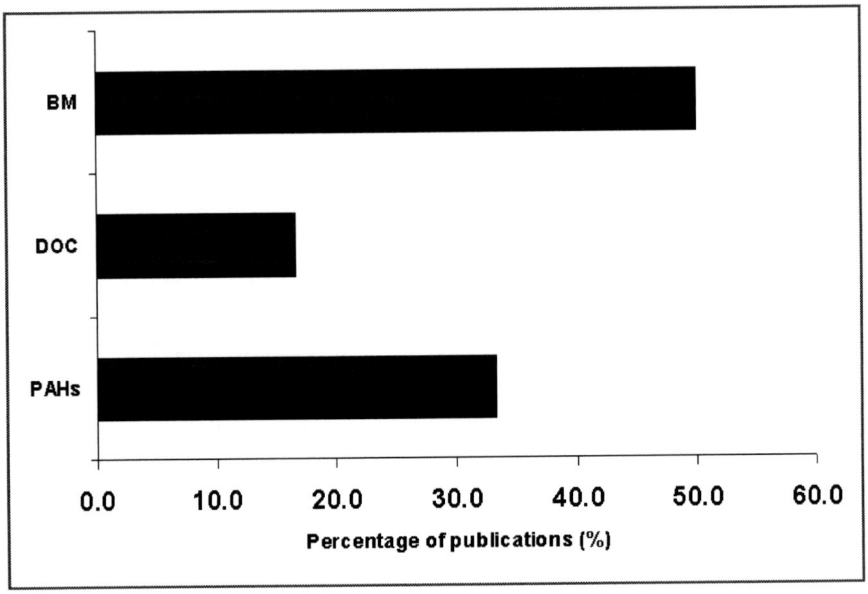

Figure 2.10. Distribution of FI determinations of organic species in marine and estuarine sediments. BM: Biomarkers; DOC: dissolved organic carbon; PAHs: polycyclic aromatic hydrocarbons.

Details and characteristics about the organic species determined by FI methodologies in marine and estuarine sediments are discussed below (only the elements that have not been commented in the previous section: marine and estuarine water).

Biomarkers

Sewage contamination has traditionally been determined by enumeration of microbiological indicators such as faecal coliforms. Many researchers have questioned the reliability of these methods due to a series of shortfalls, especially in the marine environment, where die-back/re-growth rates are essentially unknown. Thus, an alternative method of detecting sewage pollution by researchers worldwide is by the use of sterol biomarkers. Organic matter containing faecal and other sterols released into freshwater bodies, estuarine and marine waters tend to adhere to particulate organic matter. This organic matter will eventually sink to the bottom and become embedded into sediments. There, sterols are degraded at varying rates depending on a range of environmental parameters including temperature, currents, tides, total organic matter, particle size and endemic microbial populations. Knowledge about the degradation of sterols in sediments is therefore essential to maximizing their usefulness as biomarkers for tracing marine pollution and for establishing recovery/residence times for pollution monitoring programs. Sterol biomarker fingerprinting in both the water column and the sediments from marine and freshwater environments has been used with considerable success to trace sewage plumes and to identify the likely origin of sewage pollution. Compounds such as coprostanol (5β-cholestan-3-β -ol) can be used in conjunction with other sterols (e.g. cholestanol [5α(H)-cholestan-3β-ol] (Figure 2.11) to indicate the relative abundance of sewage in sediments. Additionally, many researchers utilize sterol biomarker ratios, rather than relying solely on absolute concentrations for providing pollution threshold levels. Coprostanol is produced in the digestive systems of higher animals by the microbial reduction of cholesterol and is one of the principal sterols in human and animal faeces. Coprostanol concentrations correlate well with coliform bacteria (especially *Escherichia coli*) in sewage contaminated environments. The multiplicity of sources of steroids in aquatic environments and the transformation/degradation processes that occur can, however, complicate interpretation with respect to urban sewage pollution. In theory, however, unpolluted sediments should not contain coprostanol, i.e. the compound should typically be below the limits of analytical detection. To enhance the reliability of pollution assessments using coprostanol (and other steroid compounds), some authors have derived (and tested) ratios between selected steroids [182-183].

By the other hand, distributions of lipid biomarkers in sediments can provide valuable information concerning the origin and transport pathways of organic matter and the oceanic conditions (e.g. temperature, nutrient level, stratification, etc.). Various phytoplankton groups contribute lipids to the primary produced organic matter, largely depending on the nutrient conditions of the surface waters. Allochthonous lipids are transported to the marine realm by freshwater runoff or wind and thus carry information about continental aridity and wind strength. Before final deposition in the sediments these lipid signals can be modified by zooplankton feeding, and microbial and/or chemical oxidation [184].

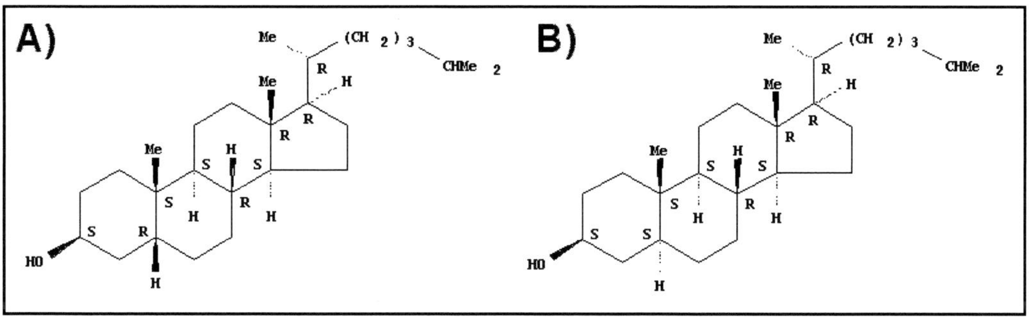

Figure 2.11. A) Coprostanol; B) Cholestanol.

SEAWEEDS

Seaweeds are traditionally used in the orient as part of the daily diet and as excellent sources of iodine. Consumption of particular seaweeds of various species of edible green, brown and red algae are high in Asia, especially in Japan, China and Korea. Seaweeds are sources of protein and carbohydrates, although these nutrient contents of some seaweed vary with species and environment. Seaweeds have also been used in the pharmaceutical, cosmetics and food industries. Seaweeds exhibit original and interesting nutritional properties. From a nutritional standpoint, the main properties of seaweeds are their high mineral (iodine, calcium) and soluble dietary fiber contents, the occurrence of vitamin B12 and specific components such as fucoxanthin, fucosterol, phlorotannin.

Several organisms have been used for monitoring heavy metal concentrations and other contaminants, for instance lichen and brown algae such as *Fucus sp*, geen algae, such *as Ulva lactuna*, red algae such as *Porphyra sp*, invertebrates such as *Mytilus edulis*, crustaceans, gastropod mollusks, phytoplanktons and other invertebrates. Thus, it has long been established that marine and estuarine macroalgae accumulate in their cell walls metals, hydrocarbons such as polychlorinated biphenyls (PCBs) and pesticides to levels many times found in the surrounding waters [185]. The use of seaweeds as monitors of pollution has increased in recent years. In this context, macroalgae have several intrinsic advantages:

- most of them are sessile in nature and can therefore be used to characterize one location over time.
- seaweeds can be collected in abundance at many coastal localities.
- they readily accumulate compounds present in their environmental waters.

Moreover, biosorption studies using living biomass including seaweed have been widely performed in large parts of the world. In this sense, the use of microorganisms as a biosorbent for metals has become a good alternative to the other preconcentration methods as regards higher recovery, economic advantages, simplicity and environmental protection. In general, microorganisms have the ability to selectively adsorb a specific element without preconcentrating the matrix. Either living or nonliving microorganisms, such as yeasts, fungi, bacteria and algae are capable of accumulating heavy metals from aqueous solutions by different chemical and biological mechanism. Microbial cell products such as metabolites, polysaccharides, and cell wall constituents are effective in metal accumulation. Biosorption process takes place on the cellular membrane. If there is no biological activity (a dead cell), the metallic species are firstly adsorbed on a cellular membrane and after passing through this membrane they are absorbed into this cell structure. Biomolecules (proteins, polysaccharides and cellulose), which contain sulfates, carboxylates, phosphates etc. in their structure, are responsible for the absorption process. It may be concluded that some processes such as absorption, cationic exchange, chelating, precipitation or crystallization take place in the cellular membranes of the microorganism [186].

The available certified reference materials (CRMs) with a seaweed matrix are the following: IAEA-140 (common brown seaweed, *Fucus* for organochlorine compounds and petroleum hydrocarbons) and IAEA-392 (algae for trace, minor and major elements) from International Atomic Energy Agency (IAEA), Austria; NIES-03 (Chlorella (green algae) for trace elements) and NIES-09 (Sargasso, for major and minor constituents and trace elements) from the National Institute for Environmental Studies, Japan; SRM4359 (seaweed, for radionuclide) from the Standard National Institute of Standards and Technology (NIST), USA; BCR-279 (*Ulva lactuca* (Sea lettuce) for trace elements) from the Institute for Reference Materials and Measurements, Belgium [187-188].

Species Determined by FI Methodologies

Over 25 papers have been published involving the determination of species in seaweeds by FI methodologies. The parameters that have been determined by using FI methodologies have been the following: As, Ge, Hg, Mo, Sn, Si, iodide/iodate, nitrate/nitrite, phosphorus/phosphate, β-dimethylsulfoniopropionate (DMSP), free amino acids (FAA) and laminarin. As can be observed in Figure 2.12, the species for which have been proposed more new FI methodologies have been iodide/iodate, phosphorus/phosphate and As with a percentage of publication of 25.8, 12.9 and 12.9%, respectively (information obtained by searching Chemical Abstracts between 1975 and September 2008 using Scifinder Scholar and the keywords "corresponding compound", seaweeds and flow injection).

Details and characteristics about the species determined by FI methodologies in seaweeds are discussed below (only the elements that have not been commented previous sections).

β-dimethylsulfoniopropionate

β-Dimethylsulfoniopropionate (DMSP) is a tertiary sulphonium compound with a low molecular mass (M_r 135), which is contained, in large and small amounts, in a variety of marine organisms: marine micro- and macroalgae cells and in only a few flowering plants such as *Spartina* spp., *Wollastonia biflord* and sugarcane. It is especially high in green sea algae (*Aonori, Anaaosa, Hitoegusa*). DMSP occurs at high intracellular concentrations (100 to 500 mmol/L) in many marine algal species [189]. In addition, a number of environmental factors, including light intensity, solar UV radiation, and availability of limiting nutrients can influence DMSP concentrations within individual algal species. DMSP was first described as a precursor of dimethylsulfide (DMS) in the red macroalga, *Polysiphonia fastigiata*. DMS is the most important volatile organic sulfur compound involved in the global cycling of sulphur, which affects global climate change, cloud formation and influences atmospheric acidity. DMSP is enzymatically cleaved to dimethyl sulfide and acrylate by the enzyme DMSP lyase, which occurs in both marine algae and bacteria. Because of its zwitter-ionic nature, DMSP appears to act as an osmoprotectant, also termed a compatible solute, to maintain turgor pressure inside plant cells against outer osmotic pressure caused by the high salinity of seawater [190].

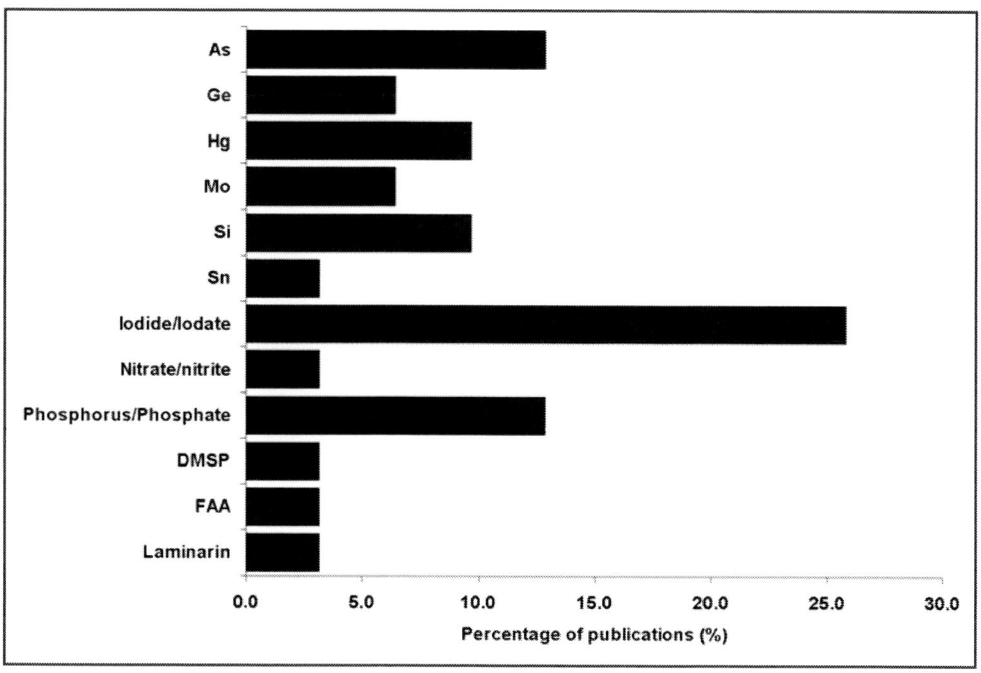

Figure 2.12. Distribution of FI determinations in seaweeds.

Free Amino Acids

Amino acids play a key role in the carbon and nitrogen metabolism. Their total concentrations and the ratios between some amino acids are modulated by the photosynthetic process and by the nitrogen assimilation. Amino acids in marine ecosystems are responsible for 13–30% of the dissolved nitrogen. Because of these factors amino acids can be used as markers to characterize distinct physiological conditions of the phytoplankton and to help the understanding of oceanic food webs. The intensity of absorption and excretion of free amino acids depend on physiological activity of algae [191].

Laminarin

In the marine environment, there are many polysaccharides, such as alginate, laminarin, chitin, and chitosan that do not exist in the terrestrial environment. Laminarin has a β -1,3-glucan structure (molecular formula: $C_{18}H_{32}O_{16}$). Brown seaweeds are a potential source of this compound [192]. In these seaweeds, soluble fiber consists of laminarin, fucans and alginates, whereas insoluble fiber is essentially cellulose.

MARINE ANIMALS/SEAFOOD

Fish and shellfish are nutritious and wholesome foods. They are perceived as an excellent source of high-quality protein, containing lipids with high levels of unsaturated fatty acids, and perhaps contributing to the enhancement of human health by reducing the risk of cardiovascular disease. Likewise, seafood is characteristically tender, easily digested, and a good source of many important minerals and vitamins. Both finfish and shellfish are subjected to contamination and cross-contamination in their natural habitat, as well as at any point during handling, processing, distribution, or preparation. Antimony, arsenic, cadmium, lead, mercury, selenium, polychlorinated biphenyls, dioxins, certain processing-related contaminants (nitrosamines and possibly products of chlorination), contaminants related to aquaculture and petroleum hydrocarbons have long been recognized as the most deleterious pollutants to biota in the world's marine, coastal and estuarine waters, and pose sufficient potential risks for consumers [193].

Trace elements from seawater and marine sediments are known to be accumulated by many species of marine invertebrates. Thus, bivalves such as oysters, mussels, clams and shells are widely used as sentinels for monitoring pollution of the marine environment, because of their ability to accumulate and tolerate exposure to heavy metals and other pollutants in water and sediments to levels well above those present in the surrounding waters or sediment, thus being able to provide information on local pollution sources. Furthermore, the subsequent analysis is more sensitive than that for water. Their usefulness as bioindicator organisms provides, ideally, an estimate of trace elements availabilities to the biomass of different areas and localities. They should be able to accumulate the pollutant in a sedentary manner without being killed. There are three possible routes by which metals can be derived namely, (1) from solution, (2) from the ingestion of food and (3) from the ingestion of particulate matter containing metals. One of the major requirements is that all species to be used as bioindicators should exhibit the same correlation in their elemental contents with those in the surrounding marine environment namely water, at all locations in the study area under all conditions. The body burdens of trace metals in most bivalves have been used to identify and map areas with exceedingly high levels of trace metals and organic pollutants, hence they can be used as biomonitors for aquatic environment. Concentrations of the various trace metals can differ in the type of tissue analyzed and so it is with variations in body size or growth rates which can influence the quantity of metals in the tissues. Mussels are sedentary, benthic and gregarious invertebrates. They filter water continuously and feed on phytoplankton. The water current is aken through the inhalant siphons that passes through the gills, labial palps, mantle of the mussels and is finally ejected through the exhalent siphon. During such processes the suspended soil particles, excess algal blooms and metal ions (Cu, Zn, Ni etc.) are removed from the water. In addition to the gills, the mantle, kidney, foot and hepatopancreas are anticipated to be major sites of metal uptake because of their large surface area, thus clearing the aquatic habitat. They accumulate both essential (Na, Ca, Mg) and non essential (Hg, Cd, Pb) metals in higher concentrations than the ambient water. Through their filter feeding and respiratory mechanisms mussels also take up other pollutants such as hydrophobic organic contaminants, poly aromatic hydrocarbons (PAHs), metallothionein and organochlorines. The accumulation of contaminants from the water column by bivalves is referred to as bioconcentration, a property that makes bivalves as the common mussel *Mytilus edulis*, potentially useful as biomonitors for seawater quality monitoring programs, and also for bioremediation to improve the quality of polluted waters [194].

Therefore, analytical determinations in seafood have interest because of three areas of concern, nutritional, toxicological and environmental. Nutritional because trace metals such as Ca, Fe, Mg, Zn, Cu, are necessary for maintenance of optimum health, toxicological since certain metals such as Pb, Cd, As and Hg and other organic pollutants are detrimental to optimum health and it is necessary effect controls to evaluate their suitability for human consumption as seafood, and environmental because seafood as marine bivalve mollusks are used as bioindicator organisms to assess bioavaility contaminant concentrations in coastal waters.

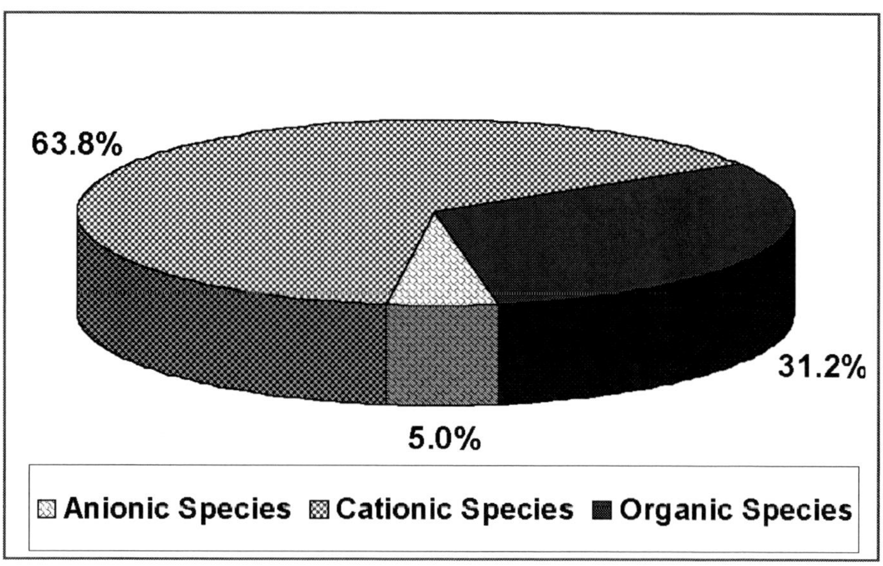

Figure 2.13. Distribution of FI determinations according to the analyte analyzed in seafood.

The available certified reference materials (CRMs) with a fish or shellfish matrix are the following: CRM627 (tuna Fish tissue for forms of arsenic), CRM422 (cod muscle for trace elements and methyl mercury), CRM463 (tuna fish for total mercury and methyl mercury), CRM464 (tuna fish for total mercury and methyl mercury), CRM349 (cod liver oil for PCBs), CRM598 (cod liver oil for organochlorine pesticides), CRM350 (mackerel oil for PCBs), CRM278R (mussel tissue for trace elements), CRM477 (mussel tissue for butyltin compounds), CRM542 (mussel for dc-saxitoxin), CRM543 (mussel for dc-saxitoxin), CRM668 (mussel for rare earth elements), CRM682 (mussel for PCBs), CRM718 (canned fresh herring for PCBs) and CRM719 (canned fresh chub for PCBs);from the Institute for Reference Materials and Measurements, Belgium; MURST-ISS-A2 (Antarctic krill for trace elements) from the Istituto Superiore di Sanità, Italy; SRM1588a (cod liver oil for PCBs, organochlorine pesticides and other organic contaminants), SRM2976 (mussel tissue for trace elements and methylmercury), SRM2977 (mussel tissue for organics and trace elements), SRM2978 (mussel tissue for organics and trace elements) and SRM1566b (oyster tissue for trace elements) from the Standard National Institute of Standards and Technology (NIST), USA; IAEA-352 (tuna fish for radioactive isotopes), IAEA-134 (cockle flesh for radioactive isotopes) and IAEA-MA-B-3/RN (fish flesh for radioactive isotopes) from International Atomic Energy Agency (IAEA), Austria; DOLT-4 (dogfish liver for trace elements), DORM-4 (dogfish muscle for trace elements); CARP-2 (fish-common carp for organochloride compounds), LUTS-1 (non-defatted lobster hepatopancreas for trace metals), TORT-2 (lobster hepatopancreas for trace metals) MUS-1B (mussel tissue for domoic acid (amnesic toxin)) and MUS-2 (mussel tissue for okadaic acid and dinophysistoxin-1 (diarrhetic shellfish toxin))from the NRC Institute for Marine Biosciences, National Research Council of Canada; LGC7101 (mackerel paste for proximates and nutrients) and LGC7160 (crab paste for trace elements) from LGC Standards, UK; PT33-K85-04 (fish based foodstuffs for proximates and trace elements) from Swedish National Food Administration; GBW08572 (prawn for trace elements) and GBW08571 (mussel for trace elements) from the Chinese National Research Center for Certified Reference Materials (Beijing, China) [195-198].

Taking into account all the analytes that have been determined by FI methodologies in seafood, the distribution of these FI determinations is shown in Figure 2.13. This Figure shows that the majority of the determinations were proposed for cationic species (63.8%), while anionic species are the analytes less determined by FI methodologies in these samples (5.0%) (information obtained by searching Chemical Abstracts between 1975 and September 2008 using Scifinder Scholar and the keywords "corresponding species", fish or shellfish and flow injection). In view of these results, it is clear that FI analytical determinations in seafood were developed practically only for inorganic analytes.

Cationic Species Determined by FI Methodologies

Over 100 papers have been published involving the determination of cationic species in seafood by FI methodologies. Thus, the elements that have been determined by using FI methodologies have been the following: Ag, Al, As, B, Ca, Cd, Co, Cr, Cu, Fe, Ga, Hg, In, Li, Mn, Mo, Ni, Pb, Sb, Se, Sn, Sr, Tl, V and Zn. As can be observed in Figure 2.14, the cationic species for those which have been proposed more new FI methodologies have been Hg, Cd, As and Pb with a percentage of publication of 22.1, 12.1, 8.7 and 8.7%, respectively (information obtained by searching Chemical Abstracts between 1975 and September 2008 using Scifinder Scholar and the keywords "corresponding species", fish or shellfish and flow injection).

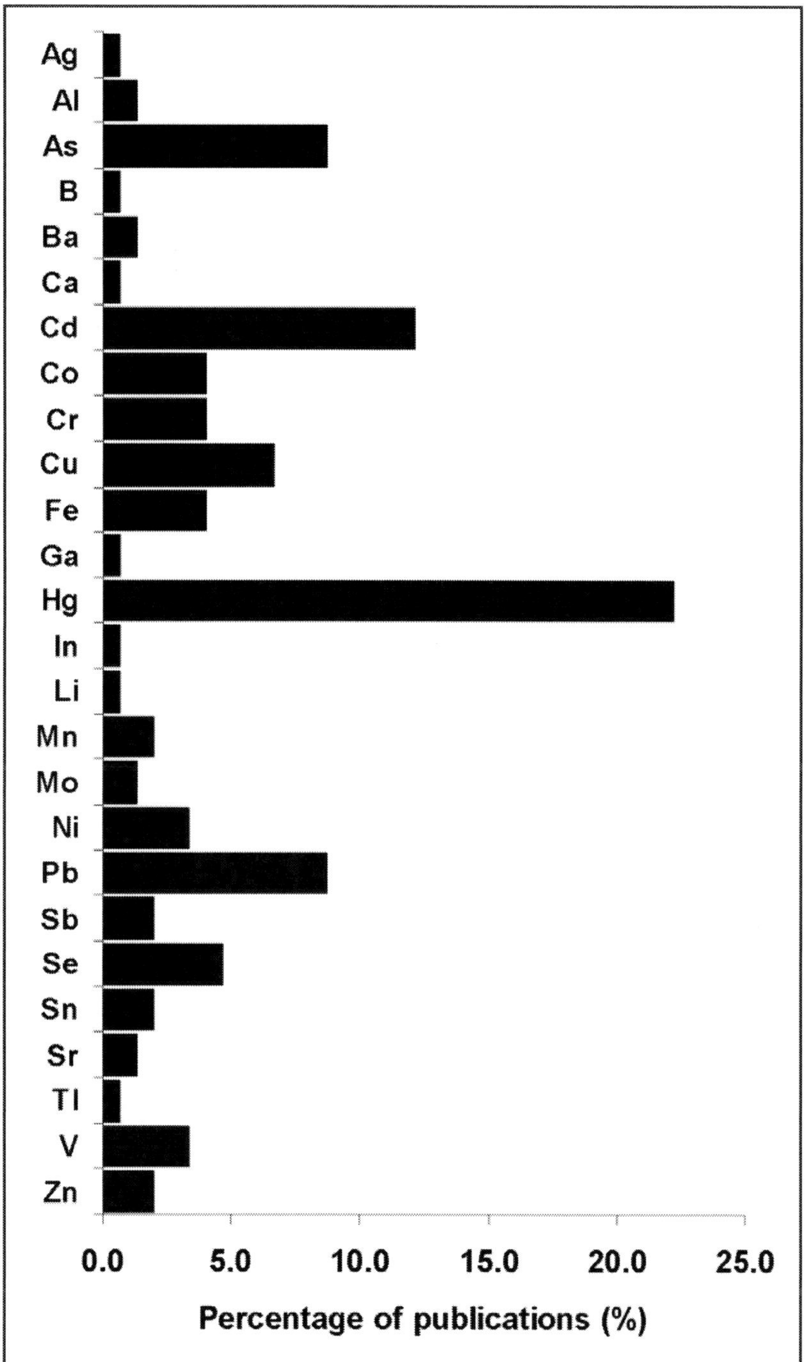

Figure 2.14. Distribution of FI determinations of cationic species in seafood.

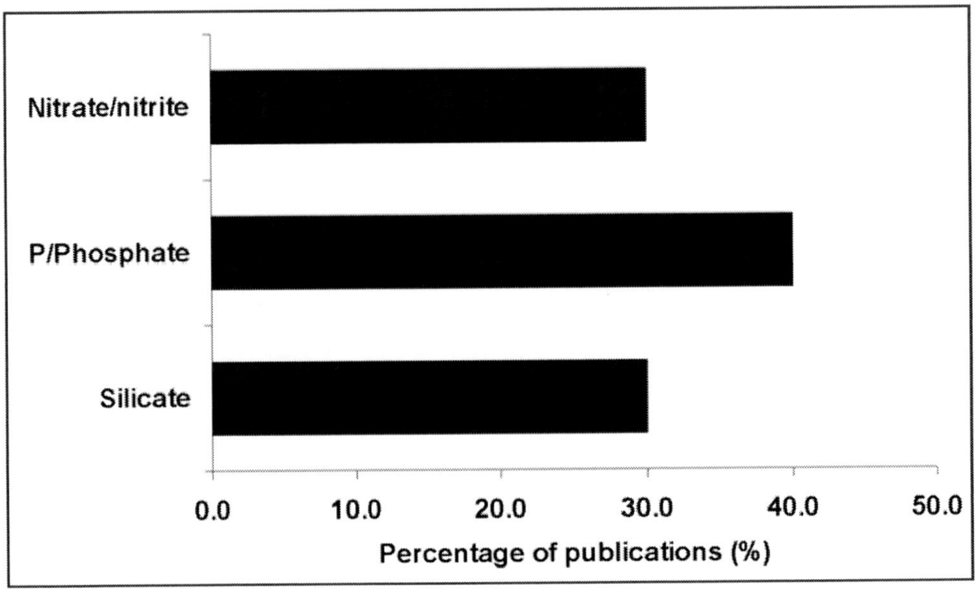

Figure 2.15. Distribution of FI determinations of anionic species seafood.

Otoliths (ear stones) are located within the inner ear of teleost fish, and are composed mainly of calcium carbonate (as aragonite). The composition of otoliths varies with ontogenic factors (e.g. size and age) and environmental factors (e.g. salinity, temperature, and water chemistry). Several recent studies have revealed the potential of otoliths for studying the structure of fish population and fish migration. Otoliths are highly complex and consist of proteins (3%), calcium carbonate (96%) and a variety of minor inorganic species [199].

Anionic Species Determined by FI Methodologies

Over 10 papers have been published involving the determination of anionic species in seafood by FI methodologies. Thus, the parameters that have been determined by using FI methodologies have been the following: nitrate/nitrite, phosphorous/phosphate and silicate. As can be observed in Figure 2.15, the anionic species for which has been proposed more new FI methodologies is phosphate with a percentage of publication of 40.0%, (information obtained by searching Chemical Abstracts between 1975 and September 2008 using Scifinder Scholar and the keywords "corresponding anionic species", fish or shellfish and flow injection).

Organic Species Determined by FI Methodologies

Over 45 papers have been published involving the determination of organic species in seafood by FI methodologies. Thus, the parameters that have been determined by using FI methodologies have been the following: amino acids (histidine, L-lysine and tyrosine), DNA/RNA, formaldehyde, histamine, hypoxanthine, polycyclic aromatic hydrocarbons (PAHs), diarrheic shellfish poisoning (DSP), paralytic shellfish poisoning (PSP), succinate/glutamate, trimethylamine, total lipid hydroperoxides (TLP), total volatile acids (TVA), total volatile basic nitrogen (TVBN), uric acid and vitamin B12. As can be observed in Figure 2.16, the organic species for those which have been proposed more new FI methodologies have been trimethylamine/TVBN, hypoxanthine and histamine with a percentage of publication of 34.1, 15.9, and 13.6%, respectively (information obtained by searching Chemical Abstracts between 1975 and September 2008 using Scifinder Scholar and the keywords "corresponding organic compound", fish or shellfish and flow injection).

Details and characteristics about the organic species determined by FI methodologies in seafood are discussed below (only the elements that have not been commented in the previous sections).

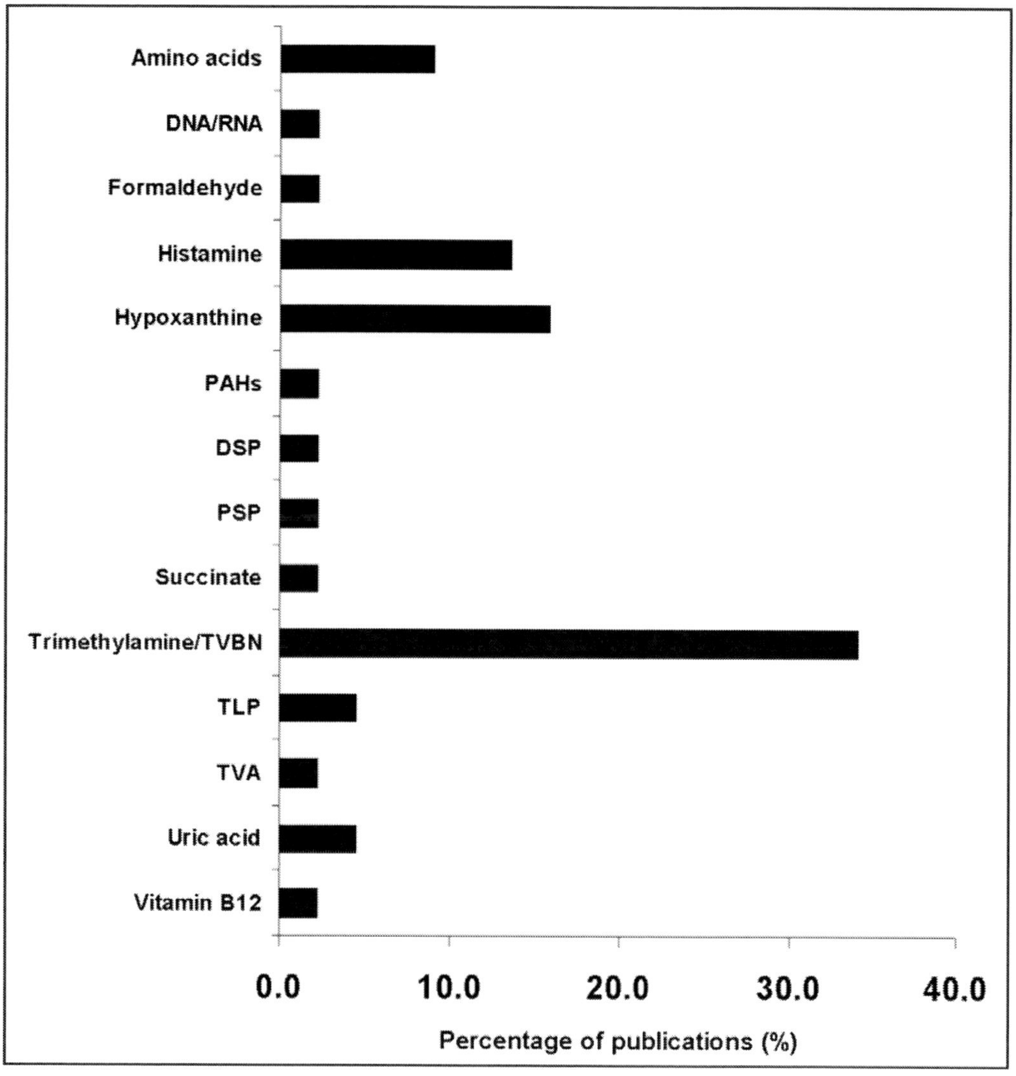

Figure 2.16. Distribution of FI determinations of organic species in seafood. PAHs: Polycyclic aromatic hydrocarbons; PSP: paralytic shellfish poisoning; DSP: diarrheic shellfish poisoning; TLP: total lipid hydroperoxides; TVA: total volatile acids; TVBN: total volatile basic nitrogen.

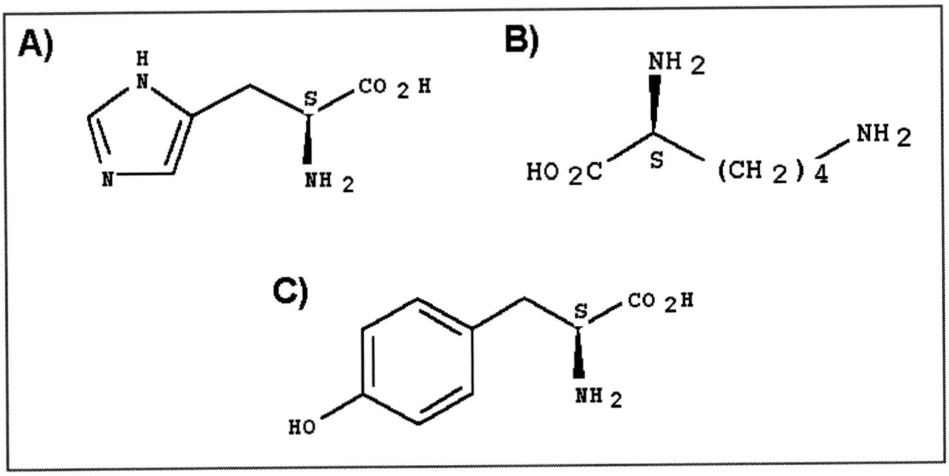

Figure 2.17. A) L-Histidine; B) L-Lysine; C) Tyrosine.

Amino Acids

Many species of fish are especially rich in the amino acid histidine. L-Histidine (His) (Figure 2.17) is an essential amino acid that can not be produced and converted by the liver, hence, it must be in the diet to be available to the body. Most His is found in meat, poultry and fish, and a small amount of free His exists in plants and fermented foods. His can degrade into the compound histamine by the bacterial decarboxylation if fish are not refrigerated promptly after catch. In the human body His is a direct precursor of histamine, and tissue histamine levels increase as the amount of dietary His increases [200].

L-Lysine (Lys) (Figure 2.17) is an essential amino acid for animal feed and human nutrition. It is in fact, one of many amino acids that the body needs for constant tissue repair and for growth as well. Lys is one of the most well known amino acids and is an essential component of all proteins and plays an important role in the metabolism of protein, carbohydrates and fatty acids. It is also a component of connective tissue and brain chemicals. Lys and another amino acid, methionine, forms carnitine. The body does not produce Lys, it must be acquired from food and/or supplementation. Thus, seafood are an excellent source of lysine especially halibut and shrimp [201].

Tyrosine (Tyr) (Figure 2.17) is a semi-essential amino acid that the body synthesizes from phenylalanine. Tyrosine is precursor of compounds with biological interest as melanin, L-dopa and dopamine, triiodo and tetraiodothyronine, norepinephrine and epinephrine. Tyr is related to several genetic metabolic disorders, including phenylketonuria, tyrosinaemia type

II, and tyrosinosis. Moreover, it is believed that several other disorders also are related to the metabolism of tyrosine, including Parkinson's disease, atherosclerosis, lung diseases, liver diseases and mental illnesses [202]. Structural analogs of tyrosine have been identified as biomarkers to provide an estimation of extent of oxidative damage to target biomolecules.

Nucleic Acids

Measurements of metabolic activity have been especially valuable as indicators of condition in studies of marine organisms, groups for which accurate determination of field metabolic rates is difficult. Molecular methods based on nucleic acid derived indices and the polymerase chain reaction has recently become an important tool in this field. One of the most widely used nucleic acid derived indices in marine ecology is the RNA:DNA ratio. Since the RNA:DNA ratio was first proposed as a biochemical indicator of the physiological and nutritional state of aquatic organisms in natural environment it has been continuously explored. These indices have been applied with success in marine ecology in microbial communities and in invertebrates and fishes. All this information can be applied in the determination of the potential survival of a captured organism in the marine environment [203]. Also RNA:DNA ratio has been used to test nutrient productivity models by demonstrating tight linkages between nearshore oceanographic processes (such as upwelling) and benthic rocky intertidal ecosystems.

Formaldehyde

In seafood and crustaceans formaldehyde is known to form post-mortem from the enzymatic reduction of trimethylamine-N-oxide (TMAO) to formaldehyde and dimethylamine. This compound accumulates during frozen storage, reacts with protein and subsequently causes protein denaturation and muscle toughness. The enzyme, trimethylamine- N-oxide demethylase, which is responsible for this process, is present in large quantities in the kidney and spleen of gadoid fish and to a less extent in the muscle tissue of these species. The determination of formaldehyde in fish is impeded by the formation of different stable chemical bondings between formaldehyde and the proteins in the fish tissue. This means that it is necessary to distinguish between free and bound formaldehyde [204]. Traditionally, the free formaldehyde is defined as the fraction, which is extractable at room temperature using, diluted trichloroacetic acid (10%) or perchloric acid (6%). If formaldehyde is extracted by steam distillation of a mixture of fish tissue and sulphuric acid (l-40% V/V) a greater content of formaldehyde is found. The difference between this acid labile amount of formaldehyde and the free formaldehyde is normally referred to as reversibly bound formaldehyde. Finally, there is an amount of formaldehyde, which is not extractable. This irreversibly bound formaldehyde fraction corresponds to the difference between the determined dimethylamine content and the sum of the free and the reversibly bound formaldehyde. Thus the total amount of formaldehyde is calculated from the amount of dimethylamine, assuming that dimethylamine and formaldehyde are formed in equimolar quantities.

Histamine

Histamine (1H-Imidazole-5-ethanamine) (Figure 2.18) is a biogenic amine produced during microbial decomposition of scombroid fish such as tuna and mackerel. Histamine has been associated with scombroid poisoning, which resembles an allergic reaction. The formation of histamine in fish and shellfish was mainly derived from decarboxylation of histidine by exogenous decarboxylase released from microflora associated with the specimens or surrounding seawater. Immediately after catching, fresh fish contains very low levels of histamine, but the content increases with the progress of fish decomposition. Therefore, histamine has also been proposed as a chemical index of freshness of fishes and poor hygienic quality of raw materials used and/or poor manufacturing conditions. According to US Food and Drug Administration (FDA), fish containing less than 10 mg/Kg histamine is of good quality, whereas a level of 30 mg/Kg indicates significant deterioration, and 50 mg/Kg is considered to be a conclusive evidence of decomposition. Because of its potential risk to health, a regulation level on histamine has been set in most of the countries or organizations. In the United States, the toxic level is set at 500 mg/Kg, and the caution level is 50 mg/Kg. The European Union has also set a caution level of 100-200 mg/Kg [205].

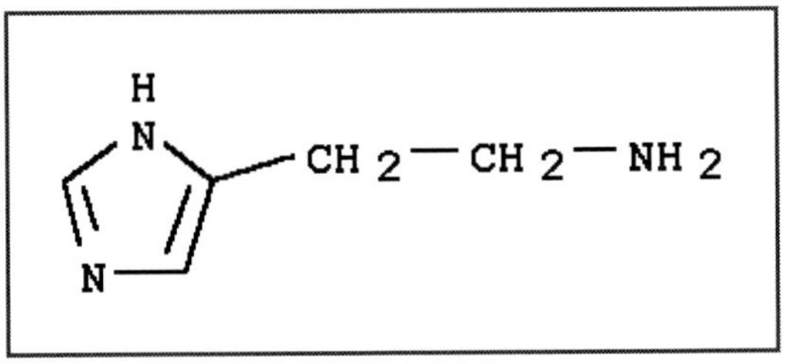

Figure 2.18. Histamine.

Hypoxanthine

Hypoxanthine (6H-Purin-6-one, 1,9-dihydro-) (Figure 2.19) is formed in dead fish by the breakdown of adenosine triphosphate (ATP), a natural biological substance found in live fish. Initially there is a build up in the level of hypoxanthine after the fish is harvested, but as bacteria begin to multiply and consume the hypoxanthine, the level drops off. Therefore, hypoxanthine levels in fish are indicative of freshness in terms of how long the fish has been dead as well as spoilage in terms of the multiplication of bacteria [206].

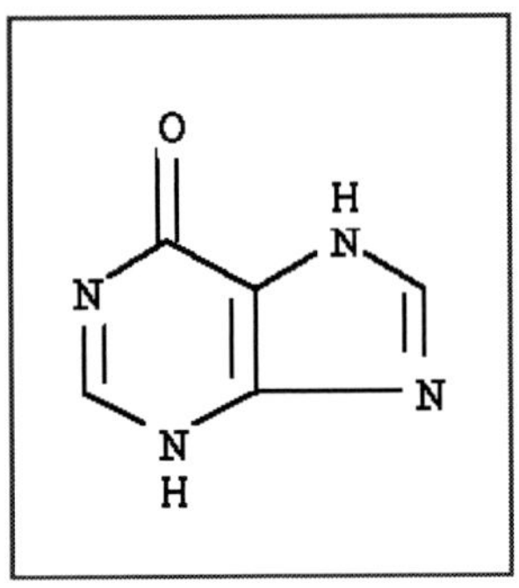

Figure 2.19. Hypoxanthine.

Diarrheic Shellfish Poisoning

Diarrhetic shellfish poisoning (DSP) is a gastrointestinal illness with symptoms such as diarrhea, nausea, vomiting, headache, chills and moderate to severe abdominal pain. DSP is usually a consequence of consuming contaminated shellfish that have ingested large quantities of toxic dinoflagellates of the genera *Dinophysis* and *Prorocentrum* through their filter feeding activities. Specifically, the main causative organisms, *Dinophysis* dinoflagellates, are known to produce at least nine toxins. Okadaic acid (1,7-Dioxaspiro[5.5]undec-10-ene-2-propanoic acid) (Figure 2.20), one of the principal causative toxins of DSP, is a complex, high molecular weight, lipophilic polyether. OA was first isolated from the sponge *Halichondria okadai*, and has been known to stimulate the phosphorylation of proteins that control sodium secretion by intestinal cells. OA may also enhance the phosphorylation of cytoskeletal or functional elements, which regulate permeability to solutes, thereby resulting in diarrheagenic effects of passive loss of fluids. OA is amongst the first recorded compounds that inhibit protein phosphatases 1 (PP1) and 2A (PP2A) [207].

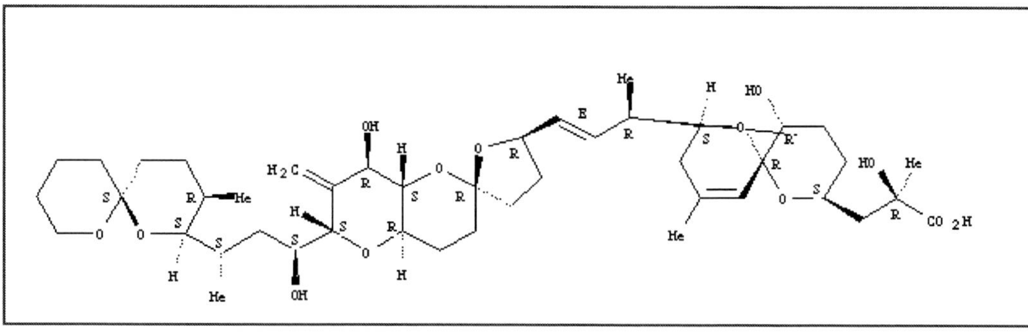

Figure 2.20. Okadaic acid.

Paralytic Shellfish Poisoning

Paralytic shellfish poisoning (PSP) is caused by a group of 26 naturally occurring potent neurotoxins. Saxitoxin (STX) (1H,10H-Pyrrolo[1,2-c]purine-10,10-diol, 2,6-diamino-4-[[(aminocarbonyl)oxy]methyl]-3a,4,8,9-tetrahydro-, (3aS,4R,10aS)-) is the most potent of the known PSP toxins (Figure 2.21). PSP toxins are biosynthesized by marine dinoflagellates of the genus *Alexandrium*, *Gymnodinium, Pyrodinium*, and some freshwater cyanobacteria like *Aphanizomenon flos-aquae, Cylindrospermopsis raciborskii, Anabaena circinalis*, respectively. These toxins induce dangerous intoxications by acting as potent sodium channel blockers. As a consequence of this reversible inhibition of the voltage-activated channels, those toxins cause paralytic symptoms such as respiratory insufficiency and they can finally lead to death. PSP toxins raise persistent problems due to their accumulation in shellfish [208].

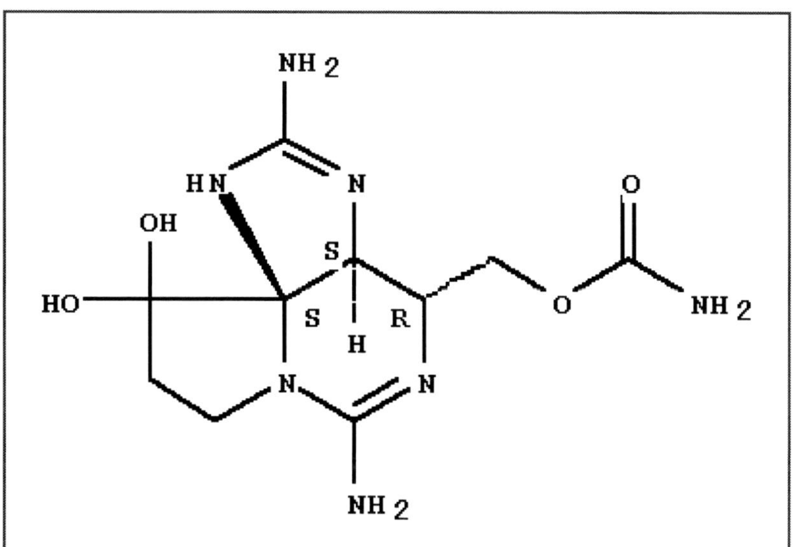

Figure 2.21. Saxitoxin.

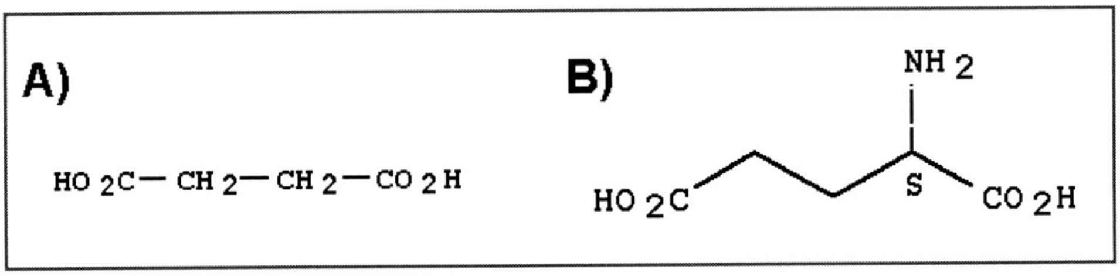

Figure 2.22. A) Succinic acid; B) L-Glutamic acid.

Succinate/Glutamate

The taste of food is composed of four basic tastes, namely sour, salty, bitter and sweet. In addition, umami taste intensifies a food flavor to make it more pleasant, more savory and more satisfying, and it is recognized as a fifth basic taste. Glutamate, 5'-inosinate, 5'-guanylate and succinate are well-known as umami components in foodstuffs. The umami taste presented by succinate differs from those obtained from glutamate, 5'-inosinate, and 5'-guanylate. Succinate assumes a peculiar umami taste accompanied with a strong sour and a slight astringency. Succinate is the main umami component in shellfishes such as corb shell and short-neck clam. These shellfishes contain large amount of succinate, which contributes a characteristic taste [209].

L-Glutamate is reported to be a potent neuroexcitatory amino acid involved in several behavior patterns. The ingestion of high concentrations of L-glutamate derived from food can induce the appearance of neurological manifestations such as Parkinson's and Alzheimer's disease [210].

Trimethylamine

Trimethylamine (TMA) (Methanamine, N,N-dimethyl-) is one of the most important organic amines, which is produced in the process of metabolism by animal organs and proteins. It is also one of the toxic gases in the biology field and foodstuff industry. It was reported that during the decay of fishes and seashells after death, some gaseous species such as TMA, dimethylamine (DMA), triethylamine (TEA), methylamine (MA) and ammonia (NH_3) are given off and the concentrations of these gases increase with the decreasing freshness of fishes. TMA is typical and common fish-odor substance in seafood, and is produced by the decomposition of trimethylamine N-oxide (TMAO) in sea creatures. The TMA measurement in seafood has been reported to be one of indicators for the evaluation of fish freshness (fresh: 0-1 mg/100g, initial corruption: 1-5 mg/100g, rotting fish: more than 6 mg/100g) [211].

Total Lipid Hydroperoxides

Lipid oxidation occurring in food is a major cause of quality deterioration in flavor, texture, consistency, and appearance. It is also a decisive factor in the causes of aging and some diseases. Lipid peroxidation is a major cause of the deterioration of fatty fish during storage [212].

Total Volatile Acids

Losses in organoleptic acceptability could be correlated with the increase in total volatile acids (TVA). Thus, TVA is a well known indicator for the decomposition of seafood products [213].

Total Volatile Basic Nitrogen

The determination of total volatile basic nitrogen (TVBN) has known to be the most common chemical parameter applied to evaluate the quality of fish and other meat products. Volatile amines such as trimethylamine (TMA), ammonia (NH_3) and dimethylamine (DMA) comprise total volatile basic nitrogen compounds (TVBN) that are released by the enzyme catalyzed decomposition of trimethylamine oxide (TMAO) by specific spoilage organisms [214]. TVBN levels have been recognized as useful indicators of seafood spoilage; under EU directive 95/149/EEC, the European Commission has specified that TVBN levels be used if sensory methods raise doubts about the freshness of seafood species. Critical limits have been set for groups of seafood species, quoted as mg TVBN per 100 g of seafood tissue, e.g. 30 mg TVBN per 100 g of tissue for cod species [215].

Uric Acid

After the death of fish, adenosine-5%-triphosphate (ATP) starts to degrade to uric acid (UA) through the following pathway: ATP→ADP→AMP→IMP→HxR→Hx→X→UA, where ADP is adenosine-5%-diphosphate, AMP is adenosine-5%-monophosphate, IMP is inosine monophosphate, HxR, inosine, Hx, hypoxanthine and UA, uric acid. To indicate fish freshness, the K value based on the degradation of these compounds in fish meat is defined as: $K=100(HxR + Hx)/(IMP + HxR + Hx)$ as ATP, ADP and AMP generally disappear around 24 h after the death. Hx and HxR concentrations depend upon the species of fish. IMP is one of the major contributing factors to the pleasant flavor of fresh fish. The accumulation of Hx and/or xanthine (X) during the storage results in an off-taste. The concentration of Hx, one of the intermediates of these reactions increases with prolonged storage and thus can be used as an indicator of fish meat freshness [216]. Therefore, simultaneous determination of these compounds is also necessary for a rapid estimation of freshness.

Vitamin B_{12}

Vitamin B_{12} (Figure 2.23), also known as cobalamin ($C_{63}H_{88}$-$CoN_{14}O_{14}P$) containing a cobalt ion within tetrapyrrole ring, is a very unusual biochemical species. Vitamin B_{12} is naturally found in animal foods including seafood. It helps maintain healthy nerve cell and red blood cell, and is also necessary to make DNA, the genetic material in all cells. The deficiency of Vitamin B_{12} may lead to fatigue, weakness, nausea, constipation, weight loss, and even as severe as addisonian pernicious anemia. The shellfishes which siphon large quantities of vitamin B_{12} synthesizing microorganisms in the sea are known to be excellent sources (>10 μg/100 g) of B_{12}. The B_{12} synthesizing microorganisms can synthesize various corrinoids (including B_{12} analogues inactive for mammals) with a different base in the α - ligand. In some shellfish, B_{12} contents determined by the *Lactobacillus leichmannii* microbiological method have been shown to be several fold greater than the values by intrinsic factor (IF)-chemiluminescence method, suggesting that most of the B_{12} found in the shellfish are inactive B_{12} analogues so that they may not be bioavailable in mammals [217].

Conclusion

Flow techniques have been widely used in the analysis of environmental, toxicological or nutritional parameters in marine samples. Seawater and estuarine water are the samples more analyzed by using FI methodologies, offering the possibility of solving a number of specific analytical problems in marine chemistry. By the other hand, inorganic compounds, above all cationic species, are the most analyzed elements in marine samples. Nevertheless, it is significant that marine samples as marine seabirds, marine mammals or marine air are not still analyzed by FI methodologies.

References

[1] Lenihan, J.M.A. &. Fletcher, W.W (1977). *The marine environment.* New York: Academic Press.

[2] Borja, A. & Collins, M. (2004). *Oceanography and marine environment of the Basque country.* Amsterdam: Elsevier.

[3] Roberts, J. (2006). *Marine environment protection and biodiversity conservation: the application and future development of the imo's particularly sensitive sea area concept.* Heidelberg: Springer.

[4] Albaigés, J. (1989). *Marine pollution.* New York: Hemisphere Publishing Corporation.

[5] Sinha, P.C. (1998). *Marine pollution.* New Delhi: Anmol Publications.

[6] Andrady, A.L. (2003). *Plastics and the environment.* New York: Wiley-IEEE Press.

[7] Sindermann, C.J. (2006). *Coastal pollution: effects on living resources and humans.* Boca-Ratón (FL): CRC Press.

[8] Harrison, R.M. (2001). *Pollution: Causes, effects and control.* Cambridge: The Royal Society of Chemistry.

[9] Kennish, M.J. (2001). *Practical handbook of marine science.* Boca-Ratón (FL): CRC Press.

[10] Calow, P. (1998). *Handbook of environmental risk assessment and management.* Oxford: Blackwell Publishing.

[11] Leonard, E.N., Ankley, G.T. & Hoke, R.A. (1996). *Environ. Toxicol. Chem.* 15, 2221-2232.

[12] Lussier, S.M., Boothman, W.S., Poucher, S., Champlin, D. & Helmstetten, A. (1999). *Environ. Toxicol. Chem.* 18, 889-898.

[13] Yap, C. K., Ismail, A., Omar, H. & Tan, S. G.(2004). *Environ.Int.* 29, 1097-1104.

[14] Manfra, L., Moltedo, G., Virno Lamberti, C., Maggi, C., Finoia, M. G., Giuliani, S., Onorati, F., Gabellini, M., Di Mento, R. & Cicero, A. M. (2007). *Arch. Environ. Contam. Toxicol.* 53, 183-190.

[15] Gonzalez, J.J., Vinas, L., Franco, M.A., Fumega, J., Soriano, J. A., Grueiro, G., Muniategui, S., Lopez-Mahia, P., Prada, D., Bayona, J. M., Alzaga, R. & Albaiges, J. (2006). *Mar.Pollut. Bull.* 53, 250-259.

[16] Perez, C., Velando, A., Munilla, I., Lopez-Alonso, M. & Oro, D. (2008). *Environ.Sci. Technol.* 42, 707-713.

[17] Sobek, A., Cornelissen, G., Tiselius, P. & Gustafsson, O. (2006). *Environ.Sci. Technol.* 40, 6703-6708.

[18] Pikkarainen, A.L. (2007). *Chemosphere* 68, 17-24.
[19] Gioia, R., Nizzetto, L., Lohmann, R., Dachs, J., Temme, C. & Jones, K.C. (2008). *Environ.Sci. Technol.* 42, 1416-1422.
[20] Jurado, E., Jaward, F.M., Lohmann, R., Jones, K.C., Simo, R. & Dachs, J. (2004). *Environ.Sci. Technol.* 38, 5505-5513.
[21] Rios, L.M., Moore, C. & Jones, P.R. (2007). *Mar.Pollut. Bull.* 54, 1230-1237.
[22] Ni, Z., Hang, H., Li, A., He, B. & Xu, F. (1991). *J. Anal. At. Spectrom.* 6, 385-387.
[23] Garaventa, F., Faimali, M. & Ceraci, S. (2003). *Pitture e Vernici, European Coatings* 79, 13-17.
[24] Atienza, J., Herrero, M. A., Maquieira, A. & Puchades, R. (1991). *Crit. Rev. Anal. Chem.* 22, 331-44.
[25] Shpigun, L.K., Kolotyrkina, I.Y. & Zolotov, Y. A. (1992). *Anal. Chim. Acta* 261, 307-314.
[26] Atienza, J., Herrero, M. A., Maquieira, A. & Puchades, R. (1992). *Crit. Rev. Anal. Chem.* 23, 1-14.
[27] Motomizu, S. & Oshima, M. (1996). *Nippon Kaisui Gakkaishi* 50, 363-377.
[28] Hirata, S. (2000). *J. Flow Injection Anal.* 17, 23-35.
[29] Wessel, J.K. (2003). *The handbook of advanced materials: Enabling new designs.* New York: John Wiley & Sons.
[30] Encyclopedia Britannica Online. http://www.britannica.com.
[31] Boyd, C.E. (2000). *Water quality: An introduction*. Heidelberg: Springer.
[32] Brown, J., Colling, A., Park, D., Phillips, J., Rothery, D. & Wright, J. (1995). *Seawater: its composition, properties and behaviour*. Oxford: Pergamon Press.
[33] Sadiq, M. (1992). *Toxic metal chemistry in marine environments.* Boca-Ratón (FL): CRC Press.
[34] Gianguzza, A., Pelizzetti, E. & Sammartono, S. (2002). *Chemistry of marine water and sediments.* Berlin: Springer Verlag.
[35] Bowen, H.J.M. (1979). Environmental chemistry of the elements. London: Academic Press.
[36] Drever, J.I. (1982). *The geochemistry of natural waters*. New Jersey: Prentice Hall
[37] Summerhayes, C. P. & Thorpe, S. A. (1996). *Oceanography: An illustrated guide*. New York: John Wiley.
[38] Kester, D. R., Duedall, I. W., Connors, D. N. & Pytkowicz, R. M. (1967). *Limnology & Oceanography* 12, 176-179.
[39] Baxter, D.C. & Frec., W. (1989). *Anal. Chim. Acta* 225, 175-183.
[40] Greenway, G.M., Nelms, S.M., Skhosana, I. & Dolman, S.J.L. (1996). *Spectrochim. Acta* 51B, 1909-1915.
[41] Liang, Y., Yuan, D., Li, Q. & Lin, Q. (2006). *Anal. Chim. Acta* 571, 184-190.
[42] http://www.sigmaaldrich.com/catalog/search/ProductDetail/ALDRICH/390542/
[43] Arancibia, V. & Muñoz, C. (2007). *Talanta* 73, 3, 546-552.
[44] http://inms-ienm.nrc-cnrc.gc.ca/calserv/crm_files_e/CASS-4_e.pdf
[45] http://inms-ienm.nrc-cnrc.gc.ca/calserv/crm_files_e/NASS-5_e.pdf
[46] http://inms-ienm.nrc-cnrc.gc.ca/calserv/crm_files_e/SLEW-3_e.pdf
[47] http://inms-ienm.nrc-cnrc.gc.ca/calserv/crm_files_e/MOOS-1_e.pdf
[48] Fukushi, K., Nakayama, Y. & Tsujimoto, J. (2003). *J. Chromatog. A* 1005, 197-205.
[49] Clancy, V.& Willie, S. (2004). *Anal. Bioanal. Chem.* 378, 1239-1242.

[50] Okamoto, T., Fukushi, K.,Takeda, S. & Wakida, S. (2007). *Electrophoresis* 28, 3447-3452.
[51] Smith, K.M. & Wells, D.E. (2007). *J.Environ. Monitor*. 9, 708-717.
[52] Lawrence, M.G. & Kamber, B.S.(2007). *Geostand. Geoanal. Res*. 31, 95-103.
[53] Ensafi, A.A. & Ghaderi, A.R. (2007). *J. Hazardous Materials* 148, 319-325.
[54] Grotti, M. & Frache, R. (2007). *J. Anal.At. Spectrom*. 22, 1481-1487.
[55] Sohrin, Y., Urushihara, S., Nakatsuka, S., Kono, T., Higo, E., Minami, T., Norisuye, K. & Umetani, S. (2008). *Anal. Chem*. 80, 6267-6273.
[56] Meeravali, N.N. & Jiang, S.J. (2008). *J. Anal.At. Spectrom*. 23, 555-560.
[57] http://www.irmm.jrc.be/html/reference_materials_catalogue/catalogue/attachements/BCR-505_cert.pdf
[58] http://www.irmm.jrc.be/html/reference_materials_catalogue/catalogue/attachements/BCR-579_cert.pdf
[59] Vereda Alonso, E.I., Palomo Gil, L., Siles Cordero, M.T.; Garcia de Torres, A., Cano Pavon, J.M. (2001). *J. Anal. At. Spectrom*. 16, 293-295.
[60] Locatelli, C. (2002). *J. AOAC Int*. 85, 233-242.
[61] Cairns, W. R. L., Ranaldo, M., Hennebelle, R., Turetta, C., Capodaglio, G., Ferrari, C. F., Dommergue, Aurelien, Cescon, P. & Barbante, C. (2008). *Anal. Chim.Acta* 622, 62-69.
[62] Directive of the European Parliament and of the Council of 6 September 2006 (2006/44/EC). Relative to the quality of fresh waters needing protection or improvement in order to support fish life. Off. J. Eur. Union. 25 November 2006, 20-31. http://eurlex.europa.eu/LexUriServ/LexUriServ.do?uri=OJ:L:2006:264:0020:0031:EN:PDF
[63] Directive of the European Parliament and of the Council of 12 December 2006 (2006/113/EC). on the quality required of shellfish waters. Off. J. Eur. Union. 27 December 2006, 14-20. http://eurlex.europa.eu/LexUriServ/LexUriServ.do?uri=OJ:L:2006:376:0014:0020:EN:PDF
[64] United States Environmental Protection Agency (EPA). Water Quality Criteria. http://www.epa.gov/waterscience/criteria/
[65] United States Environmental Protection Agency (EPA) (1982). A Compilation of Water Quality Standards for Marine Waters.
[66] Wang, T., Lee, H. K. & Li, S.F.Y. (1998). *J. Liq. Chrom. Relat. Tech*. 21, 2485-2496.
[67] Hema, S. (2007). Int. J. Chem. Sci. 5, 1160-1162.
[68] Bishop, J.K.B. (1990). *Anal. Chem*. 62, 553-557.
[69] Klinkhammer, G. P.& Chan, L. H. (1990). *Anal.Chim. Acta* 232, 323-329.
[70] Lebel, J. & Poisson, A. (1976). *Mar. Chem*. 4, 321-32.
[71] Khalifa, M. E. (1996). *Chemia Analityczna* 41, 357-362.
[72] Tsunogai, S., Hishimura, M. & Nakaya, S. (1968). *J. Oceanogr. Soc. Japan* 24, 153-159.
[73] Perry, H., Trigg, C., Larsen, K., Freeman, J., Erickson, M. & Henry, R. (2001). *Aquaculture* 198, 197-208.
[74] Shijo, Y., Shimizu, T. & Tsunoda, T. (1989). *Anal. Sci*. 5, 65-68.
[75] Ferguson, E.A.& Hogstrand, C. (1998). *Environ.Toxicol. Chem*. 17, 589-593.
[76] Miller, L.A.& Bruland, K.W.(1995). *Environ. Sci.Technol*. 29, 2616-2621.
[77] Savenko, V.S. & Tagirov, B.R. (1996). *Okeanologiya* 36, 231-234.

[78] Barriada, J.L., Tappin, A.D., Evans, E.H. & Achterberg, E.P. (2007). *Trends Anal. Chem.* 26, 809-817.

[79] Massey, R.C. & Taylor, D. (1989). Alum*inium in food and the environment.* Cambridge: The Royal Society of Chemistry.

[80] Arancibia, V. & Muñoz, C. (2007). *Talanta* 73 546–552.

[81] La Pera, L., Di Bella, G., Rando, R., Lo Turco, V. & Dugo, G. (2008). *Environ. Monit. Assess.* 145, 119-126.

[82] Neff, J. M. (1996). *Environ.Toxicol. Chem.* 16, 917-927.

[83] Gonzalez-Dávila, M. (1995). *Mar.Chem.* 48, 215-235.

[84] Falkner, K.K. & Edmond, J.M. (1990). *Earth Planet. Sci. Lett.* 98, 208-221.

[85] Vlassopoulos, D. & Wood, S.A. (1990). *Geochim. Cosmochim. Acta* 54, 3-12.

[86] Bryjak, M., Wolska, J.& Kabay, N. (2008). *Desalination* 223, 57-62.

[87] Zeebe, R.E., Sanyal, A., Ortiz, J.D. & Wolf-Gladrow, D.A. (2001). *Mar.Chem.* 73, 113-124.

[88] Ohyama, J., Maruyama, F. & Dokiya, Y. (1987). Anal. Sci. 3, 413-416.

[89] Savenko, V.S., Savenko A.V. & Shatalov I. A. (1998). *Doklady Earth Sci.* 361A, 891-893.

[90] de Baar, H. J. W., Saager, P. M., Nolting, R. F. & van der Meer, J. (1994). *Mar. Chem.* 46, 261-281.

[91] Peng, J., Liu, R. Liu, J., He, B., Hu, X.L. & Jiang, G.B. (2007). *Spectrochim. Acta* 62B, 499-503.

[92] Florence, T. M.& Batley, G. E. (1980). *Crit. Rev. Anal. Chem.* 9, 219-296.

[93] Vega, M. & van den Berg, C.M.G. (1997). *Anal. Chem.* 69, 874-881.

[94] Cosovic, B., Degobbis, D., Bilinski, H. & Branica, M.(1982). *Geochim. Cosmochim. Acta* 46, 151-158.

[95] Pan, P. & Susak, N.J. (1991). *Geochem. J.* 25, 411-420.

[96] Ellwood, M.J. &. van den Berg, C.M.G. (2001). *Mar. Chem.* 75, 33-47.

[97] Mugo, R.K. & Orians, K.J. (1992). *Anal. Chim. Acta* 271 1-9.

[98] Elderfield, H. (1970). *Earth Planet. Sci. Lett.* 9 10-16.

[99] Gardner, M. J. & Ravenscroft, J. E.(1996). *Fresenius J. Anal. Chem.* 354, 602-605.

[100] Yebra, M.C., Carro, N., Enríquez, M.F., Moreno-Cid, A. & García, A. (2001). *Analyst* 126, 933-937.

[101] Bruland, K.W., Rue, E.L., Donat, J.R., Skrabal, S.A. & Moffett, J.W. (2000). *Anal. Chim. Acta* 405, 99-113.

[102] Delgadillo-Hinojosa, F., Zirino, A. & Nasc, C. (2008). *Mar. Environ.Res.* 66, 404-411.

[103] Eric P. Achterberg, Toby W. Holland, Andrew R. Bowie, R. Fauzi C. Mantoura, Paul J. Worsfold (2001). *Anal. Chim.Acta* 442, 1-14.

[104] Roy, E.G., Jiang, C., Wells, M.L. & Tripp, C. (2008). *Anal. Chem.* 80, 4689-4695.

[105] Wu, J. (2007). *Mar. Chem.*103, 370-381.

[106] Mason, R.P., Fitzgerald, W.F. & Morel, F.M.M.(1994). *Geochim.Cosmochim.Acta* 58, 3191-3198.

[108] Whalina, L., Kimb, E.H. & Mason, R. (2007). *Mar. Chem.* 107, 278-294.

[109] Miyazaki, A., Kimura, A. & Tao, H. (2008). *Bunseki Kagaku* 57, 555-561.

[110] Doi, T., Obata, H. & Maruo, M. (2004). *Anal. Bioanal. Chem.* 378, 1288-1293.

[111] Stumm, W. & Morgan, J.J. (1996). *Aquatic Chemistry.* New York: Wiley.

[112] Sun, Y.C., Mierzwa, J. & Chi-Ren Lan, C.R. (2000). *Talanta* 52, 417-424.

[113] Li, Q.P., Zhang, J.Z., Millero, F.J. & Hansell, D.A. (2005). *Mar. Chem.* 96, 73-85.
[114] Sadiq, M. (1989). *Hydrobiologia* 176-177, 225-232.
[115] Patterson, C., Settle, D. & Glover, B. (1976). *Mar. Chem.* 4, 305-319.
[116] Byrne, R.H. (1981). *Nature* 290, 487-489.
[117] Nakata, R., Okazaki, S. & Fujinaga, T. (1982). *Bull. Chem. Soc. Japan* 55, 2283-2284.
[118] Liang, P., Liu, Y. & Guo, L. (2005). *Spectrochim. Acta* 60B, 125-129.
[119] Wang, Z.L. & Yamada, M. (2007). *Environ. Geol.* 52, 779-787.
[120] Saito, K. & Miyauchi, T. (1982). *J. Nucl. Sci. Technol.* 19, 145-150.
[121] Hirose, K. & Aoyama, M. (2002). *Anal. Bioanal. Chem.* 372, 418-420.
[122] Bertine, K.K., Koide, M. & Goldberg, E.D.(1993). *Mar. Chem.* 42, 199-210.
[123] Filella, M., Belzile, N. & Chen, Y.W. (2002). *Earth-Sci. Rev.* 57, 125-176.
[124] Cutter, G.A. & Cutter, L.S. (2001). *Deep-Sea Research II* 48, 2917-2931.
[125] Riso, R.D., Waeles, M., Garbarino, S. & Le Corre, P. (2004). *Anal. Bioanal. Chem.* 379, 1113-1119.
[126] Foti, C., Gianguzza, A., Piazzese, D. & Trifiletti, G. (2000). *Chem. Spec. Bioavailability* 12, 41-52.
[127] Zhang, Q., Minami, H, Inoue, S. & Atsuya, I. (2001). *Fresenius J. Anal. Chem.* 370, 860-864.
[128] Marina Mlakar, M. (2004). *Electroanalysis* 16, 217-223.
[129] García, E.A. &. Blanco Gomis, A. (2002). *Anal. Lett.* 35, 2337-2346.
[130] Savenko, V. S.(2001). *Geokhimiya* 1, 95-98.
[131] Rehkämper, M. & Nielsen, S.G. (2004). *Mar.Chem.* 85, 125-139.
[132] Dupont, V., Auger, Y. Jeandel, C. & Wartel, M. (1991). *Anal. Chem.* 63, 520-522.
[133] Stanley, J. K. & Byrne, R. H. (1990). *Geochim.Cosmochim. Acta* 54, 753-760.
[134] Bruland, K. W.(1989). *Limnol.Oceanogr.* 34, 269-285.
[135] Prenesti, E., Ferrara, E., Berto, S., Fisicaro, P. & Daniele, P.G. (2007). *Anal. Bioanal. Chem.*, 388, 1877-1883.
[136] Uraisin, K., Takayanagi, T., Oshima, M., Nacapricha, D. & Motomizu, S. (2006). *Talanta* 68, 951-956.
[137] Magazinovic, R.S., Nicholson, B.C., Mulcahy, D.E. & Davey, D.E. (2004). *Chemosphere* 57, 329-335.
[138] Ramos, A.A., Ohde, S., Hossain, M.M.M., Ozaki, H., Sirirattanachai, S. & Apurado, J.L. (2005). *J. Radioanal. Nucl. Chem.* 266, 19-29.
[139] Truesdale, V.W. (2007). *Mar. Chem.* 104, 266-281.
[140] Rong, L. & Takeuchi, T. (2004). *J. Chromatogr.* 1042A, 131-135.
[141] Wong, G.T.F. & Zhang, L.S. (2008). Mar.Chem. 111, 22-29.
[142] Gardolinski, P.C.F.C., David, A.R.J. & Worsfold, P.J. (2002). *Talanta* 58,1015-1027.
[143] Patey, M.D., Rijkenberg, M.J.A., Statham, P.J., Stinchcombe, M.C., Achterberg, E.P. & Mowlem, M. (2008). *Trends Anal. Chem.* 27, 169-182.
[144] Chen, G., Yuan, D., Huang, Y., Zhang, M. & Bergman, M.. (2008). Anal. Chim. Acta 620, 82-88.
[145] Calvo Quintana, J., Idrissi, L., Palleschi, G., Albertano, P., Aminec, A., El Rhazi, M. & Moscone, D. (2004). *Talanta* 63, 567-574.
[146] Floch, J., Blain, S., Birot, D. & Treguer, P. (1998). *Anal.Chim. Acta* 377, 157-166.
[147] Butler, J.N. (1991). *Carbon dioxide equilibria and their applications*. Boca-Ratón (FL): CRC Press.

[148] Al-Farawati, R. & van den Berg, C.M.G. (1997). *Mar. Chem.* 57, 277-286.
[149] http://www.epa.gov/ttn/atw/hlthef/acrolein.html.
[150] Beauchamp, R.O., Andjelkovich, D.A., Kligerman, A.D., Morgan, K.T. & Heck, H.A. (1985). *CRC Crit. Rev. Toxicol.* 14, 309-380.
[151] Gibb, S.W., Mantoura, R.F.C. & Liss, P.S. (1995). *Anal. Chim. Acta* 316, 291-304.
[152] Aminot, A. & Kérouel, R. (2006). *Mar. Chem.* 98, 223-240.
[153] Gonzalez-Mazo, E. & Gomez-Parra, A. (2003). *Anionic surfactants in marine and estuarine environments.* Comprehensive Analytical Chemistry XL. Knepper, T.P., Barcelo, D. & De Voogt, P. (Eds). Amsterdam: Elsevier.
[154] Peschka, M., Petrovic, M., Knepper, T.P. & Barceló, D. (2007). *Anal. Bioanal. Chem.* 388, 1227-1234.
[155] Liu, Y., Ji, H., Xin, H. & Liu, L. (2006). *Journal of Ocean University of China* 5, 137-140.
[156] Cappiello, A., Famiglini, G., Palma, P., Pierini, E., Trufelli, H., Maggi, C., Manfra, L. & Mannozzi, M. (2007). *Chemosphere* 69, 554-560.
[157] Santinelli, C., Gasparini, G.P., Nannicini, L. & Seritti, A. *Deep-Sea Research I* 49, 2203-2219.
[158] Nuccio, J., Seaton, P.J. & Kieber, R.J. (1995). *Limnol. Oceanogr.* 40, 521-527.
[159] Ekdahl, A., Pedersen, M. & Abrahamsson, K. (1998). *Mar.Chem.* 63, 1-8.
[160] Vong, L., Läes, A. & Blain, S. (2007). *Anal. Chim. Acta* 588, 237-244.
[161] Pandey, S., Kumar, R., Sharma, S., Nagpure, N.S., Srivastava, S. & Verma, M.S. (2005). *Ecotoxicol. Environ. Safety* 61, 114-120.
[162] Carriger, J.F. & Rand, G.M. (2008). *Ecotoxicology* 17, 680-696.
[163] Di Necio, M.S., Pistonesi, M.F., Fragoso, W.D., Pontes, M.J.C., Goicoechea, H.C., Araujo, M.C.U. & Fernández Band, B.S. (2007). *Microchem. J.* 85, 194-200.
[164] Bihari, N., Fafanđel, M. & Piškur, V. (2007). *Arch. Environ. Contam. Toxicol.* 52, 379-387.
[165] Wang, X.H., Hong, H.S., Mu, J.L., Lin, J.Q. & Wang, S.H. (2008). *J. Environ. Sci. Health* 43A, 219-226.
[166] Kim, D., Nakashima, T., Matsuyama, Y., Niwano, Y., Yamaguchi, K. & Oda, T. (2007). *J. Plankton Res.* 29, 241-247.
[167] Chester, R. (2003). *Marine geochemistry.* Oxford: Blackwell Publishing.
[168] Ross, D.A. (1982). *Introduction to oceanography.* New Jersey: Prentice Hall.
[169] Lee, S. H., Povinec, P. P., Wyse, E. & Hotchkis, M. A. C. (2008). *Applied Radiation and Isotopes* 66, 823-828.
[170] Carvalho, F.P., Villeneuve, J.P. & Cattini, Ch. (1999). *Int. J. Environ. Anal. Chem.* 75, 315-329.
[171] Waheed, S., Rahman, A., Daud, M., Zaidi, J. H. & Ahmad, S.(2004). *Radiochim. Acta* 92, 939-944.
[172] Ulberth, F. (2006). *Anal. Bioanal.Chem.* 386, 1121-1136.
[173] Giusti, L. & Zhang, H. (2002). *Environ.Geochem. Health.* 24, 47-65.
[174] Wahlen, R. & Wolff-Briche, C. (2003). *Anal. Bioanal. Chem.* 377, 140-148.
[175] Jiménez-Moreno, M., Pacheco-Arjona, J., Rodriguez-Gonzalez, P., Preud'Homme, H., Amouroux, D. & Donard, O. F. X. (2006). *J. Mass Spectrom.* 41, 1491-1497.
[176] Muñoz, J., Gallego, M. & Valcarcel, M. (2005). *Anal. Chim. Acta* 548, 66-72.

[177] Gao, E., Jiang, G., He, B., Yin, Y. & Shi, J. (2008). *J. Anal. At. Spectrom.* 23, 1397-1400.
[178] Rouxel, O., Galy, A. & Elderfield, H. (2006). *Geochim. Cosmochim. Acta* 70, 3387-3400.
[179] Colodner, D.C., Boyle, J.E.A. & Edmond, J.M. (1993). *Anal. Chem.* 65, 1419-1425.
[180] Haus, N., Zimmermann, S., Wiegand, J. & Sures, B. (2006). *Chemosphere* 66, 619-629.
[181] Ramesh, A. & Subramanian, M. S. (1997). *Analyst* 122, 1605-1610.
[182] Marklevitz, S.A.C., Almeida, E., Flemming, J. & Hellou, J. (2008). *J. Soils Sediments* 8, 86-91.
[183] Marklevitz, S.A.C., Almeida, E., Flemming, J. & Hellou, J. (2008). *J. Soils Sediments* 8, 92-97.
[184] Pratt, C., Warnken, J., Leeming, R., Arthur, M.J. & Grice, D.I. (2008). *Organic Geochem.* 39, 353-369.
[185] Vasquez, J.A & Guerra, N. (1996). *Hydrobiologia* 326/327, 327-333.
[186] Carrilho, E.N.V.M., Nobrega, J.A. & Gilbert, T.R. (2003). *Talanta* 60, 1131-1140.
[187] Okamoto, K., Yoshinaga, J. & Morita, M. (1996). *Microchim. Acta* 123, 15-21.
[188] Outola, I., Filliben, J. & Inn, K.G.W. (2008). *J. Radioanal. Nucl. Chem.* 276, 329-334.
[189] VanAlstyne, K.L. & Puglisi, M.P. (2007). *Aquat. Sci.* 69, 394-402.
[190] Sunda, W.G., Litaker, R.W., Hardison, D.R. & Tester, P.A. (2005). *Mar. Ecol. Prog.* Ser. 287, 11-22.
[191] Braven, J., Evens, R. & Butler, E.I. (1984). *Chem. Ecol.* 2, 11-21.
[192] Devillé, C., Gharbi, M., Dandrifosse, G. & Peulen, O. (2007). *J. Sci. Food Agric.* 87, 1717-1725.
[193] Ahmed, F.E. (1991). *Seafood safety*. Washington: National Academies Press.
[194] Yusof, A. M., Yanta, N. F. & Wood, A.K.H. (2004). *J. Radioanal. Nucl. Chem.* 259, 119-127.
[195] Quevauviller, P., Draaek, I., Muntau, H., Bianchi, M., Bortoli A.& Griepink B.(1996). *Trends Anal. Chem.* 15, 160-167.
[196] Caroli, S., Senofonte, O., Caimi, S., Pauwels, J., Kramer, G.N. & Robouch, P. (2001). *J. Anal. At. Spectrom.* 16, 1142-1146.
[197] Kramer, K. J. M., De Haan, E.P. M., Van Het Groenewoud, H., Dorten, W., Kramer, G.N., Muntau, H. & Quevauviller, P. (2002). *Trends Anal. Chem.* 21, 762-773.
[198] Van Leeuwen S.P.J., Van Cleuvenbergen R., Abalos M., Pasir A., Eriksson U., Cleemann M., Hajslova J. & De Boer J. (2006). *Trends Anal. Chem.* 25, 397-409.
[199] Hanson, P. J. & Zdanowicz, V. S. (2005). *J. Fish Biol.* 54, 656-668.
[200] Mackie, I.M. & Fernandez-Salguéro, J. (1977). *J. Sci. Food Agric.* 28, 935-40.
[201] García-Villar, N., Saurina, J. & Hernández-Cassou, S. (2003). *Anal.Chim. Acta* 477, 315-324.
[202] Sanfeliu Alonso, M.C., Lahuerta Zamora, L. & Martínez Calatayud, J. (2003).*Talanta* 60, 369-376.
[203] Chícharo, M.A. & Chícharo, L. (2008). *Int. J. Mol. Sci.* 9, 1453-1471.
[204] Bianchi, F., Careri, M., Musci, M. & Mangia, A. (2007). *Food Chem.*100, 1049-1053.
[205] Peng, J.F., Fang, K.T., Xie, D.H., Ding, B., Ying, J.Y., Cui, X.M., Zhang, Y & Liu, Y.Z. (2008). *J.Chromatogr. A*, 1209, 70-75.
[206] Niu, J.& Lee, J.Y. (1999). *Anal. Commun.* 36, 81-83.

[207] McNabb, P. (2008). Chemistry, metabolism, and chemical analysis of okadaic acid group toxins. In *Seafood and freshwater toxins*. Botana, L.M. (Ed.) pp. 209-228. Boca-Ratón (FL): CRC Press.

[208] Stobo, L.A., Lacaze, J.P.C.L., Scott, A.C., Petrie, J. & Turrell, E.A. (2008). *Toxicon* 51, 635-648.

[209] Tsukatani, T. & Matsumoto, K. (2000). *Anal. Chim. Acta* 416, 197-203.

[210] Oliveira, M.I. P., Pimentel, M.C., Montenegro, M.C.B.S.M., Araújo, A.N., Pimentel, M.F. & Lins da Silva, V. (2001). *Anal. Chim. Acta* 448, 207-213.

[211] Mitsubayashi, K., Kubotera, Y., Yano, K., Hashimoto, Y., Kon, T., Nakakura, S., Nishi, Y. & Endo, H. (2004). *Sensors and Actuators* 103, 463-467.

[212] Chotimarkorn, Ch., Ohshima, T. & Ushio, H. (2006). *Lipids* 41, 295-300.

[213] Quaranta, H.O. & Curzio, O.A. (1983). *Food Chem.*11, 27-30.

[214] European Union (EU). Commission Decision 95/149/EC. http://eurlex.europa.eu/LexUriServ/LexUriServ.do?uri=CELEX:31995D0149:EN:HTML

[215] Byrne, L., King T. L. & Diamond, D. (2002). *Analyst* 127, 1338-1341.

[216] Carsol, M.A., Volpe, G. & Mascini, M. (1997). *Talanta* 44, 2151-2159.

[217] Watanabe, F., Katsura, H., Takenaka, S., Enomoto, T., Miyamoto, E., Nakatsuka, T. & Nakano, Y. (2001). *Int. J. Food Sci.Nutr.* 52, 263-268.

Chapter 3

SEA AND ESTUARINE WATER. PART 1: DETERMINATION OF ORGANIC ANALYTES

ABSTRACT

This chapter summarizes and examines the manuscripts issued to date referred the application of flow injection (FI) methodologies to the determination of organic analytes in sea and estuarine water samples. Concretely, reported flow methods applied to the determination or monitoring of acrolein, amines, bentazone, chemical oxygen demand (COD), diethylene glycol, dissolved organic carbon (DOC), formaldehyde, halocarbons, iron-porphyrin-like complexes, organophosphorus pesticides, nitrophenol isomers, polycyclic aromatic hydrocarbons, red tide phytoplankton and surfactants, usually monitored to assess quality of marine waters. FI methods are described for each organic analyte. Analytical figures of merit, characteristics, features and interferences are also discussed for each organic analyte.

INTRODUCTION

The organic analytes that have been determined by using FI methodologies were the following: acrolein, amines, bentazone, chemical oxygen demand (COD), diethylene glycol, dissolved organic carbon (DOC), formaldehyde, halocarbons, iron-porphyrin-like complexes, organophosphorus pesticides, nitrophenol isomers, polycyclic aromatic hydrocarbons, red tide phytoplankton and surfactants. Below, they will be described the FI methodologies proposed for each one of these substances because they involve different analytical techniques and reagents.

Different features of FI methods for the determination of organic analytes in sea and estuarine water are illustrated in Table 3.1. In the following paragraphs, some points observed in these tables are highlighted due to their interest.

Table 3.1. Features of FI determinations of organic compounds in sea and estuarine water

Analyte		Detection	DL (µg/L)	Recovery (%)	SF (samples/h)	RSD (%)	Ref.
Acrolein		A	5.5	96	100	2.9	1
Total primary amines		F	0.8 (as glycine)	No data	150	<2	2
Total primary amines		F	<0.1µM	No data	No data	<10	3
Tertiary amines		Ch	1010	No data	No data	No data	4
MAs	MMA	C	1.9	No data	No data	1-6	5-6
	DMA		2.7				
	TMA		4.8				
	CPA		3.6				
Bentazone		A	240	97.3-101.9	100	2.1	7
COD			No data	No data	No data	<10	8
COD		PT	No data	102.9-113.6	No data	No data	9
DEG		EI-MS	31	65 ±16.3	No data	5[a] 9[b]	10
DOC		SP	No data	No data	20	2.0	11
DOC		C	800	98±9	60	No data	13
Organic pollutants		Ch	No data	No data	No data	No data	14
Formaldehyde		F	0.7	No data	No data	2	15
HC	$CHCl_3$	GC	4.8×10^{-4}	No data	4-6	1-1.3	16
	CH_3CCl_3		1.1×10^{-4}				
	CCl_4		2.0×10^{-5}				
	$CHClCCl_2$		2.5×10^{-4}				
	$CHBrCl_2$		6.0×10^{-5}				
	CCl_2CCl_2		8.0×10^{-5}				
	$CHBr_2Cl$		8.0×10^{-5}				
	$CHBr_3$		2.2×10^{-4}				
VOCs	$C_2H_2Cl_2$	MS	0.2-0.4	No data	No data	5-8	20
	C_6H_6		0.3-0.5				
	$C_6H_5CH_3$		0.6-0.9				
	C_6H_5Cl		0.3-1.1				
	CCl_4		0.7-1.0				
IPC		Ch	0.11 nM	No data	No data	No data	21
OPP and CI		B	0.5-275	No data	No data	No data	22
Malathion		B	0.05-1.3	No data	No data	No data	23
OPP		B	No data	73.2-98.3	No data	2.9	24
Methylparathion		B	No data	No data	No data	<10	25
NPD	2-NP	SP	1.2 µM	96.0	11	<5	26
	3-NP		3.2 µM	110.4			
	4-NP		0.3 µM	103.3			

Analyte		Detection	DL (µg/L)	Recovery (%)	SF (samples/h)	RSD (%)	Ref.
NPD	2-NP	DRS	0.69	109-124	7	0.9	27
	4-NP		0.42	105-112		2.8	
	2,4-DNP		0.37	92-107		1.9	
PAHs		F	No data	No data	No data	No data	28
Chattonella antiqua		Ch	No data	No data	60	No data	29
Heterosigma carterae		Ch	No data	No data	30-60	No data	30
Nonidet AT 85		Ch	5000	No data	No data	<10	31
Anionic surfactants		P	No data	No data	3	<25	33

[a] Intraday repeatability.

[b] Interday repeatability.

A: amperometry; B: biosensor; C: conductimetry; Ch: chemiluminescence; CI: carbamate insecticides (carbofuran and carbaryl); COD: chemical oxygen demand; CPA: cyclopropylamine; DEG: Diethylene glycol ; DL: detection limit; DMA: dimethylamine; DNP: dinitrophenol; DOC: dissolved organic carbon ; DRS: diffuse reflectance spectroscopy; EI-MS: electron ionization mass spectrometry; F: fluorescence; GC: gas chromatography; HC: halocarbons; IPC: iron-porphyrin-like complexes; Mas: methylamines; MMA: monomethylamine; MS: mass spectrometry; NPD: nitrophenol derivatives; OPP: organophosphorus pesticides (azinphos-Et, azinphos-Me, bromophos-Me, dichlorovos, fenitrothion, malathion, paraoxon, parathion-Et, and parathion-Me); P: photometry; PAHs: polycyclic aromatic hydrocarbons; PT: potentiometric titration; SF: sampling frequency; RSD: relative standard deviation; SP: spectrophotometry; TMA: trimethylamine; VOCs: volatile organic compounds

Acrolein

Naranjo-Rodríguez et al. [1] carried out a study of the electrochemical behavior of acrolein at a dropping mercury electrode using different polarographic techniques. Between them, described a FI method with amperometric detection. In this flow system 250 µL of a solution containing 25 mL of spiked seawater and 7 mL of 0.25 mol/L EDTA are injected into a carrier solution (a Britton-Robinson buffer solution of pH 10 containing 2.7 g of boric acid, 2.3 mL of acetic acid and 2.7 mL of orthophosphoric acid (85%) per liter), which flowing at 1.8 mL/min. The injection system was connected to the flow cell through 0.5 mm i.d. PTFE tubing. The detection was performed at a potential of -1.4 V using a mercury electrode. Before each injection, any drop hanging from the tip of the capillary needs to be dislodged and a new electrode drop manually dispensed.

Amines

Petty et al. [2] used a FI sample processing with fluorescence detection for the determination of total primary amines in seawater. Seawater samples are filtered (0.45 µm membrane), preserved with 1 drop of pentachlorophenol solution (5 mg/mL in ethanol), and stored at < 4°C. This FI procedure with a carrier stream of o-phthaldialdehyde (OPA), offers a rapid and precise method for screening large numbers of samples for total primary amines. Delmas et al. [3] measure dissolved primary by FI and fluorescence labeling OPA. This procedure is compared with high-performance liquid chromatography (HPLC). Thus, a very good agreement between both methods for most of the samples analyzed is obtained. However, for samples where ammonia greatly overbalances dissolved free amino acid (DFAA) concentrations, such as pore waters from reduced environments, the method was invalid. Tertiary amines are determined by Lancaster et. al. [4] by a method based on measurement of the chemiluminescence emission resulting from the oxidation of triethylamine by sodium hypochlorite at pH 11.0 in the presence of Rhodamine B as a sensitizer. Gibbs et al. [5] presented flow injection gas diffusion-ion chromatography (FIGD-IC), a new hyphenated technique for the simultaneous determination of nanomolar levels of ammonia (NH_3) and methylamines (MAs). This investigation team employed a three stage strategy in the development of the FIGD-IC system: (1) optimization of the IC operational conditions; (2) development and optimization of an on-line flow injection gas diffusion method (FIGD) for the selective diffusion and preconcentration of amines, and (3) coupling of the IC to the FIGD. Although seawater cannot be directly analyzed for trace NH_3, and MAs by IC, this problem was overcome through use of a flow injection system in tandem with a capillary diffusion block fitted with a gas-permeable (hydrophobic) PTFE membrane (of the type employed in the photo- and fluorometric assay of NH_3). Of the membranes evaluated, Goretex MF\OlO\PM (the thinnest and most porous material) was the most efficient for the transfer of volatile bases. Sample is pumped and treated to pH > 12.0 by addition of alkaline EDTA. Under such conditions NH_3 and MAs cations are efficiently deprotonated (> 98%) to their volatile gaseous forms, which may then undergo transmembrane diffusion and accumulate in a recirculating acidic "trapping solution" (40 mM HCl) in which they are reprotonated and trapped. In addition, the reagent chelates Mg^{2+} and Ca^{2+} to prevent their precipitation as $Ca(OH)_2$ and $Mg(OH)_2$. The acceptor solution is then injected onto an ion chromatograph (IC) where NH_4^+ and MAs cations are separated within 15 min and detected by chemically suppressed conductimetry using cyclopropylamine as an internal standard for quantification. The FIGD was coupled to the IC by transferring the enriched contents of the FIGD acceptor loop through a PTFE transmission line to charge the injection loop of the IC. From this stream a 200 μL heat-cut was injected. The method is enhanced by the same authors [6] developing automation and computer interfacing of the technique. Automation improved the analytical reproducibility and sample throughput of the system by increasing precision and accuracy valve switching and by reducing the scope for human error.

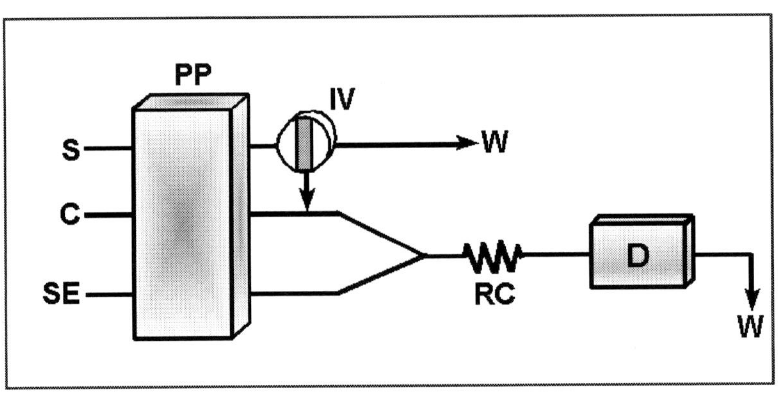

Figure 3.1. Flow-injection system for bentazone determination. C: carrier stream; D: detector; IV: injection valve; PP: peristaltic pump; RC: reaction coil; S: sample stream; SE: supporting electrolyte; W: waste.

Bentazone

Cerejeira et al. [7] proposed a new FI analytical method for bentazone determination at the low levels that the herbicide is found in water. Estuarine water samples are transported to the laboratory in dark flasks, at low temperature, and stored undisturbed under refrigeration for 24 h, after which they are analyzed. Because the need for high sensitivity, and thus low detection limits, linked with the electroanalytical oxidation of bentazone, so an amperometric detection is suggested. With the purpose of enabling the analysis of a large number of samples per hour, the detector is coupled to a flow injection analysis (FIA) system (Figure 3.1). The FIA system used is a double-channel manifold, which enabled the ionic strength and pH adjustments required by the detector. Standard solutions and samples were injected into a water carrier and combined on-line with the supporting electrolyte, before reaching the detector. This FI system allows the direct use of samples by the system, thus eliminating any sample pretreatment. Any adjustment of the samples to conform with the requirements of the detector would be made inside the manifold. This FIA method is an excellent alternative to conventional, chromatographic, or voltammetric, methods for determination of bentazone because it is considerably less polluting than liquid chromatography. Furthermore, if it is compared with other electrochemical methods reported in the literature, this FIA amperometric method has both: a low detection limit and a wide linear concentration range, and eliminates previous adsorption problems at the glassy carbon surface, which led to the need for frequent cleaning and increased the time per determination and reduced the reliability of the results.

Chemical Oxygen Demand

Jin et al. [8] presented a novel FI method for the determination of chemical oxygen demand (COD) by chemiluminescence (FI-CL). This method is based on the chemiluminescence phenomena produced by oxidation of luminol by O_3 in aqueous solution. Co^{2+} is used as masking agent to eliminate the interference of the interfering ions. $NaHCO_3$ was used as scavenger to eliminate the interference of OH free radical. The concentration range for COD determination is 0.6-24 mg/L, which was suitable for COD determination in seawater. The results obtained by this method were consistent with those obtained by permanganate method. Zhang et al. [9] determined COD with alkali potassium permanganate by FI. This method, which has many characteristics, such as rapid, successive, automatic, precision, anti-jamming and so on can be detected locally on time. Calibration graph was linear within 0-10 mg/L.

Diethylene Glycol

Diethylene glycol (DEG) is determined by Capiello et al. [10] in produced formation water (PFW) discharge and seawater. The method includes an off-line solid-phase extraction/preconcentration technique, followed by the identification and quantification of DEG exploiting an innovative analytical approach based on a quadrupole mass spectrometer coupled with a direct-electron ionization (EI) interface, a novel device for directly coupling a liquid chromatograph with an electron ionization ion source mass spectrometer (MS) (a nanoscale FI/direct EI-MS analysis). Water samples are collected in dark glass bottles to inhibit photochemical activity and saturated with mercury chloride to inhibit bacterial activity. Then the samples were refrigerated to 4 °C until the chemical analysis. The off-line extraction procedure is carried out with SPE cartridges packed with of ISOLUTE ENV+ stationary phase and using acetone as eluent, and finally, the extracts are evaporated to a volume of 200 µL. The samples were introduced into the Direct-EI interface by flow injection analysis (FIA) at nano-flow rates. No column separation is done with this sample introduction method, allowing rapid analyses. Direct-EI takes advantage of the very low operating flow rate typical of the nano-HPLC, allowing the registration of high quality EI spectra. The interface can be placed on a pre-existing GC/MS system with a simple and reversible modification, converting it into an HPLC-EI-MS instrument. Direct-EI is not influenced by mobile phase composition, analyte polarity and matrix; this is a clear advantage over atmospheric pressure ionization (API) based interfaces such as ESI, where a matrix effect is often observed when dealing with complex mixtures.

Dissolved Organic Carbon

A commercial continuous-flow method for the determination of dissolved organic carbon (DOC) is described by Aminot el al. [11]. This method is a highly improved version of the commercial continuous method proposed by Technicon [12] because avoids the following main problems: insufficient decarbonation efficiency, insufficient oxidation yield for certain organic nitrogen compounds, insufficient sensitivity and precision for marine research. The improvement is achieved by adding a hot acidic oxidation step before the photochemical UV-oxidation for complete destruction of refractory nitrogeneous compounds. The CO_2 resulting is measured colorimetrically by phenolphthalein discoloration. The calibration curve is linear from 0 to 7 mg/L. Koshy et al. [13] describe a conductimetric method for the determination of (DOC). The analysis of DOC has two parts to it. First, the inorganic carbon (IC) is determined under nonoxidizing conditions. Second, the water samples are reanalyzed with the oxidant in place. The resulting carbon dioxide is a measure of the total carbon (TC) consisting of IC naturally present and the IC generated from the oxidation of DOC. After correcting for chloride interference, the level of DOC was calculated by the difference. The method suffers from chloride interference, but the interference can be masked to levels as low as 1% with IC levels greater than 8 mg/L. Gas diffusion is achieved using a Tecator chemifold V with a TBA permatite GT gas plumbing tape (i.d. 0.2 mm). The detector consists of two square perspex blocks separated by a Teflon plate that has a passage to allow the flow of the acceptor stream. Each of the blocks had two circular platinum electrodes that detected the change in conductivity as the acceptor solution passed between them. The sample volume used is 600 µL. A chemiluminescence method with potassium permanganate at room temperature is developed by Fujimori et al. [14] for use as an indicator of organic pollutants in fresh and seawater. Potassium permanganate is chosen because this reagent does not react so rapidly with chloride ion at room temperature. The chemiluminescence intensity increased about 10% with the increase in salinity because of the production of manganese dioxide in the oxidation of chloride ions with permanganate. Thus, this method would be suitable for analyzing seawater at an estuary because the change of chemiluminescence intensity against the salinity is not very drastic.

Formaldehyde

Largiuni et al. [15] improved an analytical method for formaldehyde determination based on the Hantzsch reaction between HCHO and acetylacetone, acetic acid and ammonium acetate to form diacetyldihydrolutidine (DDL). A non-ionic surfactant (Triton X-100) was used to enhance the fluorescence signal by decreasing the quenching effects. The samples were collected and stored in polyethylene bottles, which were previously cleaned with a solution of non-ionic surfactant (Extran) for 24 h, then they were rinsed with tap water and rinsed out many times with MilliQ water. Successively, a 2% suprapur HCl solution was added and the filled bottles were stored for several days. effects. The samples were injected into the FI system, via a 4-way Teflon Rheodyne Type-50 injection valve with a loop of 250 μL, into a flow of 1.09 mL/min of ultrapure water, used as a carrier. The reagent solution (acetic acid 0.177 mol/L, acetylacetone 0.145 mol/L and ammonium acetate 2 mol/L) is added on-line at a flow-rate of 0.19 mL/min. The Hantzsch reaction takes place into the reaction coil (0.5 mm id, 200 cm long), which is warmed up to 75 °C in an in-line dry block heater. After the reaction, Triton X-100 at 10% (W/W) is continuously added to protect the fluorescence. The excitation and emission wavelengths were, respectively, 410 nm and 502 nm. The approach is simple, convenient, rapid and accurate. The interference from low molecular weight aldehydes and other organic compounds was found to be negligible. Only glyoxal shows a significant interference, but it can affect the HCHO measurements when its concentration is about 6000 times higher than that of HCHO. No interference occurred when benzaldehyde, methanol, ethanol and acetone were added at concentrations as high as 600 mg/L. The method has a very high linearity from concentrations of a few μg/L to about 1000 μg/L. The mean seawater superficial formaldehyde concentration was 15 μg/L.

Halocarbons

A mechanized system for extractive sample workup for gas chromatography coupled on-line to an on-column injector (Figure 3.2) is described by Fogelqvist et al. [16] with the aim of develop a closed system for extractive sample workup for gas chromatography that could be used for mechanized shipboard determinations of halocarbons in seawater. The technique presented is a mechanization of an earlier method for determination of halocarbons in seawater [17-18], which was modified by the use of large volumes, up to 250 μL, for on-column injection [19]. Extraction is performed in a liquid-liquid segmented flow in a glass coil internally coated with a hydrophobic layer. The extraction unit consists of a segmentor of a metal T-piece junction, which gives a regular segmentation. The extraction coil is a 5 m long glass capillary with an internal diameter of 0.7 mm. It was first silanized with diphenyltetramethyldisilazane (DPTMDS) by forcing a plug of DPTMDS in pentane slowly through and then keeping it sealed for about 12 h at 400 °C. Thereafter the column was coated with a thin layer of the gas chromatography stationary phase SE-54 by static coating and finally cross-linked. The phase separator is built with two blocks, which are machined from stainless steel instead of poly(vinylidene difluoride) (PMF) in order to avoid halocarbon contamination. The hydrophobic filter used is a commercially available fluoropore filter, and the support consists of a Teflon coated perforated steel disk. The system is closed from atmosphere and was proven to provide rapid and precise workup of seawater samples for the determination of the halocarbons chloroform ($CHCl_3$), bromodichloromethane ($CHBrCl_2$), dibromochloromethane ($CHBr_2Cl$), bromoform ($CHBr_3$), trichloroethylene ($CHClCCl_2$), tetrachloroethylene (CCl_2CCl_2), carbon tetrachloride (CCl_4), and 1,1,1-trichloroethane (CH_3CCl_3). Compared to manual extraction, smaller volumes of sample and organic phase are needed. Concentrations down to the picogram-per-liter level in water can be determined using injection volumes up to 130 μL. Kasthurikrishnan et a. [20] carried out the combination of FIA with membrane introduction mass spectrometry for analysis of volatile organic compounds (VOCs), including anthropogenic halocarbons (trans-l,2-dichloroethylene, benzene, toluene, chlorobenzene and carbon tetrachloride) in seawater. Membrane introduction mass spectrometry is performed using a benchtop ion trap mass spectrometer. This work compares the performance of a microporous (Teflon) membrane with that of an amorphous silicone membrane. Thus, the former is shown to provide lower detection limits for all VOCs tested. The microporous membrane provides faster response times by a factor of four to five for relatively more polar compounds, such as chlorobenzene. An analysis of a seven-component mixture demonstrates the ability of this on-line combination to allow multicomponent analysis of mixtures of some complexity.

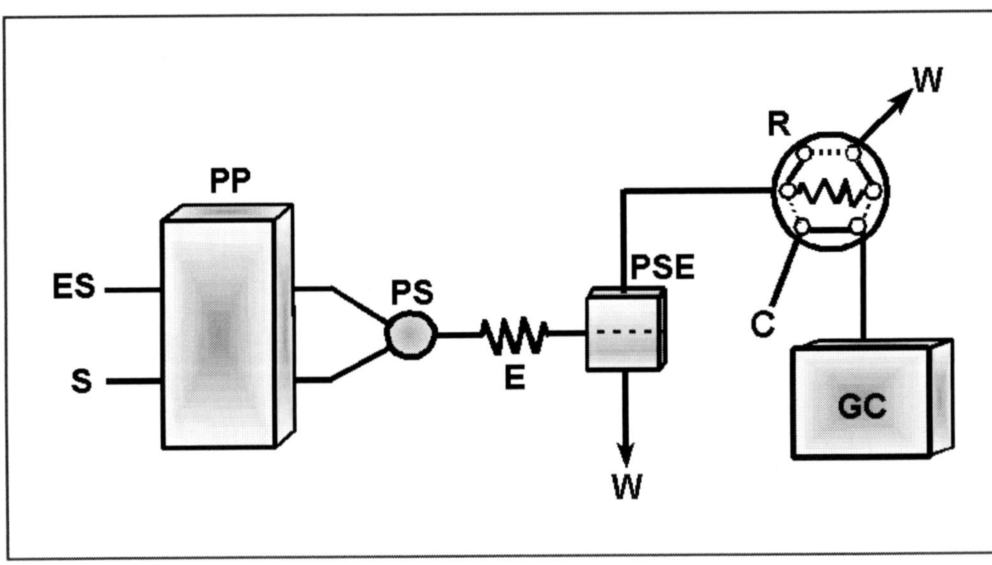

Figure 3.2. Flow-injection manifold for halocarbons determination. C: carrier gas inlet; PP: peristaltic pump; EC: extraction coil; ES: extraction solvent stream; S: sample stream; PS: Phase segmentor; PSE: phase separator; R: rotary valve; W: waste.

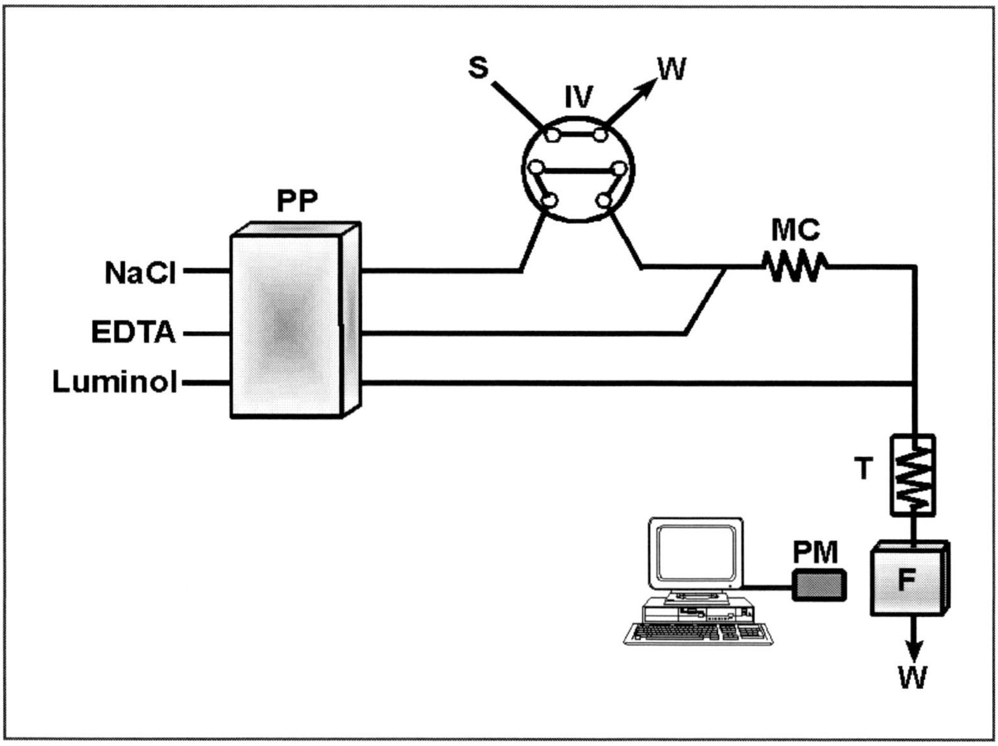

Figure 3.3. Flow-injection manifold for iron-porphyrin-like complexes determination. F: flow cell; IV: injection valve; MC: mixing coil; PP: peristaltic pump; PM: photomultiplier; S: sample; T: thermo-insulated bath; W: waste.

Iron-porphyrin-like Complexes

A new method for the non-specific determination of iron–porphyrin-like complexes in natural waters has been developed by Vong et al. [21]. It is based on the chemiluminescent oxidation of the luminol in the presence of dioxygen (O_2) at pH 13. The method has been implemented in a FIA manifold that allowed the direct injection of seawater and the determination on board (Figure 3.3). Direct injection of seawater is not possible due to the precipitation of magnesium and calcium hydroxide that would rapidly clog up the tubing. This precipitation is masked by complexing Ca^{2+} and Mg^{2+} with EDTA before mixing.

Organophosphorus Pesticides

A flow-injection system, incorporating an acetylcholinesterase (AChE) single bead string reactor (SBSR), for the determination of some organophosphorus (azinphos-Et, azinphos-Me, bromophos-Me, dichlorovos, fenitrothion, malathion, paraoxon, parathion-Et, and parathion-Me) and carbamate insecticides (carbofuran and carbaryl) is presented by Kumaran et al. [22]. The detector used is a simple pH electrode with a wall-jet entry. The measure process is based on the variations in enzyme activity due to inhibition, which produces pH changes when the substrate (acetylcholine) is injected before and after the passage of the solution containing the insecticide. The percentage inhibition of enzyme activity is correlated to the insecticide concentration. The enzyme reactor can be regenerated after inhibition and be reused for subsequent analysis. The immobilized enzyme did not lose any activity up to 12 weeks when stored at 4°C. Meng et a. [23] studied the possibility to determine malathion in seawater continuously by a biosensor based on the immobilized AChE with help of a FIA system. In this system, the carrier is seawater without malathion. The feasible work conditions for the biosensor are as follows: the flow rate was 0.39 mL/min for both the carrier and samples; the inhibition time from the samples lasted for 20 min. The concentration and injection volumes of the substrate, acetylthiocholine iodide are 41.6 mmol/L and 40 mL, respectively The sensitivity of the biosensor had a remarkable rise when the samples were oxidized previously by hypochlorous Na. The detection limits of the biosensor for malathion in seawater were 1.3 mg/L and 0.05 mg/L before and after pre-oxidation respectively. For samples with a malathion concentration from 0.1 μg/L to 10 mg/L, a good linearity between the response of biosensor and the concentration of malathion could be obtained. The inhibited activity of immobilized AChE could be reactivated completely by transporting pyridine-2-aldoxime of 0.5 mmol/L into the biosensor for 15 min after a sample containing malathion <100 μg/L is measured. As a result, the biosensor is suitable for the sensitive, accurate and continuous detection of organophosphorus pesticides in seawater. A mediator-free amperometric biosensor for screening organophosphorus pesticides with a FIA system based on anticholinesterase activity of organophosphorus pesticides to immobilized AChE is developed by Shi et al. [24]. The enzyme biosensor is prepared by entrapping AChE in Al_2O_3 sol-gel matrix screen-printed on an integrated 3-electrode plastic chip. This sol-gel matrix provides not only a friendly microenvironment for the immobilization of AChE to retain its activity for a long time, but also an effective promotion of the electron transfer between the thiocholine and the electrode. This promoting effect greatly decreased the overpotential to the detection of the thiocholine and minimized the interference from other co-existing impurities. Thus, this strategy is found not only increase the stability of the embedded AChE, but also effectively catalyze the oxidative reaction of thiocholine, making the Al_2O_3-AChE biosensor detects the substrate at 0.25 V (vs. Ag/AgCl), hundreds mini-volt lower than other reported mediator-free ones. The Al_2O_3-AChE biosensor is thus coupled to the FIA system to build up a simple and low-cost FIA electrochemical system for screening organophosphorus pesticides in real samples. A wide linear inhibition response for dichlorvos, is observed in the range of 0.1-80mM, corresponding to 7.91-84.94% inhibition for AChE. The detection limit for dichlorvos is achieved at 10 μM in the simulated seawater for 15 min inhibiting time. Meng et al. [25] design an AChE biosensor, connected with a FIA system. This is designed to detect methyl-parathion in seawater rapidly, by calorimetric determination of the activity change of the immobilized AChE from the organophosphorus pesticide. The working conditions for this system are optimized as follows: the temperature is 30 °C, the flow rate is 0.45 mL/min for both the carrier and the samples, the inhibition time from the samples lasted for 20 min, the concentration and injection volumes of the substrate and the acetylthiocholine iodide were 0.100 mol/L and 100 mL, respectively. Under these conditions, the percent inhibition of the immobilized AChE increased with the concentration of methyl-parathion in seawater. For the samples with methyl-parathion from 0.1 μg/L to 100 μg/L, there is a good linearity between the biosensor response and the logarithm of methyl-parathion concentration.

Nitrophenol Isomers

Manera et al. [26] proposed for first time a flow-through sorbent enrichment procedure with further spectrophotometric detection of nitrosubstituted phenol isomers from environmental waters. This is based on the combination of on-line disk-based solid-phase extraction for separation and enrichment with multi-syringe flow injection (MSFI) analysis. The FI manifold used is shown in Figure 3.4). The custom-built extraction unit consists of a Perspex cylindrical block (1.3 cm i.d.×2 cm long), which has been partially bored for the accommodation of the membrane holder and sorptive disk. The extraction unit is sealed with a threaded fitting ending with a conical cavity to prevent excessive pressure on the effective sorption surface of the membrane. Holes are drilled at the upper and down part of the unit to serve as a liquid inlet and outlet, respectively. Simultaneous diode-array spectrophotometric monitoring of the phenolic compounds is accomplished by application of multivariate regression modeling as a time and cost-effective alternative to chromatographic methods. The overlapping of absorbance spectra recorded during concurrent elution of the analytes is resolved by multivariate least-square regression (MLR). In spite of the intrinsic limitations of MLR, such as the requirement of a linear relationship between the analytical readout and the analyte concentration and the prior knowledge of sample composition, multivariate regression modeling via MLR is easy to handle and does not require neither training steps nor tedious calibrations with a large set of standards. Potential interferences from other phenolic derivatives that do not present any absorption band in the visible region may be minimized by monitoring the spectra above 320 nm. The coupling of time-based disk-phase MSFI extraction with multivariate analysis should regarded as a time and cost-effective alternative to conventional column separation techniques, such as liquid or gas chromatography or capillary electrophoresis, for resolution and determination of nitro-substituted phenols in complex chemical systems, due to the minimum operational maintenance of the devised analyzer and the precise metering of reagents and solutions via programmable flow. The same investigation [27] team proposed a novel flow-based optrode for automatic monitoring of phenolic derivatives (2-, 4-, and 2,4-dinitrophenol) exploiting the solid phase extraction (SPE)-optosensing alliance. The methodology is based on the uptake of the target phenolic derivatives onto an anionic exchange membrane disk, followed by their simultaneous determination via chemometric deconvolution of the diffuse reflectance spectra. Accordingly, the devised disk-based sorptive optrode is implemented in a fully automated flow network based on flow programming that ensures the performance of the analytical procedure with minimum operational maintenance. In contrast to bead-extraction optosensing, no flow impedance is observed when using disk-based sorbent materials in flow systems while better enrichment factors are obtained because of the improved specific surface area. Thus, the enrichment factors calculated as the ratio of the linear range sensitivity of the sorptive optosensor and that of direct spectrophotometric detection in aqueous medium using a 1mm lightpath cuvette turned out to be 73, 74 and 81 for 2-nitrophenol, 4- nitrophenol, and 2,4-dinitrophenol, respectively. As compared with earlier multiresidue methods for isolation and preconcentration of phenolic compounds involving liquid-liquid extraction, the proposed flow-through optosensing system should be viewed as an environmentally friendly approach because the use of hazardous solvents is overcome. Improved detectability is also assured as regards to SPE protocols as a consequence of the direct optical detection in the heterogeneous phase.

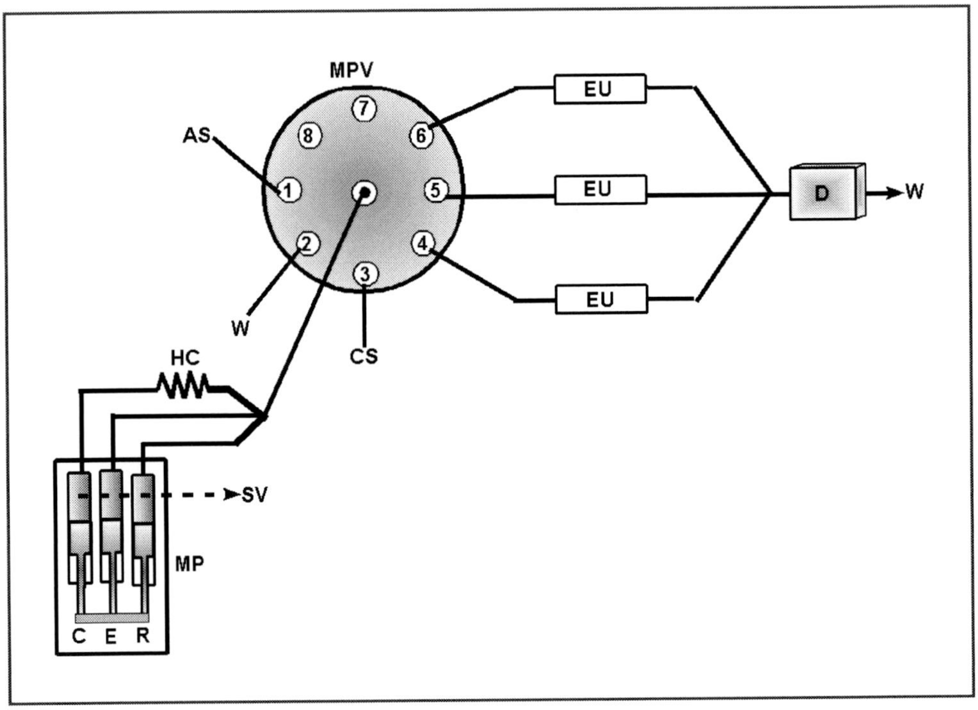

Figure 3.4. Multi-syringe flow injection system for simultaneous enrichment and determination of nitro-substituted phenols. C: carrier; CS: clean-up solution; D: detector; E: eluent; EU: extraction unit; HC: holding coil; MP: multi-syringe pump; MPV: multiposition valve; R: regeneration solution; SV: solenoid valve; W: waste.

Polycyclic Aromatic Hydrocarbons

Utsumi et al. [28] reported a FIA method with fluorescence detection as a rapid screening test of C-fuel oil contamination. This method is possible because many polycyclic aromatic hydrocarbons (PAHs) are fluorescent. Although high-performance liquid chromatography equipped with a fluorescence detector (HPLC/FLD) is a technique accurate and quantitative for determination of PAHs, it takes too much time for the rapid analysis of many oil-contaminated environmental samples. Therefore, the FIA methodology proposed in this work is a rapid screening test of C-fuel oil contamination.

Red Tide Phytoplankton

Asai et al. [29] developed a chemiluminescent FIA system for the detection of the red tide phytoplankton *Chattonella antiqua*. This system is based on system the following reactions:

MCLA + superoxide → dioxysetane → MCLA + hυ

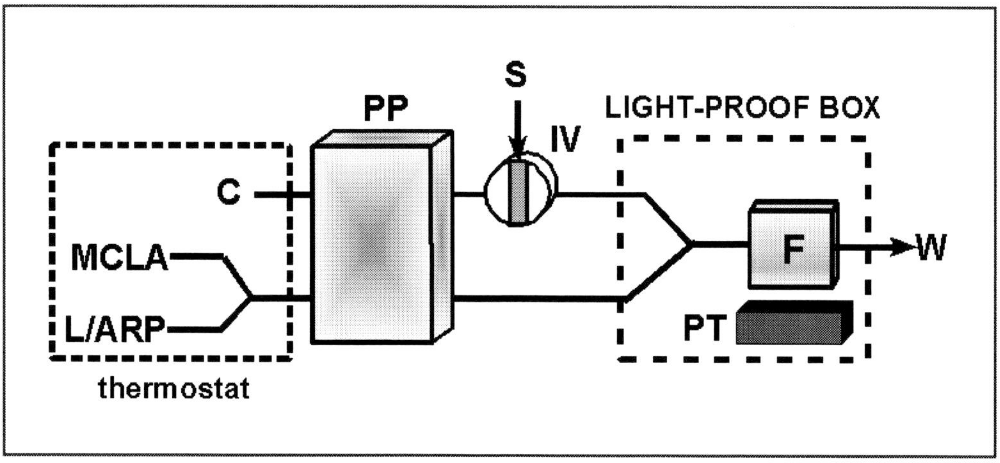

Figure 3.5. Schematic diagram of the FIA system for the detection of the red tide phytoplankton *Heterosigma carterae*. C: carrier; F: flow cell; IV: injection valve; L/ARP: luminol/ *Arthromy-ces ramosus* peroxidase solution; MCLA: 2-methyl-6-(p-methoxyphenyl)-3,7-dihydroimidazo[1,2-α]-pyrazin-3-one; PMT: photomultiplier tube; PP: peristaltic pump; S: sample; W: waste.

The Cypridina luciferin analog, 2-methyl-6-(p-methoxyphenyl)-3,7-dihydroimidazo[1,2-α]-pyrazin-3-one (MCLA), strongly emits light at 465 nm in the presence of superoxide. The system consisted of a two reagent feeding stream, a sample injector a joint for mixing MCLA and sample, a chemiluminescence reaction cell, a chemiluminescence detector and a recorder unit. The FIA system has an optimum pH of 10.7. The calibration curves for *Chattonella antiqua* displayed linearity from 2 x 10^3 to 2 x 10^4 cells/mL. The response time is approximately 1 min for one measurement cycle. Since this method is based on the superoxide released from phytoplankton, it may also be used to detect species other than *antiqua.* The same authors constructed a chemiluminescent-FIA system for the detection of O_2^- and H_2O_2 released by *H. carterae* [30], which is shown in Figure 3.5. MCLA is used as an O_2^- specific reagent. Luminol chemiluminescence, catalyzed by peroxidase from *Artheromyces ramosus* (ARP), is used to detect H_2O_2. The chemiluminescence reactions are the following:

MCLA + O_2^- → product + hυ (465 nm)
luminol + H_2O_2 + peroxidase → product + hυ (425 nm)

The resultant chemiluminescence is detected using a photomultiplier tube. A linear response was observed from 10^2 to 10^5 cells/mL *H. carterae*. The time required for one measurement cycle is ca. 2 min using MCLA-dependant luminescence or 1 min in luminol/ARP luminescence. Since these methods are based on the release of O_2^- or H_2O_2 from phytoplankton, the detection of other phytoplankton having this characteristic is also feasible.

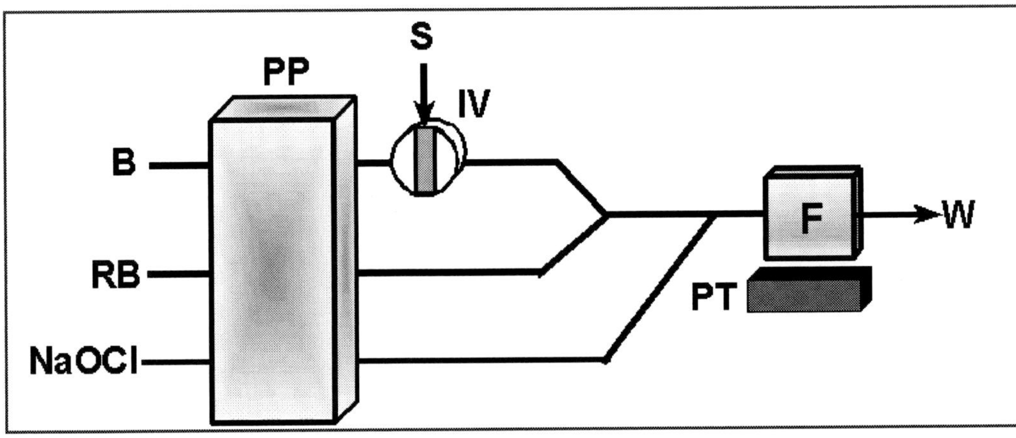

Figure 3.6. FIA manifold for the determination of non-ionic surfactant. B: borate buffer; RB: Rhodamine B solution; F: flow cell; IV: injection valve; PMT: photomultiplier tube; PP: peristaltic pump; S: sample; W: waste.

Surfactants

Lancaster et al. [31] described a FI procedure (Figure 3.6) for the determination of a commercially available fatty amine ethoxylate-based nonionic surfactant (Nonidet AT 85) in seawater over the range 0-50 mg/L. The procedure is an adaptation of a previously reported FI procedure with chemiluminescence detection for the determination of tertiary amines in aqueous media [32]. The method is based on measurement of the chemiluminescence emission resulting from oxidation of the tertiary amine group of the surfactant with hypochlorite in the presence of Rhodamine B, which acts as a sensitizer. Furthermore, in this work is discussed the design and operation of flow-through chemiluminescence detectors. In this sense, coiled tubes (made of either transparent plastic or glass) are the preferred flow cell design for monitoring chemiluminescence emission in FI manifolds. Between them glass tubes have superior optical properties and chemical resistance and facilitate the incorporation of immobilized reagents within the flow cell without increase significantly the total sample dispersion within the FI manifold. However, the major limitation of glass coils for routine use is their fragility. Therefore, a laminar flow cell is preferred. With this flow cell the maximum chemiluminescence emission intensity for the determination of the non-ionic surfactant increases with increasing cell volume. This can be explained because the surfactant gave a longer lived chemiluminescence emission than triethylamine and the larger volume flow cell is so integrating a higher fraction of the total chemiluminescence emission.

Moskvin et a. [33] developed a procedure for the FI extraction–photometric determination of anionic surfactants in natural waters in the presence of humic acids. The interference from humic acids is not observed up to concentrations of 20 mg/L. The procedure is based on the extraction-chromatographic separation of ion pairs of anionic surfactants with Methylene Blue, the elution of the separated ion pairs with chloroform, and their recovery from the aqueous phase in a chromatomembrane cell. The chromatomembrane cell with a mass-exchange layer volume of 0.5 mL was used to separate the chloroform eluate from the aqueous phase in the FI mode. The cross-section of the mass-exchange bounded by the membrane was 0.8 x 0.8 cm. The mass exchange layer is made of polytetrafluoroethylene similar to that used in extraction-chromatographic columns. The studies of the FI determination of anionic surfactants were made on a computer controlled PIAKon-01 (OOO "Rosanalit", St. Petersburg) FI analyzer with a photometric detector. A second flow switching valve was used to allow the analyzer to be used with manifolds with online preconcentration elements. The calibration range of anionic surfactants for a sample volume of 80 mL is 50-500 μg/L.

CONCLUSION

FI methodologies are shown to be a powerful utility to automate the continuous determination of organic compounds in sea and estuarine water samples. FI methodologies have several advantages to determine chemical quality of marine water: a high sample throughput, good precision, simple and relatively low cost equipment, easy automation and versatility. In fact, continuous extraction developed in FI manifolds is an interesting cost-effective and fast alternative to conventional column separation techniques, such as liquid or gas chromatography or capillary electrophoresis, for resolution and determination of organic compounds such as nitro-substituted phenols due to the minimum operational maintenance of the devised analyzer and the precise metering of reagents and solutions via programmable flow. By other hand FI methods can complement chromatographic techniques because other FI manifolds are coupled on-line to a chromatograph to automatize the sample preparation. Furthermore, FI methodologies can be a rapid and precise method for screening large numbers of samples. Most of the reported FI methods are based on the adaptation of an existing usual method to the FI mode. However, there are still not many proposed methods for the determination of organic compounds in marine waters, but it is hoped that in the future new FI methodologies will be developed to determine organic compounds in these samples, above all for those organic compounds, for which methods have not yet been proposed.

REFERENCES

[1] Naranjo Rodriguez, I., Muñoz Leyva, J.A. & Hidalgo Hidalgo de Cisneros, J.L. (1996). *Talanta* 43, 1117-1124.

[2] Petty, R.L., Michel, W.C., Snow, J. P. & Johnson, K.S. (1982). *Anal. Chim. Acta*, 142, 299-304.

[3] Delmas, D., Frikha, M.G. & Linley, E.A.S. (1990). *Mar. Chem.* 29, 145-154.

[4] Lancaster, J.S., Worsfold, P.J. & Lynes, A. (1989). *Analyst* 114, 1659-1661.

[5] Gibb, S.W., Fauzi, R., Mantoura, C. & Liss, P.S. (1995). *Anal.Chim. Acta* 316, 291-304.

[6] Gibb, S.W., Wood, J.W., Fauzi, R. & Mantoura, C. (1995). *J. Autom. Chem.* 17, 205-212.

[7] Cerejeira, R.P.A.G., Delerue-Matos, C. & Vaz, M.C.V. F. (2002). *Anal. Bioanal.Chem.* 373, 295-298.

[8] Jin, B., Zhuang, Z., Wang, X. & Lee, F.S.C.(2005). *Fenxi Ceshi Xuebao* 24, 67-70.

[9] Zhang, Y., Zhang, X., Hu, W., Jian, T., Luo, M. & Li, W. (2007). *Pige Kexue Yu Gongcheng* 17, 61-65.
[10] Cappiello, A., Famiglini, G., Palma, P., Pierini, E., Trufelli, H., Maggi, C., Manfra, L. & Mannozzi, M. (2007). *Chemosphere* 69, 554-560.
[11] Aminot, A. & Kerouel, R. (1990). *Analusis* 18, 289-295.
[12] Industrial method, n1-141-77w. (1978). Technicon, Sweden.
[13] Koshy, K. & Mataki, M. (2000). *Lab. Rob. Autom.* 12, 157-163.
[14] Fujimori, K., Ma, W., Moriuchi-Kawakami, T., Shibutani, Y., Takenaka, N., Bandow, H. & Maeda, Y. (2001). *Anal. Sci.* 17, 975-978.
[15] Largiuni, O., Becagli, S., Innocenti, M., Stortini, A.M., Traversi, R. & Udisti, R. (2005). *J. Environ. Monit.* 7, 1299-1304.
[16] Fogelqvist, E., Krysell, M. & Danielsson, L.G. (1986). *Anal. Chem.* 58, 1516-1520.
[17] Eklund, G., Josefsson, B. & Roos, C. (1978). J. *High Resolut. Chromatogr. Chromatogr. Commun* 1, 34-40.
[18] Fogelqvist, E. (1985). *J. Geophys. Res.* 90, 9181-9193.
[19] Fogelqvist, E. & Larsson, M. (1983). *J. Chromatogr.* 279, 297-306.
[20] Kasthurikrishnan, N. & Cooks, R. G. (1995). *Talanta* 42, 1325-1334.
[21] Vong, L., Laes, A. & Blain, S. (2007). *Anal. Chim. Acta* 588, 237-244.
[22] Kumaran, S. & Tran-Minh, C. (1992). *Anal. Biochem.* 200, 187-194.
[23] Meng, F., He, D., Zhu, X., Yang, Z., Ma, D. (2005). *Fenxi Huaxue* 33, 922-926.
[24] Shi, M., Xu, J., Zhang, S., Liu, B. & Kong, J. (2006). *Talanta* 68, 1089-1095.
[25] Meng, F.P., Liu, Q. & Liu, J. (2007). *Ziran Kexueban* 37, 111-115.
[26] Manera, M., Miro, M., Estela, J.M. & Cerda, V. (2007). *Anal. Chim. Acta* 582, 41-49.
[27] Manera, M., Miro, M., Estela, J.M., Cerda, V., Segundo, M.A. & Lima, J.L.F.C. (2007). *Anal. Chim. Acta* 600, 155-163.
[28] Utsumi, A., Nakashima, A., Ando, K., Kizu, R. & Hayakawa, K. (1998). *Anal. Sci.*14, 845-847.
[29] Asai, R., Matsukawa, R., Ikebukuro, K. & Karube, I. (1998). *Anal. Lett.* 31, 2279-2288.
[30] Asai, R., Matsukawa, R., Ikebukuro, K. & Karube, I. (1999). *Anal. Chim. Acta* 390, 237-244.
[31] Lancaster, J.S.& Worsfold, P. J. (1990). *Anal. Chim.Acta* 239, 189-194.
[32] Lancaster, J.S., Worsfold, P. J. & Lynes, A. (1989). *Analyst* 114, 1659-1661.
[33] Moskvin, L.N., Mikhailova, N.V. & Moskvin, A.L. (2001). *J. Anal. Chem.* 56, 763-766.

Chapter 4

SEA AND ESTUARINE WATER. PART 2: DETERMINATION OF INORGANIC ANALYTES

ABSTRACT

This chapter summarizes and examines the manuscripts issued to date referred the application of flow injection (FI) methodologies to the determination of inorganic analytes (cationic and anionic species) in sea and estuarine water samples. Thus, are reported flow methods applied to the determination of alkali metals, alkaline earth metals, silver, aluminium, arsenic, gold, boron, bismuth, cadmium, cobalt, chromium, copper, iron, mercury, indium, manganese, molybdenum, ammonium, nickel, lead, rare earths rhodium, antimony, selenium, tin, titanium, thallium, vanadium, zinc, alkalinity, hydrogen peroxide halides (bromide, chloride, fluoride and iodide), nitrate/nitrite, phosphate, silicate, sulfate and sulfide. Analytical figures of merit, characteristics, features and interferences are also discussed for each inorganic analyte.

INTRODUCTION

The determination of inorganic trace species above all trace metals in seawater represents one of the most challenging tasks in chemical analysis because the low levels of analyte (parts per billion (ppb) or sub-ppb) are very susceptible to matrix interference from alkali or alkaline-earth metals and their associated counterions. Thus, the exceptional difference between seawater and most other natural waters is its extremely high salinity and disparate levels of analyte and matrix ions that complicates direct analysis. For instance, the concentration of chloride alone makes up about 0.5 mol/L and considerably exceeds the concentrations of other anionic seawater species. The alkali metals tend to affect the atomization and the ionization equilibrium process in atomic spectroscopy, and the associated counterions such as the chloride ions might be preferentially adsorbed onto the electrode surface to give some undesirable electrochemical side reactions in voltammetric analysis. Therefore, while quantification of major ionic components, such as chloride or sodium, exhibits no significant problem, it is far more challenging to monitor anions or cations occurring in seawater at lower and especially much lower concentrations [1]. Regarding this analytical task, seawater can by all means be categorized as a complex sample [2] and hence often forces an analytical method to operate very close to the theoretical limits of performance. In consequence, most current methods for seawater analysis employ a separation technique for analyte preconcentration and/or for matrix rejection to remove interferences. These preconcentration techniques include coprecipitation, liquid-liquid extraction, solid phase extraction and electrodeposition between other, which have been implemented in a flow injection (FI) mode. Nevertheless, in addition to the very low element concentrations and a high salt content of the sample matrix, there are other factors, which also difficult seawater analysis: contamination or loss of elements from samples during sampling, preservation and pretreatment. The first two factors are dictated by the seawater medium and cannot be totally controlled, although their effects can be minimized, but any loss or contamination during sample storage and pretreatment is determined by the care taken in sample handling. The versatility of FI methodologies is such that experiments need not be confined to the laboratory and analysis and/or preconcentration methods can be accomplished in the same place of sampling. Accordingly, for a long time the determination of chemical parameters in marine waters involved traditional sampling techniques followed with laboratory measurements. However, this approach is susceptible to errors arising from contamination and/or degradation of the sample during collection, transportation and storage. In addition, to many oceanographers, such methods do not allow enough determinations to understand the spatial and temporal distribution of elements in seawater and results can not be obtained in real time excluding any possibility to optimize a strategy of study. Moreover, this approach is very time consuming and labor intensive, particularly when a large number of samples are involved. In this sense, syringe-based, fixed-site FI monitors that incorporate wet chemistries and spectrophotometric detection are commercially available for the determination of a variety of chemical species. The requirements of an on-line process monitor include contamination-free environment, high temporal and spatial resolution without the need for discrete sample collection, rapid analysis, high sampling frequency, portable with robust construction, simple design, ability to perform automated analysis, long-term stability and minimal maintenance and operating costs. Therefore, in a on-line process FI monitor, each component is controlled automatically by a simple single-board computed (Figure 4.1). This FI in-situ monitor can have incorporate a sample pretreatment step to remove suspended particulate matter, for matrix removal, and/or to increase the sensitivity of the method such as filtration, dilution and/or analyte preconcentration. Miniature solenoid-type pumps and valves for lower reagent consumption and ease of operation, can also be incorporated to switch between the sample and standards for regular self-calibration. The suitability of FI instrumentation for in-situ monitoring and study of dynamics of freshwater and marine processes has been exhaustively explored by Worsfold's group [3]. Another option, particularly useful for transient species in biogeochemical cycles is the deployment of submersible FI monitors [4]. Shipboard FI techniques have been mainly applied to the monitoring of nutrients using a solid-state photometric detector incorporate light-emitting diodes as sources (and photodiodes as detectors). So, conventional photometric assays as well as metal traces exploiting spectrophotometric or chemiluminescence catalytic methods in combination with in-line solid-phase preconcentration for sensitivity and selectivity improvement. Methods for determining other parameters in marine research (such as carbon dioxide, dissolved inorganic carbon, pH and hydrogen peroxide that require high temporal and spatial resolution have also been adapted for high-precision ship-board use in a FI system. Furthermore, the use of microelectronic sensors could be of great interest because these devices provide low energy consumption, small size, solid structure and fast response times. In addition to the fact that their solid structure is a potential advantage for pressure resistance, these devices could be integrated in multiparameter measuring systems of reduced size. Moreover, they provide the possibility of automated recalibration as the standard solution volumes needed are small. Much effort has very recently been focused on hardware and software innovations as well as set-up miniaturization. In sense, instrumentation incorporating miniature array detectors has provided enhanced information through full spectral acquisition and signal processing [5].

Since its introduction, interest has been sustained in the application of flow injection analysis (FIA) techniques to the development of automated, on-line sample pretreatment procedures Benefits are associated with flow injection separations, preconcentration techniques include high efficiency, simple on-line operation, rapidity, low sample reagent consumption, relatively simple and compact hardware, and freedom from contamination. Most FIA preconcentration methods are based on the use of a minicolumn containing a chelating resin or a solid sorbent, which can significantly reduce sample volume. Furthermore, the versatility of FI is such that experiments should not be confined to the laboratory and preconcentration methods can be accomplished in the same place of sampling. Therefore, in order to optimize and improve the first stage of the analytical process (preliminary operations, involving sampling and sample pretreatment) to be placed on the same level of the other stages, the microcolumn field sampling (MFS) technique was proposed [6]. In this technique, water samples are processed in flow systems at the sampling site (field flow preconcentation systems, Figure 4.2) and trace elements of interest are immobilized on minicolumns. The minicolumns may then be returned to the laboratory and directly inserted into a FI system for on-line elution and quantitative analysis, usually developed by atomic spectroscopic techniques. In addition to the advantages of FI methodology, the MFS technique achieved the following: high efficiency, rapidity, cost-effective and environmental-friendly sampling and analysis, and improvement in data quality in speciation and ultratrace analysis. Collection and transport of water samples (0.5-1 L) back to a laboratory was avoided, and the sampled occupied a small space for convenient storage Nevertheless, the major problem to establish a reliable methodology for trace metals determination in seawater is related to sample preservation, since most of the procedures involve several chemical stages, increasing the possibility of analyte losses and sample contamination. Thus, the accepted method for preserving water samples for trace metals determination involves the addition of a concentrated acid (nitric or hydrochloric) to reduce the pH to below 2. Nevertheless, the conventional sample preservation method has disadvantages, such as transporting and manipulating a volume of concentrated acid to each sampling site, contamination of the sample by the acid added, possible modification of chemical species to perform speciation studies, several containers containing the samples to return to the laboratory and have a place in the laboratory for sample storage. In order to avoid these drawbacks involving the sampling, storage and analyte determination, the minicolumn field sampling (MFS) technique has been proposed. In this technique flow systems are used to process the water samples at the sampling site, and trace elements of interest are immobilized on minicolumns. These minicolumns are returned to the laboratory where they are directly inserted into a FI system for on-line elution and quantitative analysis. In addition to the advantages of the FI methodology, the MFS technique achieves the following: high efficiency, rapidity, improvement in data quality and ultratrace analysis, improvements in collection, transporting and storage of seawater samples (the samples occupy a small space for convenient storage in the laboratory). However, the major limitation of FI systems for environmental monitoring is their conception as single-analyte analyzers. Fully software-controlled sequential injection discontinuous methods have proven suitable alternatives for multiparametric determinations [7]. The replacement of the injection valve of FIA manifolds with rotary multiposition valves furnished with various reagents, sample pre-treatment modules and appropriate detectors has enabled different reactions to be accommodated in a single manifold without requiring manual reconfigurations. The virtues of SIA as an automated microfluidic technique for environmental monitoring have been exploited to full potential in the design of a rugged multi-analyte monitor able to determine ten relevant parameters (namely, ammonium, nitrate, nitrite, total nitrogen, orthophosphate, organic phosphorous, chemical oxygen demand, biochemical oxygen demand, total suspended solids and total organic carbon) in wastewaters in merely three minutes [8].

The inorganic analytes that have been determined by using FI methodologies were the following: alkali metals, alkaline earth metals, silver, aluminium, arsenic, gold, boron, bismuth, cadmium, cobalt, chromium, copper, iron, mercury, indium, manganese, molybdenum, ammonium, nickel, lead, rare earths, rhodium, antimony, selenium, tin, titanium, thallium, vanadium, zinc, alkalinity, hydrogen peroxide halides (bromide, chloride, fluoride and iodide), nitrate/nitrite, phosphate, silicate, sulfate and sulfide. Below, they will be described the FI methodologies proposed for each one of these substances.

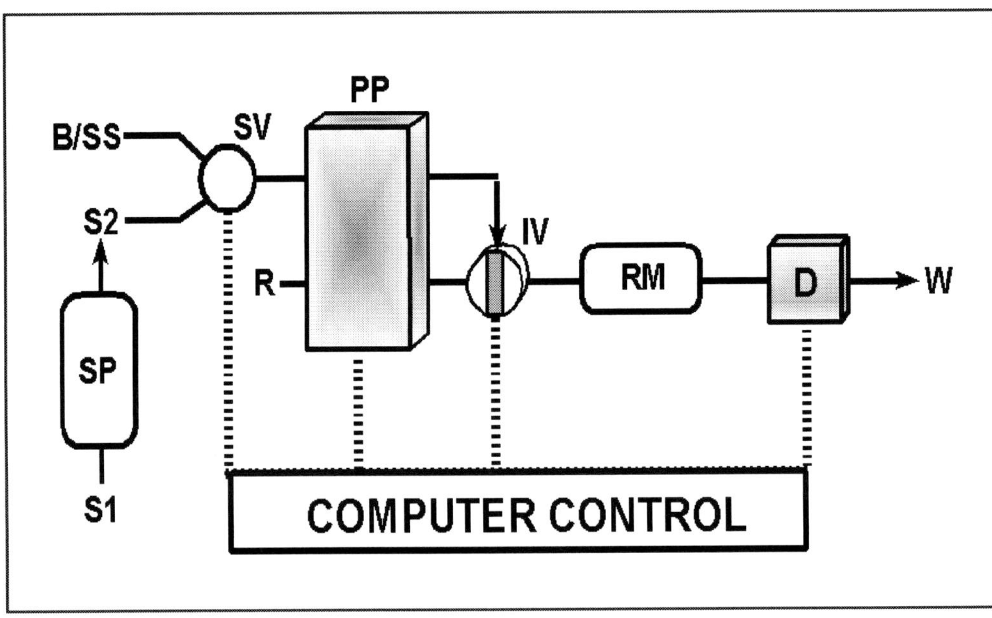

Figure 4.1. Flow-injection system for in-situ seawater monitoring. B/SS: blank and standard solutions; D: detector; IV: injection valve; PP: peristaltic pump; R: reagent(s); RM: reaction manifold; S1: seawater sample before pretreatment; S2: seawater sample after pretreatment; SP: sample pretreatment; SV: switching valve; W: waste.

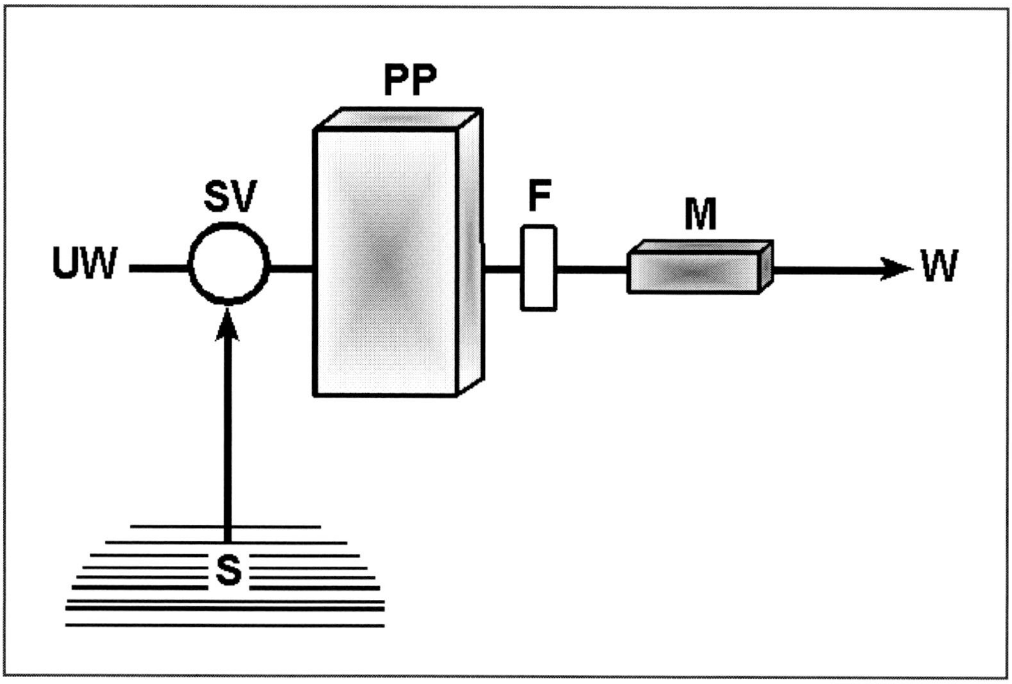

Figure 4.2. Field flow preconcentration system. F, filter; PP, peristaltic pump; S: seawater sample; SV, switching valve; M, minicolumn; UW: ultrapure water; W, waste.

Cationic Species

Measurement techniques that have been employed for the determination of trace metals by using FI methodologies include mainly spectrophotometry, atomic absorption spectrometry, anodic stripping voltammetry, differential pulse cathodic stripping voltammetry, inductively coupled plasma atomic emission spectrometry, atomic fluorescence, spectrofluorescence, chemiluminescence, liquid chromatography of the metal chelates with ultraviolet-visible absorption and, more recently, inductively coupled plasma mass spectrometry. Many of the published methods for the determination of metals in seawater are concerned with the determination of a single element. However, much of the published work is concerned not only with the determination of a single element but with the determination of groups of elements. This is particularly so in the case of techniques such as inductively coupled plasma spectrometry and inductively coupled plasma mass spectrometry. The background concentrations at which metals occur in marine water are extremely low, and much work has been done on preconcentration procedures in attempts to improve detection limits for these metals.

Different features of FI methods for the determination of cationic species in sea and estuarine water are illustrated in Table 4.1. In the following paragraphs, some points observed in this table are highlighted due to their interest.

Table 4.1. Features of FI determinations of cationic species in sea and estuarine water

Analyte	Detection	Separation Technique	DL (μg/L)	SF (s/h)	RSD (%)	Ref
Na^{+}	SP		No data	No data	No data	9
Ba^{2+}	ICP-MS	On-line Chelating resins	No data	No data	No data	10
Ba^{2+}	ICP-MS		0.0049	25	<7	11
Ca^{2+}	SP		No data	200	1.38	12
Mg^{2+}	SP		No data	200	1.38	12
Ra^{2+}	BPC	On-line adsorption: MnO_2 deposited on cotton fiber Coprecipitation with $BaSO_4$	0.05 Bq/L	4	0.4	13
Ag^{+}	ASV		0.019	No data	10.5	14
Ag^{+}	PDV		0.0054	No data	No data	15
Ag^{+}	ETAAS	On-line coprecipitation	0.0042	9-12	8.0	16
Ag^{+}	FAAS	On-line adsorption: DDTC immobilized on alumina	0.7-1.7	16	4.0	17
Ag^{+}	ICP-MS	Off-line cloud point extraction	0.004	No data	4.9	18
Ag^{+}	ICP-MS	On-line adsorption: analyte complexes with DDTP on SC18	0.15-0.8×10^{-3}	21-22	<6	19
Ag^{+}	ICP-MS	On-line anion-exchange Dowex 1-X8	6×10^{-5}	10	No data	20
Ag^{+}	ICP-MS	On-line anion-exchange Dowex 1-X8	5×10^{-5}	5	No data	21
Ag^{+}	ICP-MS	On-line anion-exchange Dowex 1-X8	8×10^{-5}	No data	< 8.0	22
Al^{3+}	F	On-line resin-immobilized 8-HQ	0.004	20	1.7	23
Al^{3+}	F		0.015-0.019	No data	No data	24
Al^{3+}	F	On-line resin-immobilized 8-HQ	0.0046	No data	2.7	25
Al^{3+}	F		0.07	No data	1.7	26
Al^{3+}	F		0.057	30	0.62	27
Al^{3+}	F	On-line commercially resin	0.0027	No data	2.5	28
Al^{3+}	ETAAS	On-line resin-immobilized 8-HQ	0.015-0.040	No data	4.0	29
As^{3+}/ As^{5+}	Ch		No data	No data	No data	30
Inorg. As	AAS	On-line gas-liquid separation	No data	150	2.5	31
As	AAS	On-line gas-liquid separation	0.15	No data	4.0	32
As As^{3+}	ETAAS	On-line solid phase extraction	0.04 0.05	22	5.5	33
As	AAS	On-line gas-liquid separation	0.14	20	2.7	34
As	AAS	On-line gas-liquid separation	0.05	65	1.8	35

Analyte	Detection	Separation Technique	DL (μg/L)	SF (s/h)	RSD (%)	Ref
As	ETAAS	Coprecipitation	0.02	9-12	8.0	16
As^{3+} As^{5+}	ETAAS	On-line adsorption on activated alumina	0.05 2.0	60	3.6 5.0	36
As^{3+} As^{3+} MMA DMA	AAS	On-line gas-liquid separation	0.005 0.002 0.0015 0.0015	No data	2.5	37
As^{3+} As^{5+} MMA DMA	AAS	On-line gas-liquid separation	0.047 0.042 0.0045 0.0063	No data	3-10	39
As^{3+}	ETAAS	On-line sorption onto a KR	0.008	No data	4.5	40
As^{3+}	AF	On-line gas-liquid separation	0.08	No data	0.5-1.0	41
As^{3+}/ As^{5+}	AF	On-line sorption onto a KR	0.023	32	1.3	42
As	ICP-MS	On-line gas-liquid separation	0.006	No data	1.6	43
As	ICP-MS		0.004-7	No data	No data	44
As	ICP-MS	On-line gas-liquid separation	0.003	No data	<5.0	45
As	ICP-MS	On-line gas-liquid separation	0.003	No data	1-4	46
As	ICP-MS	On-line gas-liquid separation	0.014	8	2.5	47
As	ICP-MS	Off-line cloud point extraction	0.006	No data	3.8	18
As^{3+}	ICP-MS	On-line adsorption: analyte complexes with DDTP on SC18	0.005-0.008	21-22	<6	19
As	ICP-MS	Solid phase filtration/concentration	1.2	No data	No data	48
As	ICP-MS		2-30	No data	No data	49
As	ICP-MS		0.003	25	<7	11
Au^{+3}	CSV		0.6-61.0	No data	6.3-7.1	50
Au	ICP-MS	On-line anion exchange	2.0×10^{-6}	No data	15	52
Au	ICP-MS	Off-line cloud point extraction	0.003	No data	4.7	18
Au	ICP-MS	On-line adsorption: analyte complexes with DDTP on SC18	0.0001-0.0006	21-22	<6	19
B	SP		20	20	0.3-0.5	53
B	SP		17	No data	<4	54
Bi	ASV	On-line immobilized 8-HQ on silica gel	No data	3-5	2.5-6.9	55
Bi^{3+}	ASV		2.0	No data	7.9	14
Bi	ETAAS	On-line coprecipitation	0.0015	9-12	8.0	16
Bi	AAS	On-line gas-liquid separation	0.07	No data	No data	56
Bi	ICP-MS	On-line adsorption: analyte complexes with DDTP on SC18	0.00006	22	<10	57

Table 4.1. Continued

Analyte	Detection	Separation Technique	DL (µg/L)	SF (s/h)	RSD (%)	Ref
Bi	ICP-MS	On-line adsorption: analyte complexes with DDTP on SC18	0.00006-0.00043	21-22	<6	19
Cd^{2+}	ASV		No data	No data	No data	58
Cd^{2+}	ASV		0.01-0.1	No data	2-5	59
Cd^{2+}	ASV	On-line immobilized 8-HQ on silica gel	No data	3-5	2.5-6.9	55
Cd^{2+}	F	On-line anion exchange resin (Dowex 1X2-400)	0.004	No data	10	60
Cd^{2+}	AF	On-line gas-liquid separation	0.0032	24	10	61
Cd^{2+}	SP		0.1	No data	1.2	62
Cd^{2+}	FAAS	On-line chelating resin (Chelex-100)	9	30-60	No data	63
Cd^{2+}	FAAS	On-line chelating resin (Chelex-100)	0.05	60	1.2-3.2	64
Cd^{2+}	ATAAS	On-line chelating resin (Chelex-100)	No data	No data	1.69	65
Cd^{2+}	FAAS	On-line chelating resin (Chelex-100)	0.0017	No data	No data	66
Cd^{2+}	FAAS	On-line adsorption: analyte complexes with 8-HQ on SC18	No data	120	1.2-2.8	67
Cd^{2+}	FAAS	On-line adsorption: analyte complexes with 8-HQ on SC18	No data	30	No data	68
Cd^{2+}	ETAAS	On-line adsorption: analyte complexes with DDTC on SC18	0.0008	22	2.0-3.3	69
Cd^{2+}	FAAS	On-line adsorption: analyte complexes with DDTC on C18	0.3	120	1.4	70
Cd^{2+}	ETAAS	On-line adsorption: analyte complexes with DDTC on SC18	0.0006-0.0008	23	2.0-3.3	71
Cd^{2+}	ETAAS	On-line adsorption: analyte complexes with DDTC on SC18	0.004	No data	2-8	72
Cd^{2+}	ETAAS	On-line adsorption: analyte complexes with DDTP on SC18	0.05	No data	No data	73
Cd^{2+}	ETAAS	On-line adsorption: analyte complexes with PAR and PADMAP on SC18	0.0017-0.004	9-12	2.7-15 1.2-8	74

Analyte	Detection	Separation Technique	DL (μg/L)	SF (s/h)	RSD (%)	Ref
Cd^{2+}	ETAAS	On-line adsorption: analyte complexes with APDC on SC18	0.005	15	3.2	75
Cd^{2+}	FAAS	On-line adsorption: analyte complexes with Phen on SC18	0.003	90	1.4	76
Cd^{2+}	ETAAS	On-line adsorption: analyte complexes with 8-HQ on XAD-2	0.004	No data	10	77
Cd^{2+}	FAAS	On-line adsorption: analyte complexes with DDTP on PCTFE	0.3	50	2.9	78
Cd^{2+}	ETAAS	On-line retention: analyte on a macrocyclic ligand immobilized on silica gel	0.0008	10	0.9-2.8	79
Cd^{2+}	ETAAS	On-line coprecipitation	0.0028	9-12	8.0	16
Cd^{2+}	FAAS	In-situ on-line retention on XAD-4 impregnated with PAN	0.0038	No data	4.1-6.5	80
Cd^{2+}	FAAS	In-situ on-line retention on XAD-4 impregnated with PAR	0.006	No data	0.8-8.9	81
Cd^{2+}	FAAS	In-situ on-line retention on chelating resin AMPA	No data	No data	No data	82
Cd^{2+}	FAAS	In-situ on-line retention on chelating resin Chelite P	0.0027	No data	0.5-9.4	83
Cd^{2+}	AAS	On-line gas-liquid separation	0.004	45	0.2-1.3	84
Cd^{2+}	ICP-AES	In-situ on-line retention on chelating resin Metpac CC-1	1.5	No data	<10	85
Cd^{2+}	ICP-AES	On-line SGF with: DPTH TS	 1.1 4.3	 40 24	 1.1 2.5	86
Cd^{2+}	ICP-MS		0.001	No data	1.1	87
Cd^{2+}	ICP-MS		0.0041	No data	No data	88
Cd^{2+}	ICP-MS	On-line chelating resin (iminodiacetate groups)	0.005	15	9	89
Cd^{2+}	ICP-MS	On-line ion-exchanger column of SO_3-oxine CM-cellulose	No data	No data	No data	90
Cd^{2+}	ICP-MS	On-line chelating resin (iminodiacetate groups)	0.0007	15	No data	91
Cd^{2+}	ICP-MS	On-line chelating resin (iminodiacetate groups)	0.0055	No data	1.2	92

Table 4.1. Continued

Analyte	Detection	Separation Technique	DL (μg/L)	SF (s/h)	RSD (%)	Ref
Cd^{2+}	ICP-MS	On-line chelating resin (iminodiacetate groups)	0.008	10	No data	93
Cd^{2+}	ICP-MS	On-line adsorption: analyte complexes with DDTP on SC18	0.0002	22	<10	57
Cd^{2+}	ICP-MS	On-line adsorption: analyte complexes with DDTP on SC18	0.2-6.7 $x10^{-3}$	21-22	<6	19
Cd^{2+}	ICP-MS	On-line SGF with 8-HQ	0.004	10	3-5	94
Cd^{2+}	ICP-MS	On-line controlled-pore glass funtionalized with 8-HQ	No data	No data	No data	95
Cd^{2+}	ICP-MS	On-line 8-HQ immobilized on fluorinated metal alkoxide glass	0.078 x 10^{-3}	No data	No data	96
Cd^{2+}	ICP-MS	8-HQ bonded ZrO_2 composition microspheres	0.01	No data	9.1	97
Cd^{2+}	ICP-MS	On-line chelating resin (Toyopearl AF-Chelate-650M)	0.0014	17	4.2	98
Cd^{2+}	ICP-MS	On-line silica gel modified with niobium(V) oxide	0.02-0.9	No data	0.8	99
Cd^{2+}	ICP-MS	On-line adsorption: analyte complexes with DDTC on C18	0.03	No data	No data	100
Cd^{2+}	ICP-MS	On-line adsorption on C18	0.03	No data	6.8	101
Cd^{2+}	ICP-MS	Off-line cloud point extraction	0.006	No data	3.4	18
Co^{2+}	Ch	On-line 8-HQ immobilized on Fractogel	0.0005	No data	5.0	103
Co^{2+}	Ch	On-line 8-HQ immobilized on silica gel	0.00062	No data	2.1	104
Co^{2+}	Ch	On-line 8-HQ immobilized on Toyopearl	0.0003	No data	No data	60
Co^{2+}	Ch	On-line 8-HQ immobilized on Toyopearl	0.0003	No data	No data	105
Co^{2+}	Ch		0.1-0.4	No data	2.0-5.5	106
Co^{2+}	AF	On-line DDTC immobilized on SC18	8 x 10^{-5} – 0.001	No data	9-13	107
Co^{2+}	SP		No data	30	3.2-4	108
Co^{2+}	SP		0.001	50	2-8	109
Co^{2+}	ETAAS	On-line adsorption: analyte complexes with DDTC on SC18	0.0017	No data	5.0	110

Analyte	Detection	Separation Technique	DL (μg/L)	SF (s/h)	RSD (%)	Ref
Co^{2+}	FAAS	On-line adsorption: analyte complexes with DDTP on SC18	0.07	No data	2-3	111
Co^{2+}	FAAS	On-line adsorption: analyte complexes with Phen on SC18	6	90	2.2	76
Co^{2+}	FAAS	On-line adsorption: analyte complexes with DCO on SC18	1.0	30	7.4	112
Co^{2+}	FAAS	On-line alumina loaded with NNDS	0.44	No data	2.3	113
Co^{2+}	FAAS	On-line polyurethane foam loaded with TAC	3.2 2.4	24 17	5 8	114
Co^{2+}	ETAAS	On-line coprecipitation	0.021	9-12	8.0	16
Co^{2+}	ICP-AES	In-situ on-line retention on chelating resin Metpac CC-1	6.8	No data	<10	85
Co^{2+}	ICP-MS		0.03-2	No data	No data	49
Co^{2+}	ICP-MS	On-line chelating resin (iminodiacetate groups)	0.06	15	7	89
Co^{2+}	ICP-MS		No data	No data	No data	115
Co^{2+}	ICP-MS	On-line chelating resins	No data	No data	No data	10
Co^{2+}	ICP-MS	On-line ion-exchanger column of SO3-oxine CM-cellulose	No data	No data	No data	90
Co^{2+}	ICP-MS	On-line chelating resin (iminodiacetate groups)	0.0005	15	No data	91
Co^{2+}	ICP-MS	On-line chelating resin (iminodiacetate groups)	0.0001	No data	3.3	92
Co^{2+}	ICP-MS	On-line chelating resin (iminodiacetate groups)	0.006	10	No data	93
Co^{2+}	ICP-MS	On-line SGF with 8-HQ	0.003	10	3-5	94
Co^{2+}	ICP-MS	On-line 8-HQ immobilized on fluorinated metal alkoxide glass	0.00002	No data	No data	96
Co^{2+}	ICP-MS	On-line silica gel modified with niobium(V) oxide	0.03-0.5	No data	3.2	99
Co^{2+}	ICP-MS	On-line adsorption: analyte complexes with DDTC on SC18	0.02	No data	No data	100
Co^{2+}	ICP-MS	On-line adsorption on C18	0.1	No data	1.3	101
Co^{2+}	ICP-MS	On-line chelating resin (iminodiacetate groups)	No data	No data	2.6-6.9	116
Co^{2+}	ICP-MS	On-line precipitation with DDTC	No data	No data	No data	117
Cr(VI)	P		1.04	40	0.6	118

Table 4.1. Continued

Analyte	Detection	Separation Technique	DL (µg/L)	SF (s/h)	RSD (%)	Ref
Cr(VI)	S	On-line Dowex 1 x 4 anion-exchange	0.0004	No data	1.67	119
Cr	Ch		0.2-0.5	No data	3.3-5.3	106
Cr^{3+} Cr(VI)	F	Off-line cloud point extraction	0.2	No data	1.8	120
Cr Cr(VI)	SP + FAAS		No data	120	0.3-0.4	121
Cr(VI)	SP		0.36	No data	No data	122
Cr^{3+}	FAAS	On-line chelating resin poly(hydroxamic acid)	100	No data	1.3	123
Cr^{3+} Cr(VI)	FAAS	On-line adsorption: analyte complexes with DDTC on SC18	0.02	10	5.1 3.5	124
Cr(VI)	ETAAS	On-line poly(styrene-divinylbenzene) beads impregnated with DPC	0.03	15	3.8	125
Cr(VI)	FAAS	On-line adsorption: analyte complexes with DDTC on PCTFE bead	0.4	30	1.8	126
Cr^{3+}/Cr	ICP-MS	On-line chelating resin (iminodiacetate groups)	0.020	12	1.9	127
Cr	ICP-MS	On-line precipitation with DDTC	No data	No data	No data	117
Cr^{3+}/ Cr(VI)/ Cr	ICP-MS		0.18	No data	5.0	128
Cu^{2+}	ASV	On-line immobilized quinolin-8-ol silica gel	No data	3-5	2.5-6.9	55
Cu^{2+}	ASV		No data	No data	No data	58
Cu^{2+}	ASV		0.01-0.1	No data	2-5	59
Cu^{2+}	P	On-line chelating resin (Chelex-100)	No data	No data	5.2	129
Cu^{2+}	Ch		5	200	2.2	130
Cu^{2+}	Ch		No data	No data	No data	131
Cu^{2+}	Ch	On-line 8-HQ immobilized on Fractogel	<0.25	No data	No data	132
Cu^{2+}	Ch		0.006	No data	No data	133
Cu^{2+}	Ch	On-line 8-HQ immobilized on Toyopearl HW-75F resin	No data	No data	<5	134
Cu^{2+}	SP		10	100	1.68	135
Cu^{2+}	SP	On-line chelating resin	0.2	No data	1.08	136
Cu^{2+}	ICP	In-situ on-line retention on chelating resin Metpac CC-1	0.5	No data	<10	85

Analyte	Detection	Separation Technique	DL (μg/L)	SF (s/h)	RSD (%)	Ref
Cu^{2+}	FAAS	On-line chelating resin (Chelex-100)	5	30-60	No data	63
Cu^{2+}	FAAS	On-line chelating resin (Chelex-100)	0.07	60	1.2-3.2	64
Cu^{2+}	ATAAS	On-line chelating resin (Chelex-100)	No data	No data	4.3	65
Cu^{2+}	ETAAS	On-line chelating resin (Muromac A-1)	0.009	14	3.8	137
Cu^{2+}	FAAS	On-line adsorption: analyte complexes with 8-HQ on SC18	No data	120	1.2-2.8	67
Cu^{2+}	FAAS	On-line adsorption: analyte complexes with 8-HQ on SC18	No data	30	No data	68
Cu^{2+}	ETAAS	On-line adsorption: analyte complexes with DDTC on SC18	0.017	22	2.0-3.3	69
Cu^{2+}	FAAS	On-line adsorption: analyte complexes with DDTC on SC18	0.2	120	1.3	70
Cu^{2+}	ETAAS	On-line adsorption: analyte complexes with DDTC on SC18	0.017-0.0085	23	2.0-3.3	71
Cu^{2+}	ETAAS	On-line adsorption: analyte complexes with DDTC on SC18	0.024	No data	2-8	72
Cu^{2+}	ETAAS	On-line adsorption: analyte complexes with DDTP on SC18	0.05	No data	No data	73
Cu^{2+}	FAAS	On-line adsorption: analyte complexes with Phen on SC18	0.3	90	3.0	76
Cu^{2+}	FAAS	On-line adsorption: analyte complexes with 1N2N on SC18	2.0	90	1.7	138
Cu^{2+}	FAAS	On-line adsorption: analyte complexes with DCOL on SC18	0.05	30	1.4	139
Cu^{2+}	ETAAS	On-line retention: analyte on a macrocyclic ligand immobilized on silica gel	0.046	10	1.8-3.9	79
Cu^{2+}	FAAS	On-line SDVB functionalized with HVPMPCEE	0.93	13	5.3	140
Cu^{2+}	FAAS	Amberlite XAD-2 functionalized with dithizone	0.019	No data	1.68	141

Table 4.1. Continued

Analyte	Detection	Separation Technique	DL (μg/L)	SF (s/h)	RSD (%)	Ref
Cu^{2+}	FAAS	On-line adsorption: analyte complexes with 8-HQ on Sep-Pak cartridge	No data	10	5.7	142
Cu^{2+}	FAAS	In-situ on-line retention on XAD-4 impregnated with PAN	0.06	No data	1.2	143
Cu^{2+}	FAAS	In-situ on-line retention on XAD-4 impregnated with PAR	0.05	No data	1.1-3.8	144
Cu^{2+}	ETAAS	On-line coprecipitation	0.015	9-12	8.0	16
Cu^{2+}	ETAAS	On-line coprecipitation	0.001	No data	3.2	145
Cu^{2+}	ETAAS	On-line sorption in a KR	0.006	26	1.5-2.5	146
Cu^{2+}	ICP-AES	In-situ on-line retention on chelating resin Metpac CC-1	0.5	No data	<10	85
Cu^{2+}	ICP-AES		0.16	30	3.2	147
Cu^{2+}	ICP-MS	On-line chelating resin (iminodiacetate groups)	0.01	15	3	89
Cu^{2+}	ICP-MS	On-line ion-exchanger column of SO_3-oxine CM-cellulose	No data	No data	No data	90
Cu^{2+}	ICP-MS	On-line chelating resin (iminodiacetate groups)	0.0012	15	No data	91
Cu^{2+}	ICP-MS	On-line chelating resin (iminodiacetate groups)	0.003	No data	1.4	92
Cu^{2+}	ICP-MS	On-line chelating resin (iminodiacetate groups)	0.05	10	No data	93
Cu^{2+}	ICP-MS	On-line chelating resin (iminodiacetate groups)	0.09	No data	2.6-6.9	116
Cu^{2+}	ICP-MS	On-line chelating resin (NTA Superflow)	0.0006	No data	2.6	148
Cu^{2+}	ICP-MS	On-line adsorption: analyte complexes with DDTP on SC18	0.005-0.033	21-22	<6	19
Cu^{2+}	ICP-MS	On-line adsorption: analyte complexes with DDTP on SC18	0.005	22	<10	57
Cu^{2+}	ICP-MS	On-line SGF with 8-HQ	0.010	10	3-5	94
Cu^{2+}	ICP-MS	On-line controlled-pore glass funtionalized with 8-HQ	No data	No data	No data	95
Cu^{2+}	ICP-MS	On-line 8-HQ immobilized on fluorinated metal alkoxide glass	0.00014	No data	No data	96

Analyte	Detection	Separation Technique	DL (μg/L)	SF (s/h)	RSD (%)	Ref
Cu^{2+}	ICP-MS	On-line silica gel modified with niobium(V) oxide	0.01-0.4	No data	1.0	99
Cu^{2+}	ICP-MS	On-line adsorption: analyte complexes with DDTC on SC18	0.04	No data	No data	100
Cu^{2+}	ICP-MS	On-line precipitation with DDTC	No data	No data	No data	117
Cu^{2+}	ICP-MS	On-line precipitation with DDTC	No data	No data	No data	149
Cu^{2+}	ICP-MS	Off-line cloud point extraction	0.03	No data	4.0	18
Fe	Ch	On-line 8-HQ immobilized on Toyopearl	0.0017-0.01	No data	3.5	24
Fe	Ch	On-line 8-HQ immobilized on Toyopearl	0.0022	No data	No data	105
Fe^{2+}	Ch	No	0.0033	No data	8.0	150
Fe	Ch	On-line 8-HQ immobilized on Toyopearl	0.0056	No data	4.0	151
Fe	Ch	On-line 8-HQ immobilized on Toyopearl	0.0022	20	3.2	152
Fe	Ch	On-line 8-HQ immobilized on Toyopearl	0.0012	No data	2-5	153
Fe^{2+}	Ch		0.001-0.007	No data	No data	154
Fe^{2+}	Ch		0.4-0.6 x 10^{-3}	No data	0.9-7.6	155
Fe^{2+}	Ch		0.006	300	1.0-3.7	156
Fe/ Fe^{2+}/ Fe^{3+}	Ch	On-line 8-HQ immobilized on a vinyl copolymer resin	0.001	20	2.7-9.1	157
Fe^{2+}	Ch	No On-line 8-HQ immobilized on a vinyl copolymer resin	0.002 0.0005	120 24	2.3 6.6	158
Fe	Ch		0.006	No data	No data	159
Fe	Ch		5.1-28	No data	0.5-3.72	160
Fe^{2+}	Ch		No data	No data	No data	161
Fe^{2+}	Ch	On-line 8-HQ immobilized on Fractogel	0.025	12	2-5	162
Fe^{2+} Fe	Ch	Amberlite XAD-4 functionalized with HEED groups	0.045 0.020	8	<4.0	163
Fe	SP		0.0014	No data	2.5	26
Fe	SP	No data	0.05 0.03	60	1.8-8.1	164

Table 4.1. Continued

Analyte	Detection	Separation Technique	DL (µg/L)	SF (s/h)	RSD (%)	Ref
Fe	SP	On-line 8-HQ immobilized on a vinyl copolymer gel	0.0014	No data	2.5	165
Fe	SP	On-line 8-HQ metal chelating resin column	0.0009	No data	2.5	166
Fe	SP	No On-line 8-HQ metal chelating resin	0.089 0.008	No data	7.0 6.0	167
Fe	SP	On-line NTA chelating resin	No data	No data	No data	168
Fe	SP	On-line NTA chelating resin	No data	No data	No data	169
Fe^{3+}	SP		5	30	<1.5	170
Fe^{2+}/ Fe^{3+}	SP		80	No data	1.7-3.4	171
Fe	SP	On-line chelating disks (iminodiacetic groups)	0.5	7	2.2-3.6	172
Fe^{2+}/ Fe^{3+}	SP	On-line 8-HQ immobilized on Fractogel	2.3	No data	0.88	173
Fe^{3+}	SP		0.1	60	2.4	174
Fe^{2+} Fe	SP		3.4 3.9	No data	0.2-0.4 0.8	175
Fe^{3+}	SP		1.0	No data	3.5	176
Fe	FAAS	On-line adsorption: analyte complexes with DCO on silica C18	4.0	30	9.1	112
Fe^{2+}/ Fe^{3+}	FAAS	On-line adsorption of the iron species in a tubular reactor	1.8	No data	3-10	177
Fe	ICP-MS	On-line chelating resin (iminodiacetate groups)	11	No data	1.1	92
Fe	ICP-MS	On-line precipitation with DDTC	No data	No data	No data	117
Fe	ICP-MS	On-line NTA chelating resin	0.0011	No data	4.4	148
Hg^{2+}	ASV		0.38	No data	5.77	14
Hg	P		0.25	No data	1.9	178
Hg^{2+}	Ch	On-line gas-liquid separation	0.8	12	3.1	179
HgCl2 MHgCl	AF	On-line gas-liquid separation	0.025 0.023	No data	3.8 1.0	180
Hg	AF	On-line gas-liquid separation	0.00001	3	<5	181
Hg	AF	On-line gas-liquid separation	0.003	No data	4-10	182
Hg Hg inorg.	AAS	On-line gas-liquid separation	0.09 0.06	No data	0.7 2.1	183
Hg	AAS	On-line gas-liquid separation	0.1	90	2.0	184

Analyte	Detection	Separation Technique	DL (μg/L)	SF (s/h)	RSD (%)	Ref
Hg Hg inorg.	AAS	On-line gas-liquid separation	0.45 0.47	No data	2.2 1.5	185
Hg	AAS	On-line gas-liquid separation	0.0002	No data	5-8.2	186
Hg	AAS	On-line gas-liquid separation	0.4	No data	No data	187
Hg	AAS	On-line chelating resins On-line gas-liquid separation	0.23	No data	0.85	188
Hg	AAS	On-line resin Chelex-100 On-line gas-liquid separation	0.0001	No data	No data	189
Hg inorg. Methyl Hg	AAS	On-line adsorption: analyte complexes with DDTC on SC18 On-line gas-liquid separation	0.016	No data	3.4	190
Hg inorg. Methyl Hg	AAS	On-line anion exchange On-line gas-liquid separation	0.4	No data	<3	191
Hg inorg. Methyl Hg	AAS	On-line L-L extraction On-line gas-liquid separation	2 4	10-12	3 5	192
Hg inorg. Methyl Hg	AAS	HPLC On-line gas-liquid separation	0.0034 0.0017	12	5.7 2.9	193
Hg^{2+}	ICP-AES	On-line SGF with TS On-line gas-liquid separation	5	40	2.1	194
Hg^{2+}	ICP-AES	On-line SGF with TS On-line gas-liquid separation	No data	30	<3	195
Hg	ICP-MS	On-line gas-liquid separation	0.004	No data	1.3	43
Hg	ICP-MS	On-line SGF with DPTH	1	40	0.5-1.3	196
Hg	ICP-MS	On-line gas-liquid separation	0.07	No data	2.2-2.3	197
Hg inorg. Methyl Hg	ICP-MS	Solid phase microextraction Gas chromatography	0.00011 0.00035	No data	4.8 2.4	198
In	ASV	On-line immobilized quinolin-8-ol silica gel	No data	3-5	2.5-6.9	55
In	ICP-MS	Solvent extraction and back extraction	$1.1x10^{-6}$ $2.3x10^{-6}$	No data	No data	199
In^{3+}	ICP-MS	PTFE filter impregnated HDEHP	0.0098	10	1.9	200
Mn^{2+}	Ch	On-line 8-HQ immobilized on Fractogel	0.0055	10	3	201

Table 4.1. Continued

Analyte	Detection	Separation Technique	DL (μg/L)	SF (s/h)	RSD (%)	Ref
Mn^{2+}	Ch	On-line 8-HQ immobilized on Fractogel	No data	No data	No data	202
Mn^{2+}	SP		No data	No data	No data	24
Mn^{2+}	SP		No data	No data	1.3-13	203
Mn^{2+}	SP	On-line 8-HQ immobilized on vinyl polymer gel	0.002	11	5	204
Mn^{2+}	SP	On-line chelating resin (iminodiacetate groups)	0.00165	No data	3.2-6.2	205
Mn^{2+}	SP		0.02	30	1.2-1.6	206
Mn^{2+}	SP		0.073	60	0.5	207
Mn^{2+}	SP		No data	25	No data	129
Mn^{2+}	SP		No data	No data	0.19	208
Mn^{2+}	SP		0.01	15	5-8	209
Mn^{2+}	SP		No data	4 - 12	5 - 8	210
Mn^{2+}	SP		0.0022	No data	No data	211
Mn^{2+}	SP		0.0012	No data	No data	212
Mn^{2+}	ETAAS	On-line sorption in a KR	0.029	26	2.9	146
Mn^{2+}	ETAAS	On-line coprecipitation	0.003	9-12	8.0	16
Mn^{2+}	FAAS	On-line adsorption: analyte complexes with DDTP on SC18	0.07	No data	2-3	111
Mn^{2+}	FAAS	On-line adsorption: analyte complexes with DCO on SC18	0.5	30	7.0	112
Mn^{2+}	ETAAS	On-line adsorption: analyte complexes with Eriochrome Black T on anion exchange resin	0.002	No data	3.0	213
Mn^{2+}	ATAAS	On-line chelating resin (Chelex-100)	No data	No data	3.16	65
Mn^{2+}	ICP-MS		0.7	No data	No data	88
Mn^{2+}	ICP-MS		0.0053	25	<7	11
Mn^{2+}	ICP-MS	On-line Chelating resins	No data	No data	No data	10
Mn^{2+}	ICP-MS	On-line ion-exchanger column of SO3-oxine CM-cellulose	No data	No data	No data	90
Mn^{2+}	ICP-MS	On-line chelating resin (iminodiacetate groups)	No data	15	No data	91
Mn^{2+}	ICP-MS	On-line chelating resin (iminodiacetate groups)	0.093	10	No data	93
Mn^{2+}	ICP-MS	On-line SGF with 8-HQ	0.028	10	3-5	94

Analyte	Detection	Separation Technique	DL (μg/L)	SF (s/h)	RSD (%)	Ref
Mn^{2+}	ICP-MS	On-line chelating resin (Toyopearl AF-Chelate-650M)	0.0086	17	2.2	98
Mn^{2+}	ICP-MS	On-line chelating resin (iminodiacetate groups)	No data	No data	2.6-6.9	116
Mn^{2+}	ICP-MS	On-line precipitation with DDTC	No data	No data	No data	117
Mo	FAAS	On-line adsorption: analyte complexes with 8-HQ on C18	1.5	30	5.0	214
Mo	ETAAS	On-line adsorption: analyte complexes with DDTC on C18	0.027	No data	No data	215
Mo	ETAAS	On-line coprecipitation	0.02-0.04	3.1	No data	216
Mo	ETAAS	On-line chelating resin (Muromac A-1)	0.06	14	3.8	137
Mo	ICP-MS		0.011	25	<7	11
Mo	ICP-MS		1.4	No data	No data	49
Mo	ICP-MS		0.7	No data	No data	88
Mo	ICP-MS	On-line Chelating resins	No data	No data	No data	10
Mo	ICP-MS	On-line chelating resin (iminodiacetate groups)	0.40	10	No data	93
Mo	ICP-AES	On-line adsorption on activated alumina	0.2	No data	2-5	217
NH_4^+	C	On-line gas-liquid separation	100	No data	<3	218
NH_4^+	C	On-line gas-liquid separation	No data	60	1.4-3.9	219
NH_4^+	F		0.018	No data	No data	220
NH_4^+	F		0.54	9	1	221
NH_4^+	F		0.126	30	5.7	222
NH_4^+	F		0.0198	3600	2.2-6.7	223
NH_3	SP	On-line gas-liquid separation	0.008	10	1	224
NH_4^+	SP		No data	No data	0.35	225
NH_4^+	SP		19.8	No data	3.6	226
NH_4^+	SP	On-line gas-liquid separation	0.9	No data	<2	227
NH_4^+	SP	On-line gas-liquid separation	No data	No data	No data	228
NH_4^+	SP	On-line gas-liquid separation On-line ion-exchange	0.2-0.5	No data	0.68	229
NH_4^+	SP		2.34	60	2.2	230
NH_4^+	SP		26	10	4.9	231
NH_4^+	SP		5×10^6	No data	0.36	232
Ni^{2+}	ETAAS	On-line coprecipitation	0.0038	9-12	8.0	16
Ni^{2+}	ETAAS	On-line sorption in a KR	0.0076	26	2.5	146
Ni^{2+}	ETAAS	On-line adsorption: analyte complexes with DDTC on SC18	0.036	22	2.0-3.3	69

Table 4.1. Continued

Analyte	Detection	Separation Technique	DL (μg/L)	SF (s/h)	RSD (%)	Ref
Mn^{2+}	ICP-MS	On-line chelating resin (iminodiacetate groups)	No data	15	No data	91
Ni^{2+}	ETAAS	On-line adsorption: analyte complexes with DDTC on SC18	0.021-0.036	23	2.0-3.3	71
Ni^{2+}	FAAS	On-line adsorption: analyte complexes with 8-HQ on SC18	No data	30	No data	68
Ni^{2+}	FAAS	On-line adsorption: analyte complexes with DDTP on SC18	0.07	No data	2-3	111
Ni^{2+}	FAAS	On-line adsorption: analyte complexes with DCO on SC18	1.0	30	7.2	112
Ni^{2+}	FAAS	On-line adsorption: analyte complexes with DMG on SC18	3.0	90	1.7	233
Ni^{2+}	ICP-AES	In-situ on-line retention on chelating resin Metpac CC-1	7.6	No data	<10	85
Ni^{2+}	ICP-MS	On-line Chelating resins	No data	No data	No data	10
Ni^{2+}	ICP-MS	On-line chelating resin (iminodiacetate groups)	0.006	15	4	89
Ni^{2+}	ICP-MS	On-line ion-exchanger column of SO3-oxine CM-cellulose	No data	No data	No data	90
Ni^{2+}	ICP-MS	On-line chelating resin (iminodiacetate groups)	0.005	15	No data	91
Ni^{2+}	ICP-MS	On-line chelating resin (iminodiacetate groups)	0.011	No data	0.8	92
Ni^{2+}	ICP-MS	On-line chelating resin (iminodiacetate groups)	0.034	10	No data	93
Ni^{2+}	ICP-MS	On-line chelating resin (Toyopearl AF-Chelate-650M)	0.028	17	3.3	98
Ni^{2+}	ICP-MS	On-line silica gel modified with niobium(V) oxide	0.02-0.5	No data	3.6	99
Ni^{2+}	ICP-MS	On-line SGF with 8-HQ	0.054	10	3-5	94
Ni^{2+}	ICP-MS	On-line 8-HQ immobilized on fluorinated metal alkoxide glass	0.0009	No data	No data	96
Ni^{2+}	ICP-MS	On-line adsorption: analyte complexes with DDTC on C18	0.32	No data	No data	100

Analyte	Detection	Separation Technique	DL (µg/L)	SF (s/h)	RSD (%)	Ref
Ni^{2+}	ICP-MS	On-line adsorption on C18	0.1	No data	2.6	101
Ni^{2+}	ICP-MS	On-line precipitation with DDTC	No data	No data	No data	117
Ni^{2+}	ICP-MS		0.03-20	No data	No data	44
Pb^{2+}	ASV		No data	No data	No data	58
Pb^{2+}	ASV		0.01-0.1	No data	2-5	59
Pb^{2+}	ASV P		No data	6	4.91	234
Pb^{2+}	ASV	On-line immobilized 8-HQ silica gel	No data	3-5	2.5-6.9	55
Pb^{2+}	Ph	On-line anion exchanger resin (Dowex 1X2-200)	0.1	No data	3	235
Pb^{2+}	Ph	On-line anion exchange resin (Dowex 1X2-400)	0.104	12	3	60
Pb^{2+}	AF	On-line gas-liquid separation	0.48	No data	2.4	236
Pb^{2+}	SP	On-line ion-exchange Chelex 100 Dowex 1-X8	 3.6 0.7-1.4	 30 6-30	2-12	129
Pb^{2+}	SP	On-line chelating resin (Chelex 100)	5	36	9	237
Pb^{2+}	SP	On-line adsorption: analyte complexes with DMS on C18	0.1	24	1.1-2.5	238
Pb^{2+}	ETAAS	On-line chelating resins On-line gas-liquid separation	0.25	No data	0.70	188
Pb^{2+}	ETAAS	On-line solid sorbent	0.0045	No data	No data	239
Pb^{2+}	FAAS	On-line chelating resin (Chelex-100)	32	30-60	No data	63
Pb^{2+}	FAAS	On-line chelating resin (Chelex-100)	0.5	60	1.2-3.2	64
Pb^{2+}	ATAAS	On-line chelating resin (Chelex-100)	No data	No data	1.69	65
Pb^{2+}	FAAS	On-line chelating resin (Chelex-100)	No data	25	No data	240
Pb^{2+}	ETAAS	On-line retention: analyte on a macrocyclic ligand immobilized on silica gel support	0.020	10	1.2-8.3	79
Pb^{2+}	FAAS	In-situ on-line retention on chelating resin AMPA	No data	No data	No data	82
Pb^{2+}	FAAS	In-situ on-line retention on chelating resin AMPA	0.0028	No data	2.5-4.3	241
Pb^{2+}	FAAS	In-situ on-line retention on XAD-4 impregnated with PAN	0.005	No data	3.1-4	242

Table 4.1. Continued

Analyte	Detection	Separation Technique	DL (µg/L)	SF (s/h)	RSD (%)	Ref
Ni^{2+}	ICP-MS	On-line adsorption: analyte complexes with DDTC on C18	0.32	No data	No data	100
Pb^{2+}	FAAS	On-line silica gel chemically modified with niobium(V) oxide	0.35	12	1.6	243
Pb^{2+}	FAAS	On-line adsorption: analyte complexes with 8-HQ on SC18	No data	120	1.2-2.8	67
Pb^{2+}	ETAAS	On-line adsorption: analyte complexes with DDTC on SC18	0.0065	22	2.0-3.3	69
Pb^{2+}	FAAS	On-line adsorption: analyte complexes with DDTC on C18	3	120	1.0	70
Pb^{2+}	ETAAS	On-line adsorption: analyte complexes with DDTC on SC18	0.004-0.0065	23	2.0-3.3	71
Pb^{2+}	ETAAS	On-line adsorption: analyte complexes with DDTC on SC18	0.003	24	1.0	244
Pb^{2+}	ETAAS	On-line adsorption: analyte complexes with DDTP on SC18	0.04	No data	No data	73
Pb^{2+}	ETAAS	On-line adsorption: analyte complexes with 8-HQ on XAD-2	0.01	No data	10	77
Pb^{2+}	FAAS	On-line adsorption: analyte complexes with DDTC on PCTFE bead	1.2	30	2.1	126
Pb^{2+}	ETAAS	On-line coprecipitation	0.032	9-12	8.0	16
Pb^{2+}	ETAAS	On-line coprecipitation with APDC	0.005	No data	4.8	145
Pb^{2+}	ETAAS	Coprecipitation with $Fe(OH)_3$ On-line retention on Pb-Spec	No data	No data	No data	245
Pb^{2+}	ETAAS	On-line gas-liquid separation	1.5-4.0	100-180	2.5-4.0	246
Pb^{2+}	ETAAS	On-line gas-liquid separation	0.3	No data	1.7	247
Pb^{2+}	ETAAS	On-line gas-liquid separation	173.3	No data	No data	248
Pb^{2+}	ICP-MS	On-line adsorption: analyte complexes with DDTC on SC18	0.05	No data	No data	100

Analyte	Detection	Separation Technique	DL (µg/L)	SF (s/h)	RSD (%)	Ref
Pb^{2+}	ICP-MS	On-line adsorption: analyte complexes with DDTP on SC18	0.0003	22	<10	57
Pb^{2+}	ICP-MS	On-line adsorption: analyte complexes with DDTP on SC18	0.0003-0.0045	21-22	<6	19
Pb^{2+}	ICP-MS	On-line adsorption: analyte complexes with DDPA on SC18	$2.8x10^{-5}$ - 0.00016	21-31	No data	249
Pb^{2+}	ICP-MS	On-line adsorption on C18	0.09	No data	3.4	101
Pb^{2+}	ICP-MS	On-line retention on Pb-Spec	0.5	No data	No data	250
Pb^{2+}	ICP-MS	On-line ion-exchanger column of SO_3-oxine CM-cellulose	No data	No data	No data	90
Pb^{2+}	ICP-MS	On-line adsorption: analyte complexes with 8H5S on florisil	0.22	No data	No data	251
Pb^{2+}	ICP-MS	On-line adsorption: analyte complexes with 8H5S on florisil	0.204		0.98	252
Pb^{2+}	ICP-MS	On-line SGF with 8-HQ	0.003	10	3-5	94
Pb^{2+}	ICP-MS	On-line controlled-pore glass funtionalized with 8-HQ	No data	No data	No data	95
Pb^{2+}	ICP-MS	On-line 8-HQ immobilized on fluorinated metal alkoxide glass	0.00028	No data	No data	96
Pb^{2+}	ICP-MS	8-HQ bonded ZrO_2 composition microspheres	0.01	No data	9.1	97
Pb^{2+}	ICP-MS	On-line silica gel modified with niobium(V) oxide	0.01-0.4	No data	4.1	99
Pb^{2+}	ICP-MS	On-line 8-HQ immobilized on SiO_2 surface	0.010	No data	9.1	253
Pb^{2+}	ICP-MS	On-line chelating resin (iminodiacetate groups)	0.001	15	10	89
Pb^{2+}	ICP-MS	On-line chelating resin (iminodiacetate groups)	0.0012	15	No data	91
Pb^{2+}	ICP-MS	On-line chelating resin (iminodiacetate groups)	0.001	No data	0.1	92
Pb^{2+}	ICP-MS	On-line chelating resin (iminodiacetate groups)	No data	No data	2.6-6.9	116
Pb^{2+}	ICP-MS	Off-line cloud point extraction	0.004	No data	4.8	18
Pb^{2+}	ICP-MS		0.001	No data	1.1	87
Pb^{2+}	ICP-MS		0.0008	No data	No data	88
Pb^{2+}	LEI	On-line adsorption: analyte complexes with DDPA on SC18	0.0033-0.011	9-18	No data	254

Table 4.1. Continued

Analyte	Detection	Separation Technique	DL (µg/L)	SF (s/h)	RSD (%)	Ref
^{139}La	ICP-MS	On-line chelating resin (iminodiacetate groups)	0.00005	5	5-6	89
^{140}Ce	ICP-MS	On-line chelating resin (iminodiacetate groups)	0.00004	5	5-6	89
^{141}Pr	ICP-MS	On-line chelating resin (iminodiacetate groups)	0.00004	5	5-6	89
^{143}Nd	ICP-MS	On-line chelating resin (iminodiacetate groups)	0.00029	5	5-6	89
^{147}Sm	ICP-MS	On-line chelating resin (iminodiacetate groups)	0.00006	5	5-6	89
^{152}Sm	ICP-MS	On-line chelating resin (iminodiacetate groups)	0.00008	5	5-6	89
^{151}Eu	ICP-MS	On-line chelating resin (iminodiacetate groups)	0.00003	5	5-6	89
^{151}Eu	ICP-MS	On-line adsorption: analyte complexes with 8-HQ on Amberlite XAD-7	1.6x10-5	No data	1.5	255
^{153}Eu	ICP-MS	On-line chelating resin (iminodiacetate groups)	0.00002	5	5-6	89
^{159}Tb	ICP-MS	On-line chelating resin (iminodiacetate groups)	0.00002	5	5-6	89
^{159}Tb	ICP-MS	On-line adsorption: analyte complexes with 8-HQ on Amberlite XAD-7	$2.3x10^{-6}$	No data	2.8	255
^{155}Gd	ICP-MS	On-line chelating resin (iminodiacetate groups)	0.00002	5	5-6	89
^{157}Gd	ICP-MS	On-line chelating resin (iminodiacetate groups)	0.00008	5	5-6	89
^{160}Gd	ICP-MS	On-line chelating resin (iminodiacetate groups)	0.00004	5	5-6	89
^{161}Dy	ICP-MS	On-line chelating resin (iminodiacetate groups)	0.00012	5	5-6	89
^{163}Dy	ICP-MS	On-line chelating resin (iminodiacetate groups)	0.00006	5	5-6	89
^{165}Ho	ICP-MS	On-line chelating resin (iminodiacetate groups)	0.00002	5	5-6	89
^{165}Ho	ICP-MS	On-line adsorption: analyte complexes with 8-HQ on Amberlite XAD-7	$1.7x10^{-6}$	No data	1.2	255
^{166}Er	ICP-MS	On-line chelating resin (iminodiacetate groups)	0.00004	5	5-6	89
^{167}Er	ICP-MS	On-line chelating resin (iminodiacetate groups)	0.00006	5	5-6	89

Analyte	Detection	Separation Technique	DL (μg/L)	SF (s/h)	RSD (%)	Ref
^{169}Tm	ICP-MS	On-line chelating resin (iminodiacetate groups)	0.00002	5	5-6	89
^{169}Tm	ICP-MS	On-line adsorption: analyte complexes with 8-HQ on Amberlite XAD-7	3.5×10^{-6}	No data	2.3	255
^{171}Yb	ICP-MS	On-line chelating resin (iminodiacetate groups)	0.00016	5	5-6	89
^{173}Yb	ICP-MS	On-line chelating resin (iminodiacetate groups)	0.00016	5	5-6	89
^{174}Yb	ICP-MS	On-line chelating resin (iminodiacetate groups)	0.00009	5	5-6	89
^{175}Lu	ICP-MS	On-line chelating resin (iminodiacetate groups)	0.00002	5	5-6	89
^{175}Lu	ICP-MS	On-line adsorption: analyte complexes with 8-HQ on Amberlite XAD-7	1.5×10^{-6}	No data	2.0	255
^{239}Pu	ICP-MS	Off-line coprecipitation Off-line anion-exchange	5×10^{-9}	No data	12.0	256
^{239}Pu	ICP-MS	On-line retention on alumina	No data	No data	No data	257
U	Ch		0.2	No data	1.0	258
U	Ch		0.2	No data	2.5	259
^{238}U	ICP-MS	On-line Chelating resins	No data	No data	No data	10
U	SWASV	Wall jet Hg-film electrode with cupferron as complexing reagent	No data	No data	8.3	260
U	ICP-MS	On-line retention on alumina	No data	No data	4.5	261
U	ICP-MS		No data	No data	No data	262
U	ICP-MS		0.10	No data	No data	88
$^{234}U/^{238}U$	ICP-MS		No data	60	5.0	263
^{238}U	ICP-MS	On-line chelating resin (iminodiacetate groups)	0.007	10	No data	93
^{238}U	ICP-MS	On-line retention on chitosan resin	No data	No data	2	264
U	ICP-MS		0.00038	25	<7	11
U	ICP-MS	On-line retention on chitosan resin	No data	6	1-1.9	265
Rh	ICP-MS		0.006-0.1	No data	2-7	266
Sb^{3+}/Sb^{5+}	AF	On-line sorption onto a KR On-line gas-liquid separation	0.0023	24	4.5	267
Sb	AAS	On-line gas-liquid separation	0.24	No data	10	32
Sb^{3+}/Sb^{5+}	AAS	On-line gas-liquid separation	0.007	No data	0.4	268
Sb^{3+}	AAS	On-line sorption onto a KR On-line gas-liquid separation	0.021	42	2.9	269

Table 4.1 Continued

Analyte	Detection	Separation Technique	DL (µg/L)	SF (s/h)	RSD (%)	Ref
Sb	AAS	On-line gas-liquid separation	0.02	No data	<6	270
Sb	ETAAS	On-line coprecipitation	0.022	9-12	8.0	16
Sb^{3+}/Sb	AAS	On-line gas-liquid separation	0.005 0.01	15 10	No data	271
Sb	AAS	On-line gas-liquid separation	0.37-1.3	60	1-8.1	272
Sb^{3+}/ Sb^{5+}	ETAAS	Off-line adsorption on nanometer size TiO2	0.05 0.06	No data	2.1 2.4	273
Sb	ICP-MS	On-line gas-liquid separation	0.0013-0.004	No data	1.1	43
Se^{4+}	ASV		0.068	No data	1.9-4.1	274
Se^{4+} Se	AF	On-line gas-liquid separation	0.005 0.004	No data	5.6 3.3	275
Se^{4+}	SP		2	7	3.4-2.7	276
Se^{4+} Se^{6+}	AAS	On-line gas-liquid separation	0.7-1.5 0.7-1.5	No data	3-5	277
Se^{4+}	ETAAS	On-line coprecipitation	0.001	33	0.7	278
Se^{4+}/ Se^{6+} org. Se(II)	AAS	On-line gas-liquid separation	0.00015	12	No data	279
Se^{4+}/ Se^{6+}	ETAAS	On-line adsorption: analyte complexes with DDTC on SC18	0.0045	11	3.8	280
Se	ICP-MS	On-line gas-liquid separation	0.005	No data	<2	281
Se	ICP-MS	On-line gas-liquid separation	0.007	5	10	47
Se^{4+}	ICP-MS	On-line adsorption: analyte complexes with DDTP on SC18	0.005	22	<10	57
Se	ICP-MS	Off-line cloud point extraction	0.02	No data	5.3	18
Sn	ETAAS	On-line gas-liquid separation	0.130	No data	2.0-9.0	282
Sn org.	ICP-MS	Micellar liquid chromatography	No data	No data	No data	283
Sn	ASV	Nonelectrolytic preconcentration with an epoxy-carbon powder 8-HQ composite electrode	0.055	No data	3.4-9.7	284
^{47}Ti ^{48}Ti	ICP-MS	On-line ion-exchanger column of SO_3-oxine CM-cellulose	0.02 0.01	No data	<10	285
Tl	ICP-MS	On-line gas-liquid separation	0.01	No data	2.4-2.8	286
Tl	ICP-AES	On-line minicolumn of ZrO_2	0.09	No data	6.3	287

Analyte	Detection	Separation Technique	DL (μg/L)	SF (s/h)	RSD (%)	Ref
V	SP	On-line minicolumn of immobilized 8-HQ	0.01	No data	0.2-0.7	288
V	ICP-MS	On-line Chelating resins	No data	No data	No data	10
V	ICP-MS		0.01-1	No data	No data	44
V	ICP-MS	On-line chelating resin (iminodiacetate groups)	0.0043	15	No data	91
V	ICP-MS	On-line chelating resin (iminodiacetate groups)	0.042	10	No data	93
V	ICP-MS		0.0018	25	4.4-5.4	11
V	ICP-MS	On-line silica gel modified with niobium(V) oxide	0.03-0.8	No data	4.5	99
V	ICP-MS	On-line ion-exchanger column of SO_3-oxine CM-cellulose	0.001	No data	<10	285
Zn^{2+}	ASV	On-line immobilized 8-HQ on silica gel	No data	3-5	2.5-6.9	55
Zn^{2+}	ASV		No data	No data	No data	58
Zn^{2+}	F	On-line minicolumn of 8-HQ immobilized on Fractogel	0.007	No data	6	289
Zn^{2+}	F	On-line anion exchange resin (Dowex 1X2-400)	0.26	No data	8	60
Zn^{2+}	SP		1.5	No data	1.85	290
Zn^{2+}	FAAS	On-line chelating resin (Chelex-100)	10	30-60	No data	63
Zn^{2+}	FAAS	On-line chelating resin (Chelex-100)	0.03	60	1.2-3.2	64
Zn^{2+}	FAAS	On-line chelating resin (Chelex-100)	0.5	No data	2.7	291
Zn^{2+}	FAAS	On-line chelating resin (Chelite P)	0.02	26	0.4-2.3	292
Zn^{2+}	FAAS	On-line adsorption: analyte complexes with 8-HQ on SC18	No data	30	No data	68
Zn^{2+}	FAAS	On-line adsorption: analyte complexes with DCO on SC18	0.5	30	6.5	112
Zn^{2+}	ICP-AFS	On line chelating resin, (CMPEPMI)	<0.1	No data	No data	293
Zn^{2+}	ICP-AES	In-situ on-line retention on chelating resin Metpac CC-1	0.7	No data	<10	85
Zn^{2+}	ICP-AES	On-line SGF with 1,5-bis(di-2-pyridyl)methylene thiocarbohydrazide	1.7	40	0.8-1.5	294

Table 4.1. Continued

Analyte	Detection	Separation Technique	DL (μg/L)	SF (s/h)	RSD (%)	Ref
Zn^{2+}	ICP-MS	On-line Chelating resins	No data	No data	No data	10
Zn^{2+}	ICP-MS	On-line chelating resin (iminodiacetate groups)	0.028	15	16	89
Zn^{2+}	ICP-MS	On-line chelating resin (iminodiacetate groups)	0.009	15	No data	91
Zn^{2+}	ICP-MS	On-line chelating resin (iminodiacetate groups)	0.012	No data	1.8	92
Zn^{2+}	ICP-MS	On-line chelating resin (Toyopearl AF-Chelate-650M)	0.046	17	4.4	98
Zn^{2+}	ICP-MS	On-line chelating resin (iminodiacetate groups)	No data	No data	2.6-6.9	116
Zn^{2+}	ICP-MS	On-line ion-exchanger column of SO_3-oxine CM-cellulose	0.014	No data	0.7-1.7	295
Zn^{2+}	ICP-MS	On-line SGF with 8-HQ	0.022	10	3-5	94
Zn^{2+}	ICP-MS	On-line 8-HQ immobilized on fluorinated metal alkoxide glass	0.0053	No data	No data	96
Zn^{2+}	ICP-MS	On-line silica gel modified with niobium(V) oxide	0.02-0.3	No data	2.2	99
Zn^{2+}	ICP-MS	On-line adsorption: analyte complexes with DDTC on C18	0.11	No data	No data	100
Zn^{2+}	ICP-MS	On-line adsorption on C18	0.3	No data	0.5	101

1N2N: 1-nitroso-2-naphthol; AAS: atomic absorption spectrometry; AF: atomic fluorescence; AMPA: aminophosphonic acid ; APDC: 1-pyrrolidinecarbodithioic acid ; ASV: anodic stripping voltammetry; ATAAS: atom trapping atomic absorption spectrometry; BPC: background proportional counter; C: conductimetry; Ch: chemiluminescence; CMPEPMI: carboxymethylated polyethylenimine-polymethylene- polyphenylene isocyanate; CSV: cathodic stripping voltammetry; DCO: 5,7-dichlorooxine ; DCOL: 5,7-dichloroquinoline-8-ol; DDPA: ammonia diethyl dithiophosphate; DMS: dibromo-methyl sulfonazo; DDP: diethyldithiophosphoric; DDTC: diethyldithiocarbamate; DDTP: salts of diethyldithiophosphoric acid; DL: detection limit; DMA: dimethylated arsenic; DMG: dimethylglyoxime; DPC: 1,5-diphenylcarbazide; DPTH: 1,5-bis(di-2-pyridyl) methylene thiocarbohydrazide ; ETAAS: electrothermal atomic absorption spectrometry; F: fluorescence; FAAS: flame atomic absorption spectrometry; HDEHP: bis(2-ethylhexyl) hydrogen phosphate; HEED: N-hydroxyethylethylenediamine; HPLC: high-performance liquid chromatography; 8-HQ: 8-hydroxyquinoline; 8H5S: 8-hydroquinoline-5-sulfonic acid; HVPMPCEE: (S)-2-[hydroxy-bis-(4-vinyl-phenyl)-methyl]-pyrrolidine-1-carboxylic acid ethyl ester; KR: knotted reactor; ICP-AES: inductively coupled plasma-atomic emission spectrometry ; ICP-AFS: inductively-coupled plasma atomic fluorescence spectrometry; ICP-MS: inductively coupled plasma-mass spectrometry; LEI: flame laser-enhanced ionization; L-L: liquid-liquid; MMA: monomethylated arsenic; NNDS: 1-nitroso-2-naphthol-3,6-disulfonate; NTA:

nitrilotriacetic acid; P: potentiometry ; PADMAP: 2-(2-pyridylazo)-5-dimethylaminophenol ; PAN: 1-(2-pyridylazo)-2-naphthol ; PAR: 4-(2-pyridylazo)resorcinol ; PCTFE: poly-chlorotrifluoro-ethylene; PDV: pulse differential voltammetry; PH: phosphorescence; Phen: 1,10-Phenanthroline ; PTFE: polytetrafluoroethylene; RSD: relative standard deviation; S: sensor; SC18: silica C18; SDVB: styrene divinylbenzene resin; SF: sampling frequency; SGF: silica gel funtionalized; SP: spectrophotometry; SWASV: square wave adsorptive stripping voltammetry; TAC: with 2-(2-thiazolylazo)-p-cresol ; TS: methylthiosalicylate

Alkali Metals (Sodium)

Uehara et al. [9] presented a flow injection (FI) system for monitoring sodium ion in brine based on a catalytic decomposition of a chelate-type chromoionophore with highly concentrated sodium ions. Bis(1,4,7,10,13-pentaoxa-16-aza-cyclooctadecane-N-carbodithioato) cobalt(II) (Co-A18CC) is newly synthesized and examined as a chelate-type chromoionophore. The catalytic decomposition of Co-A18CC is highly specific for sodium ion. A linear calibration is obtained for sodium ion in the range of 0-5 mol/L. Minor components of brine, such as magnesium, bromide, sulfate, and carbonate, did not interfere with the decomposition reaction.

Alkaline Earth Metals (Barium, Calcium, Magnesium and Radium)

Ebdon et al. [10] developed a continuous matrix elimination procedure for the determination of barium and other trace metals by inductively coupled plasma-mass spectrometry (ICP-MS). So, elements of interest are concentrated quantitatively and isolated from the sodium chloride matrix using a chelating ion-exchange column. For this, two columns are evaluated, an inhouse dynamically coated chelating exchange resin (Xylenol Orange coated onto an anion-exchange resin, Dowex 1 X8) and a commercially available chelating resin (Metpac CC-1). Although the two columns have essentially the same functional group (iminodiacetic acid), they behave somewhat differently from one another. This might be due in part to the environment of the functionality itself, in the case of Xylenol Orange this is adjacent to a benzene ring. The other factor, which effects the performance, is the distribution of the functionality across the resin. Thus, both column are found to be efficient to retain alkaline earth elements. Linear calibrations are obtained in the concentration range of interest (<200 μg/L). Field et al. [11] added methane (CH_4) to the sample gas flow to improve the sensitivity of non-metals with high first ionization potentials. Thus, can be obtained a reduction of both chlorine (Cl) and oxygen (O) based interferences. Using direct flow injection and CH_4 addition the method can perform >700 continuous determinations of 10-fold diluted seawater in <30 h by sector field inductively coupled plasma mass spectrometer (SF-ICP-MS) analysis. Yuan et al. [12] established a FI method for simultaneous determination of Ca and Mg by spectrophotometry by using Xylenol Orange and cetyltrimethylammonium bromide. The total amount of Ca and Mg is determined at 585 nm by using a NH_3-NH_4Cl buffer at pH 10.5 containing triethanolamine. Ca is determined at 605 nm by using a NH_3-NH_4Cl-sodium citrate solution. An automatic method for preconcentration and separation of radium in water samples has been developed by Fajardo et al. [13]. This method combines both multisyringe (MSFIA) and multi-pumping (MPFS) flow analysis techniques. Ra is adsorbed on-line on MnO_2, deposited on cotton fiber, is eluted with hydroxylamine and subsequently coprecipitated with $BaSO_4$. ^{226}Ra activity is determined off-line by using a low background proportional counter.

Silver

Voltammetric Detection

Ye et al. [14] studied the feasibility of an epoxy-graphite tube impregnated in the bulk with 2-mercaptobenzoxazole, in conjunction with a wall-jet cell, for the continuous flow and FI stripping determination of silver, mercury and bismuth. Shpigun et al. [15] presented a procedure including Ag accumulation on the C compositional electrode (CCE) followed by anode current recording in the pulse-differential voltammetry mode in conditions of FI system. The CCE is made from the mixture of C powder silopren and hydrophobic Si carrier inoculated with 1,4,7,10-tetrathiacyclododecane. For the period of accumulation (30 min) the anode peak height linearly depends on the Ag ion concentration at 0.1-1.0 nM.

Atomic Absorption Spectrometric Detection

A flow injection online preconcentration system coupled to an electrothermal atomic absorption spectrometer for the determination of trace metals in saline media is described by Sella et col. [16]. The filterless, magnetic collection of coprecipitated analytes is based on the tetrahydroborate reductive precipitation of added iron and palladium in alkaline medium. The precipitate is dissolved in a 20 μL volume of mixed acid and transported direct to the graphite furnace. With the exception of Cr (33% recovery in seawater), recoveries of Ag, As, Bi, Cd, Co, Cu, Mn, Ni, Pb, Sb and Tl averaged 84% from deionized water and 67% from coastal seawater. The sensitivity of the graphite furnace technique can be enhanced over 400-fold (for an 11 mL sample vol.) compared to a standard 20 μL injection volume. Dadfarnia et al. [17] proposed a FI system incorporating a microcolumn of immobilized diethyldithiocarbamate (DDTC) on surfactant-coated alumina combined with flame atomic absorption spectrometry (FAAS).

Inductively Coupled Plasma Mass Spectrometric Detection

A preconcentration method for low Ag, As, Au, Cd, Cu, Pb, and Se using cloud point extraction is proposed by Mesquita da Silva et al. [18]. For this, these analytes are complexed with ammonium-O,O-diethyl-dithiophosphate (DDTP). Complexation allows analyte separation from alkali, alkaline earth, and other elements, which are not complexed. After phase separation, the enriched analytes were determined by inductively coupled plasma-mass spectrometry (ICP-MS), using ultrasonic nebulization. Pozebon et al. [19] studied and compared the performance of two FI for on line separation, preconcentration and determination by ICP-MS of Cu, Cd, Pb, Bi, Au, Ag, As(III) and Se(IV). One system is coupled to the nebulizer of the spectrometer while the other one is coupled to an electrothermal vaporizer also coupled to the same spectrometer. The matrix separation and analyte preconcentration are accomplished by retention of the analytes complexed with DDTP on C18 immobilized on silica in a column coupled directly to the pneumatic nebulizer or to the auto-sampler arm of an electrothermal vaporizer. The eluate is on line diluted with water and introduced directly to the pneumatic nebulizer or vaporized after being automatically injected into the graphite tube, prior to the vaporization of the analytes. Yan et al. [20] described an online FI method for the direct determination of silver using isotope dilution inductively coupled plasma mass spectrometry (ID-ICP-MS). A mini-column packed with Dowex 1-X8 anion exchange resin is used to separate and concentrate Ag. This method achieves recoveries of 100 ± 1% and 96 ±1%. Barriada et al. [21] proposed an automated FI system for the determination of dissolved silver at ultratrace concentrations by using a magnetic sector inductively coupled plasma mass spectrometry (MS-ICP-MS) instrument controlled under LabVIEW. This FI approach with incorporation of an anion exchange minicolumn allows ready removal of interferences caused by the saline matrix. The software allowed full control of all FI components (valves and pumps) and removed manual time control and, therefore, operator errors. Ndung'u et al. [22] studied interference effects on the analysis of Ag in estuarine and oceanic waters using on-line ICP-MS. A mini-column packed with a strong anion exchange resin (Dowex 1-X8) is used to separate Ag from the saline samples prior to on-line determination by ICP-MS.

Aluminium

Fluorescence Detection

A highly sensitive method for the shipboard determination of Al by FIA has been developed by Resing et al. [23]. The method employs on-line preconcentration of Al onto a column of resin-immobilized 8-hydroxyquinoline. The Al is subsequently eluted into the FIA system from the resin with acidified seawater. The eluted Al reacts with lumogallion to form a chelate, which is detected by its fluorescence. The fluorescence is enhanced approximately 5-fold by the addition of a micelle-forming detergent, Brij-35. An overview of sampling techniques and FIA methods to determine low concentrations of Fe, Mn, and Al in filtered seawater is given by de Jong et al. [24]. Two different types of surface water sampling are tested and compared: conventional hand filling of a sample bottle from a rubber dinghy away from the ship; and underway pumping of seawater using a tow fish. The latter method gave the best results. Also, conventional membrane filtration and cartridge filtration for large volumes filtration were compared using Fe and Al data from water column samples. Good agreement was observed for both filter types, although to define dissolved metal species, the latter filter type was preferred. Aluminium is detected by fluorimetry measuring the Al-lumogallion complex. Obata et al. [25] developed a flow-through method using lumogallion fluorometric detection for onboard ship determination of aluminium. Because iron was selectively removed by passing the sample through an 8-quinolinol immobilized chelating resin column (MAF-8HQ), this method could be applied to samples collected at the hydrothermal oceanic regime. Several known organic complexing agents are studied as a model group to examine the effect of naturally occurring organic ligands on the detection of aluminum in seawater. A towed surface sampling device coupled to two automated FIA systems is described by Vink et al. [26]. The towed system permits uncontaminated sampling of seawater from research vessels while underway at full speed. Coupling the sampler to the FIA systems permits automatic determination of Al and Fe in surface waters at natural levels at 5 min intervals. Al is determined by spectrofluorimetric detection of the Al-lumogallion chelate. A FI spectrofluorimetric method to directly determine Al in water is described by

Kara et al. [27]. This method is based on the reaction of Al with N-o-vanillidine-2-amino-p-cresol (OVAC) in acidic medium at pH 4.0, forming a water-soluble complex. Excitation and emission wavelengths used were 423.0 and 553.0 nm, respectively. Brown et al. [28] improved a FI analytical method for the determination of dissolved aluminum in seawater, based on the work of Resing and Measures [23]. The most significant modification to this earlier work is the use of on-line preconcentration of seawater onto a commercially available resin, eliminating the resin-synthesis step. Other modifications include the addition of a column-conditioning step prior to sample loading that increases the %retention and sensitivity, on-line buffering of samples prior to column loading to achieve optimal sample pH, an increased acid-eluent concentration to efficiently remove Al from the resin column, and an increased reaction buffer concentration.

Atomic Absorption Spectrometric Detection

Yuan et al. [29] developed a method for the determination of endogenous levels of aluminium in water samples using an automated on-line preconcentration system with FI coupled directly to an electrothermal atomic absorption spectrometer. Two preconcentration materials, 8-quinolinol immobilized on controlled-pore glass (8-Q-CPG) and Amberlite XAD-2, poly(styrene/divinyl benzene) copolymer (XAD-2) were investigated and compared. Both systems were found to be suitable for preconcentration. However, the sampling flow-rate for the 8-Q-CPG system was found to be much lower than that of the XAD-2 system, relative to the same magnitude of preconcentration. The chelating kinetics of the 8-Q-CPG system were less favorable than the adsorption kinetics of the XAD-2 system.

Arsenic

Chemiluminescence Detection

Fujiwara et al. [30] proposed a chemiluminescence method based on a combination of AsH_3 generation from arsenite and/or arsenate, and the measurement of the chemiluminescence, which occurs, upon mixing O_3 and AsH_3.

Atomic Absorption Spectrometric Detection

Yamamoto et al. [31] examined the performance of a gas-liquid separator with microporous PTFE tubing for use with a FI manifold for hydride-generation atomic absorption spectrometry (HG-AAS). Arsine, generated from ppb levels of As, is separated with a tube length of 25 cm. Lu et al. [32] proposed an online cold-trap hydride collection, FI system for determination As and Sb in seawater. The system consist of a hydride generator, a nebulizer for gas/liquid separation and liquid nitrogen bath hydride collector. Hydrides are generated by the reactions of samples with $NaBH_4$; after passing through the system. Arsenic(III) is extracted using sodium diethyldithiocarbamate as the complexing agent and C18 reversed phase packing as the column material for solid phase extraction. Arsenic(V) is on-line reduced to its trivalent oxidation state prior to extraction with a mixture of sodium sulfite, hydrochloric acid, sodium thiosulfate, and potassium iodide. Thus, arsenic(III) and total arsenic can be determined sequentially by graphite furnace atomic absorption spectrometry (GFAAS) [33]. Willie [34] described a simple method to distinguish between

As species that react with sodium tetrahydroborate (III) to form AsH_3 and the naturally occurring As species that are unreactive. Organoarsenic species that do not react with sodium borohydride under acidic conditions such as arsenobetaine, arsenocholine and tetramethylarsenic, are converted to As(V) by on-line photo-oxidation or microwave heating. The sample is subsequently acidified, reduced with sodium borohydride and the generated arsine is trapped in a heated graphite furnace prior to atomization. Xu et al. [35] proposed a method for the determination of trace arsenic with L-cysteine as prereductant using FI-HG-AAS. A flow injection online preconcentration system coupled to an electrothermal atomic absorption spectrometer for the determination of trace metals in saline media is described by Sella et col. [16] described a FI method including a separation by coprecipitation and graphite furnace atomic absorption spectrometric detection. Karthikeyan et al. [36] developed a rapid and sensitive sorbent extraction hydride generation-FI atomic absorption spectrometric (HG-AAS) method for the determination of As(III) and As(V) based upon online preconcentration on a microcolumn packed with activated alumina. In these procedure arsenicals compounds are complexed on-line with quinolin-8-ol-5-sulfonic acid and adsorbed on the column. The retention efficiency is better than 98% with sensitivity enhancement of 12 and 10 for As(III) and As(V), respectively, for a 20 s preconcentration period. The potential of the continuous FI HG-AAS is studied for the speciation of major As species in seawater by Cabon et al. [37-38]. For this, are used two different techniques. As is determined by AAS after hydride generation and collection in a graphite tube coated with Ir. Thus, it is possible to determine As(III), total As, hydride reactive As and by difference non-hydride reactive As. By cryogenically trapping hydride reactive species on a chromatographic phase, followed by their sequential release and AAS in a heated quartz cell, inorganic As, monomethylated arsenic (MMA), dimethylated arsenic (DMA) and eventually non-hydride reactive organic compounds can be determined. An innovative analytical system based on the combination of cryogenic trapping and unambiguous gas chromatographic separation performed within a packed cold finger trap (PCFT) has been developed by Tung-Ming et al. [39]. Herbello-Hermelo et al. [40] presented a FI online sorption preconcentration system using a polytetrafluoroethylene (PTFE) knotted reactor coupled with electrothermal atomic absorption spectrometry (ETAAS) to determine selectively As(III). In this method, separation of As inorganic species is achieved by the online formation of As(III)-pyrrolidine dithiocarbamate (PDC) complex in an acid medium, its sorption onto the inner walls of the knotted reactor (KR) and elution with ethanol.

Atomic Fluorescence Detection

Hou et al. [41] developed a method for direct determination of As(III) and As(V) in seawater with flow injection-hydride generation-non dispersion atomic fluorescence spectrometry. This method have no matrix effects of NaCl, Na_2SO_4, $MgCl_2$, and $CaCl_2$ on As(III) determination and no interference of concomitant elements, such as Cd, Cu, Mn, Zn, and Pb. Yan et al. [42] proposed a FI sorption preconcentration and separation method including a KR coupled to hydride generation atomic fluorescence spectrometry (HG-AFS) for speciation of inorganic arsenic. The method involved online formation of the As(III)-pyrrolidinedithiocarbamate (PDC) complex, its adsorption onto the inner walls of the KR, elution with HCl, and detection by HG-AFS. Total inorganic arsenic was determined after prereduction of As(V) to As(III) with L-cysteine. The concentration of As(V) is calculated by the difference of the total inorganic arsenic and As(III).

Inductively Coupled Plasma Mass Spectrometric Detection

Stroh et al. [43] used vapor generation FI inductively coupled plasma mass spectrometry to determine As, Sb, and Hg at ultratrace levels. Alves et al. [44] used an Ar-H_2 mixed-gas plasma to improve the sensitivity obtained with cryogenic desolvation with inductively-coupled plasma mass spectrometry (ICP-MS). The polyatomic ions ClO^+, CaO^+, and $ArCl^+$, which normally cause severe overlap interferences, areattenuated to manageable levels by cryogenic desolvation. The samples are introduced by FI to minimize clogging of the sampling orifice. Huang et al. [45] used a simple and very inexpensive in situ nebulizer-hydride generator with ICP-MS. The application of hydride generation (HG)-ICP-MS alleviated the spectral interferences and sensitivity problems of arsenic determinations encountered when conventional pneumatic nebulization is used for sample introduction. The sample is introduced by FI to minimize deposition of solids on the sampling orifice. The arsenic in the sample is reduced to As(III) with L-cysteine before being injected into the HG system.

Chen et al. [46] used a simple and very inexpensive in-situ nebulizer/hydride generator with ICP-MS for the determination of As, Sb, Bi and Hg in water samples. The application of hydride generation ICP-MS alleviated the sensitivity problem of As, Sb, Bi and Hg determinations encountered when the conventional pneumatic nebulizer is used for sample introduction. The sample was introduced by flow injection to minimize the deposition of solids on the sampling orifice. The elements in the sample were reduced to the lower oxidation states with L-cysteine before being injected into the hydride generation system. Arsenic and selenium (IV) are determined in seawater following hydride generation (HG) by reaction with $NaBH_4$ in an automated FI system. The hydrides are trapped on reduced palladium in a graphite furnace and subsequently vaporized into an ICP for MS detection [47]. Arsenic and other trace metals are preconcentrated using cloud point extraction previous ICP-MS determination [18]. Pozebon et al. [19] studied and compared the performance of two FI for on line separation by retention of the analytes complexed with DDTP on C18 immobilized on silica in a column, and determination by ICP-MS of Cu, Cd, Pb, Bi, Au, Ag, As(III) and Se(IV). Lee et al. [48] developed a rapid and simple analytical method for the determination of transition metals and rare earth metals at sub-ppb (ppb = ng/L) levels in the presence of alkali and alkaline earth metal matrices. This method is based on a solid phase filtration/concentration technique for the matrix separation and analyte preconcentration. For this, the flow injection system used is equipped with a chelating disk made of an iminodiacetate resin and detection of the metals is accomplished by ICP-MS. Holliday et al. [49] described a method that allows the direct analysis of seawater by ICP-MS using a simple external calibration without internal standardization. This method provides robust operating conditions that greatly reduce non-spectroscopic interferences (i.e. matrix effects) while still maintaining sufficient sensitivity for analyte quantitation. Field et al. [11] added methane (CH_4) to the sample gas flow to improve the sensitivity of non-metals with high first ionization potentials to reduce chlorine (Cl) and oxygen (O) based interferences for the determination of several elements including arsenic by sector field inductively coupled plasma mass spectrometer (SF-ICP-MS) analysis.

Gold

Voltammetric Detection

Ye et al. [50] employed an epoxy-impregnated graphite tube composite electrode bulk modified with 2-mercaptobenzoxazole for the determination of gold by a differential-pulse cathodic stripping procedure in conjunction with either continuous flow or FI. Both open and closed circuit accumulation of Au(III) were studied. The addition of Rhodamine 6G to the accumulation medium enhanced the sensitivity.

Inductively Coupled Plasma Mass Spectrometric Detection

Falkner et al. [51-52] presented and applied a method for the determination of Au in seawater at 10-15 mol/L levels. The technique involves preconcentration by anion exchange of Au as a cyanide complex, using 195Au radioatracer (tl/2 = 183 days) to monitor recoveries. Samples are then introduced by FI into an inductively coupled plasma quadrupole mass spectrometer. Gold and other metals are also preconcentrated using cloud point extraction [18] or on-line complexed with DDTP on C18 immobilized on silica in a column previous determination by ICP-MS [19].

Boron

Nose et al. [53] developed a spectrophotometric method based on the complexation reaction between D-sorbitol and boric acid followed by the acid-base reaction of methyl orange. Under the optimum conditions, the calibration graph was essentially linear up to 1.2 mg/L of boron (120 µL injections). Nishioka et al. [54] determined trace B in seawater using azomethine H (415 nm).

Bismuth

Voltammetric Detection

Daih et al. [55] developed procedures based on the application of immobilized quinolin-8-ol silica gel and stripping voltammetry for the determination of Cu, Pb, Cd, Bi, In and Zn in seawater. Ye et al. [14] studied the feasibility of an epoxy-graphite tube impregnated in the bulk with 2-mercaptobenzoxazole, in conjunction with a wall-jet cell, for the continuous flow and FI stripping determination of silver, mercury and bismuth.

Atomic Absorption Spectrometric Detection

A flow injection online preconcentration system coupled to an electrothermal atomic absorption spectrometer for the determination of trace metals in saline media is described by Sella et col. [16]. Moscoso-Perez et al. [56] optimized a hydride generation procedure, via FI, coupled to ETAAS for Bi determination. These investigators studied the bismuthine trapping and atomization efficiency from graphite tubes permanently treated with uranium, tantalum, lanthanum oxide, niobium, beryllium oxide, chromium oxide and tantalum carbide.

Inductively Coupled Plasma Mass Spectrometric Detection

Pozebon et al. [57] proposed a procedure for the determination of Cu, Cd, Pb, Bi and Se(IV) in seawater by electrothermal vaporization inductively coupled plasma mass spectrometry, after online separation using a FI system. Matrix separation and analyte preconcentration is accomplished by retention of the analytes complexed with the ammonium salt of O,O-diethyldithiophosphoric acid on C18 immobilized on silica in a minicolumn coupled directly to the autosampler arm of the vaporizer. The methanol used as eluent is vaporized after being automatically injected into the graphite tube, prior to the vaporization of the analytes.

Cadmium

Voltammetric Detection

Ang et al. [58] determined the levels of Cd, Cu, Pb, and Zn in coastal seawater samples by FI anodic stripping voltammetry (ASV). The ASV is performed in the differential pulse mode at a glassy C-based Hg film electrode. Tay et al. [59] developed a simple method of oxygen removal for a FI anodic stripping voltammetry system, based on an electrochemical flow-cell design, which incorporates effective nitrogen purging. The analyses were performed in the differential-pulse mode at a renewable preplated glassy carbon based mercury film wall-jet electrode. Daih et al. [55] developed procedures based on the application of immobilized quinolin-8-ol silica gel and stripping voltammetry for the determination of Cu, Pb, Cd, Bi, In and Zn in seawater.

Fluorescence Detection

Worsfold et al. [60] described an integrated luminometer able to perform fluorescence, room temperature phosphorescence and chemiluminescence measurements on seawater samples. For cadmium determination, is used a fluorescence FI manifold and the method is based on the fluorescence modification resulting from the binding of this metal to a polyazacrown covalently linked to a luminophore via one methylene group. The FI manifold includes a Dowex 1X2-400 column not only for preconcentration of Cd and Zn on the column, but also to separate other trace metals to prevent their fluorescence quenching. Wu et al. [61] developed a simple FI separation system coupled to hydride generation atomic fluorescence spectrometry for ultra-trace cadmium determination. The preconcentratioin of cadmium is carried out on the inner walls of the knotted reactor, this procedure is based on the retention of Cd complex with 1-phenyl-3-methyl-4-benzoylpyrazol-5-one, which is eluted with hydrochloric acid.

Spectrophotometric Detection

Ma et al. [62] established an updated method with high sensitivity to determine trace cadmium based on the system Cd (II)-potassium iodide-ethyl violet.

Atomic Absorption Spectrometric Detection

On-line preconcentration systems involving a commercial chelating resin with iminodiacetic groups (Chelex-100) has been proposed by several authors to preconcentrate cadmium previous flame (FAAS) or atom trapping atomic absorption spectroscopic determination (ATAAS) of cadmium [63-66]. Bonded silica with octadecyl functional groups packed in micro or minicolumns has been used to adsorb on-line cadmium complexes with 8-hydroxyquinoline[67-68], diethylammonium or sodium diethyldithiocarbamate [69-72], diethyldithiophosphate [73], 4-(2-pyridylazo)resorcinol (PAR) or 2-(2-pyridylazo)-5-dimethylaminophenol [74], pyrrolidinecarbodithioic acid [75] and 1,10-Phenanthroline [76] in FI systems involving flame or graphite furnace atomic absorption detection. Minicolumns packed with XAD-2 resin and poly-chlorotrifluoroethylene (PCTFE) as sorbent material have been employed to retain cadmium complexes between 8-hydroxyquinoline [77] and diethyl-dithiophosphate (DDPA) [78], respectively. Other material for Cd preconcentration used in FI systems coupled with atomic absorption detectors is a macrocyclic ligand immobilized on a silica gel support [79]. Sella et col. [16] described a FI method including a separation by coprecipitation and graphite furnace atomic absorption spectrometric detection.

Yebra et al. proposed several in-situ minicolumn preconcentration systems. These systems employ a field flow preconcentration technique involving a minicolumn containing Amberlite XAD-4 impregnated with the complexing agents 1-(2-pyridylazo)-2-naphthol (PAN) [80], 4-(2-pyridylazo) resorcinol (PAR) [81], a chelating resin with aminophosphonic acid groups [82] and the commercial available chelating resin Chelite P (aminomethylphosphonic groups) [83]. Cd-loaded minicolumns are returned to the laboratory where they are incorporated into a FI elution-detection system.

Bermejo-Barrera et al. [84] developed methods for the determination of ultratrace amounts of cadmium by cold vapor/trapping and atomization in a graphite furnace. Thus, studied iridium-, tungsten- and zirconium-coated graphite tubes for the in situ preconcentration of cadmium cold vapor. Cobalt, gallium and silicon are used as catalysts for cadmium species generation. The iridium-coated graphite tube gives the best analytical performance.

Inductively Coupled Plasma Atomic Emission Spectrometric Detection

Nickson et al. [85] described a simple but robust battery-powered field sampling unit for the selective in situ preconcentration of trace elements (Cd, Co, Cu, Mn, Ni, Pb, and Zn) from natural waters on a commercially available iminodiacetate resin (Metpac CC-1). This system allows duplicate samples to be preconcentrated in parallel. The microcolumns are then taken to the laboratory and placed in a FI manifold coupled to an inductively coupled plasma-atomic emission spectrometer (FI-ICP-AES) for elution and quantitation. Zougagh et al. [86] proposed two FI-ICP-AES methods for the preconcentration and determination of trace amounts of cadmium. These are based on the adsorption of the metal ion on a micro-column placed in the injection valve of the FI manifold and packed with silica gel functionalized with 1,5-bis(di-2-pyridyl) methylene thiocarbohydrazide (DPTH-gel) and silica gel functionalized with methylthiosalicylate (TS-gel), respectively.

Inductively Coupled Plasma Mass Spectrometric Detection

Multielement determination of trace metals in seawater is accomplished by inductively coupled plasma-mass spectrometry (ICP-MS) directly [87-88] or with an on-line preconcentration step. This last approach is developed by using chelating materials for metal preconcentration: SO_3-oxine CM-cellulose [90], chelating resins with iminodiacetate groups [89, 91-93], diethyldithiophosphoric acid groups on C18 immobilized on silica [19, 57], 8-

hydroxyquinoline immobilized onto silicone tubing (Sil-8-HQ) [94], immobilized on controlled-pore glass [95] or on fluorinated metal alkoxide glass (MAF-8HQ) [96] and bonded ZrO_2 composition microspheres [97], Toyopearl AF-Chelate-650M resin [98], silica gel modified with niobium(V) oxide [99], bonded SiO_2 with octadecyl functional group C18 packed in a microcolumn to collect diethyldithiocarbamate complexes [100], adsorption on C18 [101] or including off-line cloud point extraction [18].

Ndung'u et al. [102] presented results on the effect of organic complexation on the on-line solvent-extraction and chelating resin column partitioning of estuarine water samples.

Cobalt

Chemiluminescence Detection

FIA is used to automate the determination of cobalt in seawater by the Co-enhanced chemiluminescent oxidation of gallic acid in alkaline hydrogen peroxide. The method includes an on-line preconcentration step involving a minicolumn of immobilized 8-hydroxyquinoline and measurements made shipboard [103] or at laboratory [104]. In order to enhance the sensitivity obtained with the traditional reagent gallic acid, cobalt(II) is determined by means of a pyrogallol-hydrogen peroxide–sodium hydroxide reaction in the presence of methanol and the surfactant cetyltrimethylammonium bromide (CTAB) [60,105]. Tortajada-Genaro et al. [106] proposed two stopped-flow manifolds for individual or simultaneous determination of chromium and cobalt. Automated procedures based on multicommutation systems have emphasized the differences of their catalytic effect in luminol–hydrogen peroxide chemiluminescence reaction.

Atomic Fluorescence Detection

Yuzefovsky et al. [107] developed a method involving laser-excited atomic fluorescence spectrometry in an electrothermal atomizer (ET-LEAFS). This method is integrated with semi online FI microcolumn preconcentration.

Spectrophotometric Detection

FI spectrophotometric techniques for the determination of cobalt in seawater are based on the catalytic effect of cobalt(II) on the oxidation of N,N'-diethyl-p-phenylenediamine by hydrogen peroxide in the presence of Titron as an activator [108-109].

Atomic Absorption Spectrometric Detection

On-line preconcentration systems for ETAAS or FAAS involving the use of a sorbent material as C18-bonded silica and chelating agents such as Na diethyldithiocarbamate [110], dialkyldithiophosphates [111], 5,7-dichlorooxine [112] and 1,10-Phenanthroline [76]. Also, was investigated the possibility of using alumina loaded with 1-nitroso-2-naphthol-3,6-disulfonate [113] and polyurethane foam loaded with 2-(2-thiazolylazo)-p-cresol (TAC) [114] as chelating materials to preconcentrate cobalt in FI systems. Sella et col. [16] described a FI method including a separation by coprecipitation and graphite furnace atomic absorption spectrometric detection.

Inductively Coupled Plasma Atomic Emission Spectrometric Detection

Nickson et al. [85] described a simple but robust battery-powered field sampling unit for the selective in situ preconcentration of trace elements (Cd, Co, Cu, Mn, Ni, Pb, and Zn) from natural waters on a commercially available iminodiacetate resin (Metpac CC-1).

Inductively Coupled Plasma Mass Spectrometric Detection

Determination of trace cobalt and other trace metals in seawater is accomplished by inductively coupled plasma-mass spectrometry (ICP-MS) directly [49, 115] or with an on-line preconcentration/separation step. This separation/preconcentration step carried out by using several chelating materials for metal preconcentration: SO_3-oxine CM-cellulose [90], chelating resins [10] and chelating resins with iminodiacetate groups [89, 91-93, 116], 8-hydroxyquinoline immobilized onto silicone tubing (Sil-8-HQ) [94], 8-hydroxyquinoline immobilized on fluorinated metal alkoxide glass (MAF-8HQ) [96], silica gel modified with niobium(V) oxide [99], bonded SiO_2 with octadecyl functional group C18 packed in a microcolumn to collect diethyldithiocarbamate complexes [100], adsorption on C18 [101], or group-selective precipitation with diethyldithiocarbamate [117].

Chromium

Electrochemical Detection

Ohura et al. [118] proposed a sensitive and rapid potentiometric method for the determination of Cr(VI) using both a redox potential electrode and an Fe(III)-Fe(II) potential buffer containing bromide. The method is based on the detection of a large transient potential change due to bromine generated by an oxidation reaction between Br^- and Cr(VI) in the presence of Fe(II). Yang et al. [119], by taking advantage of the electrocatalytic effect of polyaniline, prepared and used a polyaniline/polystyrene composite electrode as a sensitive FIA detector for the analysis of Cr(VI).

Chemiluminescence Detection

Tortajada-Genaro et a. [106] proposed automated procedures based on multicommutation systems. These emphasized the differences of their catalytic effect in luminol–hydrogen peroxide chemiluminescence reaction.

Fluorescence Detection

Paleologos et al. [120] applied a new method based on the cloud point extraction technique for the differentiation and the selective determination of Cr species. Cr(III) reacts with 8-hydroxyquinoline (8-HQ) in a surfactant solution yielding a hydrophobic complex, which then is entrapped in situ in the surfactant micelles. The Cr(VI) assay is based on its reduction to Cr(III) by sulfite, which subsequently reacts with 8-HQ in a similar manner.

Spectrophotometric Detection

Lynch et al. [121] illustrated the application of a method for the determination of Cr(VI) and total Cr. It employs the use of sequential spectrophotometric (using diphenylcarbazide as reagent) and atomic absorption detectors in a FI system. Wang et al. [122] studied a method to determine trace chromium(VI) in sea water by reversed-phase FIA. Chromium (VI) can be determined with diphenylcarbazide as a cabling reagent, and acetone as a carrier.

Atomic Absorption Spectrometric Detection

Cr(III) is preconcentrated by using an on-line FI system with a column packed with poly(hydroxamic acid) resin previous its determination by FAAS. For this purpose, is designed a three-way FI manifold [123]. Prasada Rao et al. [124] developed a rapid, sensitive and selective method for the speciative determination of Cr(VI) and Cr(III) by FAAS, using online preconcentration on a micro-column packed with C18 bonded silica gel. This is based on the selective formation of diethyldithiocarbamate complexes of Cr(VI) in the 1-2 pH range and Cr(III) in the 4-9 pH range in the presence of Mn(II) (which enhances the Cr(III) signal 10-fold). Long et al. [125] presented a new concept for selective and sensitive determination of trace metals via electrothermal atomic absorption spectrometry based on the principle of bead injection (BI) with renewable reversed-phase surfaces in a sequential injection-lab-on-valve (SI-LOV) mode. The methodology involves the use of poly(styrene-divinylbenzene) beads containing pendant octadecyl moieties (C18-PS/DVB), which are preimpregnated with a selective organic metal chelating agent prior to the automatic manipulation of the beads in the microbore conduits of the LOV unit. The potential of the SI-BI-LOV scheme is demonstrated by taking Cr(VI) as a model analyte, using a 1,5-diphenylcarbazide (DPC)-loaded bead column as the active microzone. Anthemidis et al. [126] developed a time-based FI online solid phase extraction method to determine Cr(VI) and Pb^{2+} by FAAS. In this work, is evaluated the use of hydrophobic poly-chlorotrifluoroethylene (PCTFE) beads as absorbent for online preconcentration. Thus, effective formation of ammonium pyrrolidine dithiocarbamate complexes and subsequent retention in a PCTFE packed column, is achieved at pH 1.0-1.6 and 1.5-3.2 for Cr(VI) and Pb^{2+}, respectively. Yebra et al. [82] proposed an in-situ minicolumn preconcentration system involving a chelating resin with aminophosphonic acid groups for Cr(III) detemination in seawater.

Inductively Coupled Plasma Mass Spectrometric Detection

Hirata et al. [127] described an automated, low-pressure FI method with online column preconcentration using ICP-MS described to determine Cr^{3+} and total Cr in seawater. For this purpose, a home-made column of commercially available iminodiacetate resin, Muromac A-1, is used to concentrate Cr^{3+}. Total Cr is determined after reducing Cr^{6+} to Cr^{3+} with hydroxylamine. Sun et al. [128] developed a simple method for the determination of Cr(III), Cr(VI), and total chromium that use a flow injection online desalter-inductively coupled plasma-mass spectrometry system. To overcome the spectral interference caused by the matrix chloride ion, the authors employed cool plasma to successfully suppress chloride-based molecular ion interference during ICP-MS measurement. Chen et al. [117] used a group-selective on-line precipitation with diethyldithiocarbamate precipitation coupled to ICP-MS.

Copper

Electrochemical Detection

Voltammetric detection of copper is carried out by anodic stripping voltammetry (ASV): Daih et al. [55] developed procedures based on the application of immobilized quinolin-8-ol silica gel and ASV. Ang et al. [58] performed ASV in the differential pulse mode at a glassy C-based Hg film electrode. Tay et al. [59] developed a simple method of oxygen removal, based on an electrochemical flow-cell design, which incorporates effective nitrogen purging. The analyses were performed in the differential-pulse mode at a renewable preplated glassy carbon based mercury film wall-jet electrode. Zolotov et al. [129] proposed a potentiometric FI system with a Cu ion-selective electrode after online preconcentration of sample on Chelex 100 resin.

Chemiluminescence Detection

Reported FI systems involving chemiluminescence detection include the following reagents and reactions: β-nitrostyrene/NaOH/hexadecyltrimethylammonium bromide sensitized with fluorescein [130], formation of a complex between Cu and 1,10-phenanthroline and the subsequent chemiluminescence during the oxidation of the complex by hydrogen peroxide at alkaline pH [131-134].

Spectrophotometric Detection

Yuan et al. [135] established a rapid method for the determination of trace copper, based on the catalytic action on the decoloration reaction of ferric ion and thiosulfate. Ma et al. [136] developed an online preconcentration FI-spectrometric method by using a new preconcentration column, mixing solutions of oxalic acid and citric acid, ethyl violet reagent.

Atomic Absorption Spectrometric Detection

The determination of copper by ETAAS or FAAS includes an on-line preconcentration systems involving iminodiacetate chelating resins (Chelex-100 or Muromac A-1) [63-65, 137], adsorption of analyte complexes on silica C18 column with 8-hydroxyquinoline [67-68], with diethyldithiocarbamate [69-72], with diethyldithiophosphate [73], with 1,10-Phenanthroline [76], with 1-nitroso-2-naphthol [138], with 5,7-dichloroquinoline-8-ol [139], a complexing reagent (macrocyclic material) immobilized on silica gel [79], a styrene-divinylbenzene resin functionalized with (S)-2-[hydroxy-bis-(4-vinyl-phenyl)-methyl]-pyrrolidine-1-carboxylic acid ethyl ester [140], Amberlite XAD-2 functionalized with dithizone for minicolumn field sampling [141], Sep-Pak cartridge [142], adsorbent supports (Amberlite XAD-4) impregnated with 1-(2-pyridylazo)-2-naphthol (PAN) [143] and 4-(2-pyridylazo) resorcinol (PAR) [144] used in field flow preconcentration systems, precipitation [16, 145] and on-line sorption/preconcentration with ammonium pyrrolidinedithiocarbamate or 8-hydroxyquinoline (HQ) in a knotted reactor [146].

Inductively Coupled Plasma Atomic Emission Spectrometric Detection

Nickson et al. [85] described a simple but robust battery-powered field sampling unit including a commercially available iminodiacetate resin (Metpac CC-1). FI sample introduction is coupled with an atomic emission source, the microwave plasma torch (MPT).

The use of FI in conjunction with an ultrasonic nebulizer is compared with direct nebulization. This FI approach offers several advantages over continuous sample introduction: rapid sample throughput and a reduction of memory effects without a loss in sensitivity or precision. Furthermore, by appropriate choice of sample dispersion, a significant reduction of the Na and K interferences can be achieved without a substantial deterioration in sensitivity [147].

Inductively Coupled Plasma Mass Spectrometric Detection

Determination of trace copper and other trace metals in seawater is accomplished by inductively coupled plasma-mass spectrometry (ICP-MS) with a previous on-line preconcentration/separation step. This separation/preconcentration step is carried out by using several chelating materials for metal preconcentration: SO_3-oxine CM-cellulose [90], chelating resins with iminodiacetate groups [89, 91-93, 116], a commercially available Nitriloacetic Acid (NTA) Superflow resin [148], ammonium salt of O,O-diethyldithiophosphoric acid on C18 immobilized on silica [19, 57], 8-hydroxyquinoline immobilized onto silicone tubing (Sil-8-HQ) [94], controlled-pore glass [95] or immobilized on fluorinated metal alkoxide glass (MAF-8HQ) [96], silica gel modified with niobium(V) oxide [99], bonded SiO_2 with octadecyl functional group C18 packed in a microcolumn to collect diethyldithiocarbamate complexes [100], group-selective precipitation with diethyldithiocarbamate [117, 149] and off-line cloud point extraction [18].

Iron

Current oceanographic studies require the determination of iron at sea in real time and this necessitates the use of portable, shipboard instrumentation, for which FI techniques are ideally suited. Thus, most of the FI methods proposed for iron determination in seawater samples by chemiluminescence and spectrophotometric detection involve shipboard analysis.

Chemiluminescence Detection

Chemiluminescence detection of iron takes place with and without prior a preconcentration step exploiting the reaction between Fe(II) and luminol or Brilliant Sulfo Flavin. Luminol chemiluminescence is induced following the oxidation of Fe(II) by oxygen, hydrogen peroxide, both or without them [105, 150-161]. This reaction creates superoxide and hydoxyl radicals, which initiate the three-step oxidation of luminol. Using this reaction, Fe(II) determination is achieved by the detection of light emitted during the final oxidation step in the pH range 9-11. The chemiluminescent reaction resulting from the addition of Fe(II) to a solution of hydrogen peroxide and alkaline brilliant sulfoflavin (BSF; 4-amino-N-(p-toly1)- naphthalimide-3-sulfonate) is also used to chemiluminescence detection of iron [162-163].

Iron(II) can be determined directly by its enhancing effect on the luminol or Brilliant Sulfo Flavin reactions and "total" iron(II+III) can be determined after acidification and sample reduction steps, usually employing sulfite.

Spectrophotometric Detection

Regarding the spectrophotometric determination of iron, catalytic methods based on iron-catalyzed oxidation of p-phenetidine by periodate [164] and based on the influence of Fe(III) on the oxidation rate of N,N-dimethyl-p-phenylenediamine by hydrogen peroxide in neutral or slight acidic medium are carried out in FI mode [26, 164-170]. The method is highly sensitive and is successfully applied to the analysis of natural waters with or without a prior step of concentration/separation step. It is however susceptible to pronounced variations in ionic strength, therefore its applicability could be limited when it was applied to the analysis of sample lots with high variability in salinity. Alternatively, other spectrometric reagents used in FI systems are thiocyanate [171-172], 1,10-phenanthroline [173], N,N-dimethylformamide [174], ferrozine [175] and diphenylamine-4-sulfonic acid sodium salt [176]. To discriminate between Fe^{3+} and Fe^{2+}, a redox minicolumn can be included in the FI manifold (in the loop of one of the valves) to develop on-line the iron oxidation [171] or carry out the on-line reduction including a channel of a reducing reagent as ascorbic acid in the FI manifold [173, 175].

Atomic Absorption Spectrometric Detection

Tony el al. [112] described a rapid, sensitive FI atomic absorption spectrometric procedure to determine Fe, Co, Ni, Mn, and Zn based on on-line preconcentration on a microcolumn packed with C18 material. These metals are complexed with 5,7-dichlorooxine from weakly acidic or neutral solutions in the flow system and adsorbed on the column. Preconcentrated elements are eluted with acidified methanol and injected directly into the nebulizer for atomization in an air-acetylene flame for measurement. Andersen [177] investigates the performance of a FIA manifold that is optimized for preconcentration without coprecipitation. The preconcentration of iron is accomplished on-line in a tubular reactor of the FIA manifold where the iron species are adsorbed directly to the walls. This is the first example of an analytical method that does not involve steps of extraction, co-precipitation or complexing in the preconcentration protocol.

Inductively Coupled Plasma Mass Spectrometric Detection

Determination of trace iron and other trace metals in seawater is realized by inductively coupled plasma-mass spectrometry (ICP-MS) with a previous on-line preconcentration/ separation step. This separation/preconcentration step is carried out by using chelating resins with iminodiacetate [92] or nitrilotriacetic acid [148] groups, and by group-selective precipitation with sodium diethyldithiocarbamate as precipitant reagent [117].

Bowie et al. [178] presented the results from a laboratory blind intercomparison exercise with two widely used analytical methods for the determination of iron in seawater. The two methods used are coprecipitation followed by isotope dilution inductively coupled plasma mass spectrometry (ICP-MS) and chemical reduction to iron(II) followed by FI with chemiluminescence detection (FI-CL). The results show that there is no statistical difference ($P = 0.05$) between the shipboard FI-CL method and the directly traceable, low blank, isotope dilution ICP-MS method for the determination of iron in surface South Atlantic seawater.

Mercury

Electrochemical Detection

Ye et al. [14] utilized the epoxy-impregnated graphite tube electrode bulk-modified with 2-mercaptobenzoxazole, employed in a wall-jet configuration for the continuous flow and FI stripping voltammetric determination of Hg^{2+}. Richter et al. [179] develop a new method to construct very efficient flow cells for Hg detection by potentiometric stripping analysis using the thin Au layer of recordable compact disks as working electrode. This new source of electrodes (CDtrodes) exhibited very attractive performance, similar to that obtained with common Au electrodes, with superior versatility. The low cost of this new source of Au electrodes allows a frequent replacement of the electrode, avoiding cumbersome clean-up treatments

Chemiluminescence Detection

Amini et al. [179] developed a gas-diffusion FI method for the chemiluminescence detection of Hg^{2+} based on the luminol-H_2O_2 reaction. Hg(II) is reduced on-line to metallic mercury. The gas-diffusion cell is immersed in a water bath thermostated at 85 ºC to enhance the evaporation of elemental mercury. The mercury vapor thus generated diffused across the Teflon membrane of the gas-diffusion cell into an acceptor stream containing $KMnO_4$ in H_2SO_4 where it is oxidized to Hg(II). This methodology offers the possibility for the construction of portable low-cost online automatic analyzers for mercury monitoring.

Atomic Fluorescence Detection

Bloxham et al. [180] described a FI atomic fluorescence method incorporating an on-line bromide-bromate oxidation step to determine mercury in filtered sea-water samples. A heated reaction coil is incorporated in the FI manifold to increase the conversion of organic mercury into inorganic mercury(II) chloride from 50 to approaching 100%. Tseng et al. [181] developed a semi-automatic dissolved elemental Hg analyzer (DEMA) to determine dissolved elemental Hg in natural water. This on-line set-up couples the main analytical steps from sample introduction, gas-liquid separation, and Au amalgamation/separation to final detection/data acquisition using FI techniques. It provides ease of operation, high analytical performance, and is suitable for shipboard use. Leopold et al. [182] proposed an automated FI system utilizing the extraordinary oxidation power of bromine monochloride for the transformation of dissolved mercury species to Hg(II) and oxidation of dissolved organic carbon. This system has been coupled to cold vapor atomic fluorescence spectrometry for highly sensitive mercury detection.

Atomic Absorption Spectrometric Detection

Owing to the simplicity, high sensitivity and relative freedom from interferences, cold vapor generation-trapping and atomization in a graphite furnace by selective reduction with $NaBH_4$ or $SnCl_2$, without [183-187] or with a preconcentration step [188-189] before the mercury cold vapor generation, has generally been used for the determination of total and inorganic mercury in seawater. Total mercury is determined after on-line oxidation with potassium peroxodisulfate and sulfuric acid as oxidizing agents for decomposition of organomercury compounds [185]. Capelo et al. [187] proposed a new oxidation method based on room temperature ultrasonic irradiation for degradation of organomercurials and subsequent mercury determination by FI cold vapor atomic absorption spectrometry.

Methyl and inorganic mercury species are on-line preconcentrated and separated by sorbent [190-191] or liquid-liquid extraction [192]. Mercury speciation is also carried out by flow injection high-performance liquid chromatography cold vapor atomic absorption spectrometry (FI-HPLC-CVAAS) [193].

Inductively Coupled Plasma Atomic Emission Spectrometric Detection

Cañada-Rudner et al. [194] depicted a novel and expeditious approach to the determination of mercury based on a separation/preconcentration system coupled on-line to a FI manifold. The method employs on-line preconcentration of mercury on a minicolumn placed in the injection valve of the FI manifold. Zara et al. [195] described a FI system including a mercury(II) preconcentration step using a resin Chelite-S packed minicolumn.

Inductively Coupled Plasma Mass Spectrometric Detection

Stroh et al. [43] used vapor generation FI-ICP-MS to determine Hg without preconcentration step. Cañada-Rudner et al. [196] described an automated preconcentration system coupled to ICP-MS. The preconcentration step is performed on a chelating resin microcolumn placed in the injection valve of a simple flow manifold. Wei et al. [197] used a simple and inexpensive laboratory built vapor generator with ICP-MS. The applications of vapor generation ICP-MS alleviated the nonspectroscopic interferences and the sensitivity problem of mercury determination encountered when the conventional pneumatic nebulizer is used for sample introduction. The concentration of mercury is determined by isotope dilution method. Bravo-Sanchez et al. [198] developed an approach to the speciation analysis of Hg by the hyphenation of solid phase microextraction to gas chromatography-inductively coupled plasma mass spectrometry.

Indium

Daih et al. [55] proposed procedures based on the application of immobilized quinolin-8-ol silica gel and stripping voltammetry. Alibo et al. [199] developed two methods to measure indium in natural waters by FI inductively coupled plasma mass spectrometry (ICP-MS). One is the isotope dilution technique using an ^{113}In enriched spike and the other utilizes natural yttrium present in the sample as an internal standard. In the former, optimization of the ^{113}In spike to minimize error is often difficult for samples in which indium concentration are variable, whereas in the latter method, a separate determination of Y in the sample is necessary and hence more sample is required. Murakami et al. [200] studied and established that a preconcentration method involving porous polytetrafluoroethylene (PTFE) filter tube impregnated bis(2-ethylhexyl) hydrogen phosphate (HDEHP) as a sorbent to establish is a practical preconcentration method for ultra trace analysis with ICP-MS.

Manganese

Chemiluminescence Detection

Chapin et al. [201] used FIA with chemiluminescence detection to determine manganese. The oxidation of 7,7,8,8-tetracyanoquinodimethane in an alkaline solution produces light. Mn(II) catalyzes this reaction and its concentration can be determined by measuring the rate of photon emission. A column containing 8-hydroxyquinoline immobilized on a solid support is used in-line in the FIA system to preconcentrate Mn. Von Langen et al. [202] modified the procedure proposed by Chapin et al. [201] to reduce blanks in the determination of Mn to perform the study of the oxidation kinetics of manganese (II) in seawater.

Spectrophotometric Detection

FI spectrophotometric procedures have been developed based on the catalytic properties of manganese in the oxidation of an organic substrate (leucomalachite green [24, 203-205], 2,2'-azinobis(3-ethylbenzothiazoline-6-sulfonic acid) [206], 4,4'-bis(dimethylamino) diphenyl- methane [207], N,N-diethylaniline [129, 208-210]) with potassium periodate or tiron with hydrogen peroxide [205, 211]. Chin et al. [212] proposed a spectrophotometric method using the complexing reagent 1-(2-pyridylazo)-2-naphthol (PAN). Color formation was rapid, which permitted the technique to be used in FI and continuous-flow systems.

Atomic Absorption Spectrometric Detection

Atomic absorption detection with flame or electrothermal atomization for manganese determination includes a preconcentration step involving on-line preconcentration in a knotted reactor [146], by precipitation [16], on-line adsorption of manganese complexes with dialkyldithiophosphates [111] and 5,7-dichlorooxine [112] on silica C18, on-line adsorption of manganese complexes with 1-(1-hydroxy-2-naphthylazo)-6-nitro-2-naphthol-4-sulfonic acid (Eriochrome Black T) on an anion exchange resin [213] and by utilization of a commercial chelating resin (Chelex-100) [65].

Inductively Coupled Plasma Atomic Emission Spectrometric Detection

Nickson et al. [85] described a field sampling unit for the selective in situ preconcentration of trace elements (Cd, Co, Cu, Mn, Ni, Pb, and Zn) from natural waters on a commercially available iminodiacetate resin (Metpac CC-1) and their subsequent determination at the laboratory by FI-ICP-AES.

Inductively Coupled Plasma Mass Spectrometric Detection

Manganese is determined directly by ICP-MS [88,11]. Nevertheless, most of the proposed methods involving ICP-MS detection including a previous separation step for preconcentration and/or matrix elimination [10, 88, 90-91, 93-94, 98, 116-117].

Molybdenum

Atomic Absorption Spectrometric Detection

On-line sorption of complexes with 8-hydroxyquinoline [214], dithiocarbamate [215]on microcolumn packed with C18 material, coprecipitation [216] and a chelating Muromac-A1 resin are the separation strategies used to preconcentrate molybdenum from seawater samples for flame or electrothermal atomic absorption spectrometric detection.

Inductively Coupled Plasma Atomic Emission Spectrometric Detection

Furuta et al. [217] used a FIA system incorporating a microcolumn of activated alumina for the determination of molybdenum by ICP-AES. The transient signal resulting from the injection of molybdenum is determined by a spectrally segmented photodiode-array spectrometer.

Inductively Coupled Plasma Mass Spectrometric Detection

Molybdenum is determined directly by ICP-MS being seawater samples introduced by FI [11, 49, 88] or utilizing FIA manifolds comprising a chelating ion-exchange minicolumn filled with a chelating resin with iminodiacetic groups [10, 93] or activated alumina [217].

Ammonium/Ammonia

Electrochemical Detection

Oms et al. [218] described a simple sequential injection method for the determination of ammonium with conductometric detection. The sample and alkaline solution are sequentially aspirated into a single tubular channel and mixed by flow reversal. The formed ammonia permeates through a gas permeable membrane and is collected on a static acceptor solution. The change in conductance of the acceptor solution is related to the concentration of ammonium ions in the sample. A similar method, but developed in FIA mode is used by Hall et al. [219] to determine total dissolved inorganic carbon and ammonia in seawater samples.

Fluorescence Detection

Masserini et al. [220] developed a sensor package for the simultaneous determination of nanomolar concentrations of nitrite, nitrate, and ammonia in seawater by fluorescence detection. Aminot et al. [221] presented a version of the o-phthaldialdehyde-fluorescence ammonium determination for FIA with a view to its use for in situ, low-power consumption systems. Thus, the reaction temperature was limited to 30°C and FIA is used in stop-flow mode. Watson et al. [222] established a method using FI, gas-diffusion, derivatization, and then fluorescent detection. The fluorescent derivative formed by reacting o-phthaldialdehyde and sulfite with ammonia gives high sensitivity while removing potential interferences. Amornthammarong et al. [223] proposed a ship-board fluorometric flow analyzer for near real-time, high-resolution underway measurement of ammonia in seawater. The fluorometric method is based on the reaction of ammonia with o-phthaldialdehyde and sulfite.

Spectrophotometric Detection

Yamane et al. [224] presented a FIA system for simple and rapid determination of ammonia at sub-ppm levels. The system consists of a direct online coupling of the gas diffusion separation using porous polytetrafluoroethylene (PTFE) membrane tubing and subsequent spectrophotometric detection of ammonia by the formation of indophenol blue dye with sodium hypochlorite and 1-naphthol in a continuous flow system. Spectrophotometric determination of ammonium ion is based on sodium salicylate-hypochlorite reaction [225] and on the formation of indophenol blue [226]. Several methods for the determination of ammonium by FI methodologies are based on the conversion of ammonium ions in the sample to ammonia by reaction with sodium hydroxide and the subsequent diffusion of NH_3 across a hydrophobic membrane. Generated ammonia reacts with an acid-base indicator (as phenol red) in an acceptor stream [227-229] or the FI manifold includes a derivatization reaction based on indophenol blue with salicylate [230]. Ammonium ion can be oxidized to nitrite by hypochlorite in the presence of large amounts of potassium bromide. These methods are based on absorbance measurement of azo dye obtained by the reaction of nitrite formed with N-(1-naphthyl)ethylenediamine dihydrochloride and sulfanilamide [231-232].

Nickel

Atomic Absorption Spectrometric Detection

Nickel has been detected by atomic absorption spectrometry (FAAS and ETAAS) by using FI manifolds including a separation step for preconcentration and matrix elimination purposes. These FI systems are based on precipitation [16], sorption in a knotted reactor [146], adsorption of nickel complexes with diethyldithiocarbamate [69,71], 8-hydroxyquinoline [68], dialkyldithiophosphates [111], 5,7-dichlorooxine [112] and dimethylglyoxime [233] on octadecyl functional group (C18) bonded silica gel as sorbent

Inductively Coupled Plasma Atomic Emission Spectrometric Detection

Nickson et al. [85] utilized a field sampling unit for the selective in situ preconcentration of trace elements (Cd, Co, Cu, Mn, Ni, Pb, and Zn) from natural waters on a commercially available iminodiacetate resin (Metpac CC-1) and their subsequent determination at the laboratory by FI-ICP-AES.

Inductively Coupled Plasma Mass Spectrometric Detection

Preconcentration and matrix elimination is usually accomplished when nickel is determined by ICP-MS. Thus, have been used mini or microcolumns packed with chelating resins [10] above all iminodiacetate resins [89-93, 98], silica gel modified with niobium(V) oxide [99], 8-hydroxyquinoline immobilized onto silicone tubing [94], 8-hydroxyquinoline immobilized on fluorinated metal alkoxide glass [96] and bonded SiO_2 with octadecyl functional group C18 to collect diethyldithiocarbamate complexes [100] or seawater samples [101]. Continuous precipitation is also used with the same purposes [117]. By the other hand, nickel is determined directly by using a small dose (2%) of H_2 to the aerosol gas flow enhanced analyte signals by a factor of 2-3, which compensated for the loss of analyte signal that accompanied earlier efforts at cryogenic desolvation with ICP-MS. The samples are simply diluted with 1% nitric acid so that the chloride could be removed as hydrochloric acid [44].

Ndung'u et al. [102] compared the results obtained for estuarine water for total dissolved trace metal concentrations using online chelating resin column partitioning ICP-MS with those obtained by ETAAS after liquid-liquid extraction using a combination of 1-pyrrolidinedithiocarbamate/diethyldithiocarbamate. ICP-MS determinations are 10-20% lower than those achieved by solvent extraction ETAAS.

Lead

Electrochemical Detection

FI potentiometry and anodic stripping voltammetry performed in the differential pulse mode at a glassy carbon based mercury film wall-jet electrode are used for lead determination in seawater samples without [58-59, 234] or including a preconcentration step involving immobilized quinolin-8-ol silica gel [55].

Phosphorescence Detection

San Vicente de la Riva et al. [235] studied the on-line formation, in a FI system, of chelates formed between the heavy metal ion Pb(II) and the reagents 8-hydroxy-5-quinolinesulfonic acid, 8-hydroxy-7-quinolinesulfonic acid and 8-hydroxy-7-iodo-5-quinolinesulfonic acid, which exhibit strong room temperature phosphorescence retained on the surface of anion exchange resin beads. For this, they reported three flow-through optosensing systems. Worsfold et al. [60] describe an integrated luminometer able to perform room temperature phosphorescence on seawater samples.

Atomic Fluorescence Detection

Cheng et al. [236] developed a method based on hydride-generation atomic fluorescence spectrometry using potassium ferricyanide in alkaline solution as oxidant and nitric acid as reaction medium for the determination.

Spectrophotometric Detection

FI spectrophotometric methods for lead determination in seawater samples are based on detection of the Pb and 4-(2-pyridylazo)resorcinol [129], dithizone [237] and dibromo-methyl sulfonazo [238] complexes.

Atomic Absorption Spectrometric Detection

FI-FAAS and FI-ETAAS methodologies proposed to determine lead in seawater samples include a preconcentration step. This preconcentration step can incorporate a mini or microcolumn [188, 239] filled with a commercially available chelating resin as Chelex-100 [63-65, 240], a macrocyclic ligand immobilized on a silica gel support [79], a poly(aminophosphonic acid) chelating resin field sampling-preconcentration [82, 241], Amberlite XAD-4 impregnated with the complexing agent 1-(2-pyridylazo)-2-naphthol for a field flow preconcentration [242], silica gel chemically modified with niobium(V) oxide

[243], silica C18 for on-line adsorption of lead complexes with 8-hydroxyquinoline [67], diethyldithiocarbamate [69-71, 244] and diethyldithiophosphate [73], XAD-2 resin for on-line adsorption of lead complex with 8-hydroxyquinoline [77], poly-chlorotrifluoroethylene beads as absorbent for online adsorption of lead complex with ammonium pyrrolidine dithiocarbamate [126]. Precipitation or coprecipitation FI systems are also proposed for on-line lead preconcentration [16, 145, 245].

The attractiveness of combining FI with lead hydride generation atomic absorption spectrometry to improve the selectivity and sensitivity of lead determination is studied. It is proven that the use of FI alleviates or even eliminates two of the major shortcomings of lead hydride generation: slow reaction kinetics and interferences caused by transition metals [246-248].

Inductively Coupled Plasma Atomic Emission Spectrometric Detection

Lead an other metal ions are selectively preconcentrated by using a robust battery-powered field sampling unit on a commercially available iminodiacetate resin (Metpac CC-1) [85].

Inductively Coupled Plasma Mass Spectrometric Detection

Matrix separation and analyte preconcentration is accomplished by on-line retention of lead complexed with diethyldithiocarbamate [100], O,O-diethyldithiophosphoric acid [57] and ammonia diethyl dithiophosphate [19, 249] on C18-bonded silica [101], Pb-Spec® (an immobilized crown ether with a cavity size selective for Pb^{2+}) [250], SO_3-oxine CM-cellulose [90], 8-hydroquinoline-5-sulfonic acid on florisil [251-252], 8-hydroxyquinoline immobilized onto silicone tubing [94], hydroxyquinoline immobilized on controlled-pore glass [95], 8-hydroxyquinoline immobilized on fluorinated metal alkoxide glass [96], hydroxyquinoline bonded ZrO_2 [97], silica gel modified with niobium(V) oxide [99], 8-hydroxyquinoline bonded silica [253], iminodiacetate-based resins [89, 91-92, 116] and cloud point extraction [18]. Kikuchi et al. [87] used a pump and an injector of an ion chromatograph system coupled with an ICP-MS as a sample flow injector to avoid interferences arising from the injection of samples by the reduction of sample volume and to increase the sensitivity. Rosland et al. [88] described the direct determination of trace metals in seawater by ICP-MS based on sample introduction by FIA and electrothermal vaporization.

Flame Laser-enhanced Ionization Technique

Ke et al. [254] were the first to couple a FI manifold with flame laser-enhanced ionization (LEI) detection. The result is the sum of the merits of FI separation and preconcentration and of extreme sensitivity and selectivity for the LEI detector. FI is incorporated with a microcolumn packed with a C18 bonded silica. The chelating agent ammonia diethyl dithiophosphate (DDPA) is used to form the Pb-DDPA complex, which may be sorbed in the microcolumn and then eluted with methanol. The preconcentrated Pb is then detected by the LEI technique with either single-step or two-step excitation.

Rare earths, Uranium and Plutonium

Willie et al. [89] presented a method for the quantitation of rare earth elements (REE) by inductively coupled plasma orthogonal acceleration time-of-flight mass spectrometry. Online separation and preconcentration is achieved using a timed FI system incorporating a column containing a iminodiacetate-based resin.

Vincente et al. [255] studied an on-line Eu, Tb, Ho, Tm and Lu preconcentration and determination system implemented with inductively coupled plasma mass spectrometry associated to a FI method. Quinolin-8-ol and Amberlite XAD-7 are used for the retention of Eu, Tb, Ho, Tm and Lu, at pH 10.0.

Eroglu et al. [256] developed a FI-inductively coupled plasma-mass spectrometric (FI-ICP-MS) procedure, utilizing ultrasonic nebulization with membrane desolvation (USN/MD), for the determination of plutonium in seawater at fg/L concentration levels. Seawater samples after filtration, are subjected to co-precipitation with NdF_3, followed by ion exchange to enrich plutonium and to reject seawater matrix ions and co-existing uranium. The same authors [257] used a basic alumina microcolumn in combination with a FI system with ICP mass spectrometry for the preconcentration and determination of plutonium at the pg/L level.

Zhu et al. [258] reported a new FI method for the determination of uranium based on the chemiluminescence reaction of the U(II)-lucigenin with a Jones reductor. Yang et al. [259] designed a FIA manifold for the chemiluminescence determination of uranium in seawater, composing the process of online pre-reduction of the chemiluminescently inactive U (VI) to its trivalent state by passing through a self-made constant potential flow-electrolytic cell with spectrographically pure, hollow carbon electrode as working electrode at -0.70 V (vs. Ag/AgCl electrode) and the process of chemiluminescence reaction of U(III) with luminol in an alkaline solution.

Ebdon et al. [10] developed an on-line preconcentration-matrix elimination procedure for the analysis of concentrated brines by ICP-MS. First row transition metals are concentrated quantitatively and isolated from the sodium chloride matrix using a chelating ion-exchange column. Economou et al. [260] determined U(VI) by square wave adsorptive stripping voltammetry on a rotating-disk Hg-film electrode using a wall jet Hg-film electrode with cupferron as complexing agent (flow-through mode). Dadfarnia et al. [261] combined a FI system incorporating a microcolumn of activated alumina with ICP-MS for on-line trace enrichment and determination of U in surface waters and seawater. Karpas et al. [262] proposed a method based on FI-ICP-MS for determination of uranium. Rosland et al. [88] described the direct determination of uranium and other trace elements in seawater by ICP-MS based on sample introduction by FIA and electrothermal vaporization (ETV). Halicz et al. [263] proposed the determination of uranium by inductively coupled plasma quadrupole mass spectrometry because can yield reliable results for the $^{234}U/^{238}U$ ratio without any treatment of the sample (except acidification). In addition, they show that combining a FI sample introduction system with an ultrasonic nebulizer, significantly increases the sensitivity and enhances the ability to cope with a relatively high salt content in non-saline water samples. Lee et al. [93] developed a new on-line FI pre-treatment system using a disk-type chelating resin with iminodiacetate functional group for the simultaneous multi-element determination of trace metals (including U) in sea-water samples by ICP-MS. Oshita et al. [264] established an online FI pretreatment system using a newly synthesized chitosan resin for the determination of U by ICP-MS. Field et al. [11] described a rapid, high-throughput method to determine trace elements (P, V, Mn, As, Mo, Ba, U) in seawater with application to tracing ballast water exchange in ocean-going vessels and detection by SF-ICP-MS. Oshita et al. [265] synthesized a cross-linked chitosan resins possessing serine moiety (serine-type chitosan resins) for the collection of trace uranium and the separation of matrixes in seawater prior to FI-ICP-MS measurement.

Rhodium

Jiménez et al [266] studied the direct FIA-ICP-MS determination of Al, As, Cr, Mn, Mo, Ni, Pb, Rh, Sb, Te, and V in saline matrixes with different total dissolved solids content (TDS) (0.03-30%). Sample introduction by FIA permits the nebulization of matrixes with higher TDS in the ICP-MS.

Antimony

Atomic Fluorescence Detection

Wu et al. [267] described a non-chromatographic approach for the determination of Sb(III) and Sb(V) in natural water samples using FI on-line sorption preconcentration coupled with hydride generation atomic fluorescence spectrometry. With the sample pH kept at 1.0, only Sb(III) forms complexes with ammonium pyrrolidine dithiocarbamate (APDC) and this is retained on the inner walls of the knotted reactor in the presence of Sb(V).

Atomic Absorption Spectrometric Detection

Lu et al. [32] propose an on-line cold-trap hydride collection, FI system to determine As and Sb in seawater. The system consists of a hydride generator, a nebulizer for gas/liquid separation, and liquid nitrogen bath hydride collector. De la Calle Guntinas et al. [268] determined antimony(III) and antimony(V) by FI-hydride generation AAS and by continuous-flow-hydride generation and transport. Both methods are more sensitive than the batch system. Yan et al. [269] synchronously coupled a FI on-line sorption preconcentration system to an ETAAS system for the selective determination of (ultra)trace amounts of Sb(III). The determination is achieved by selective complexation of Sb(III) with ammonium pyrrolidine dithiocarbamate, sorption of the complex onto the inner walls of a knotted reactor, elution with ethanol and subsequent ETAAS detection. Ding et al. [270] developed a continuous-flow (timed injection) electrochemical hydride generation system. Lead is used as cathode material for the production of stibine. Both Sb(III) and Sb(V) are equally converted into their hydrides by electrochemical means. The hydride is trapped in a Pd-coated graphite furnace prior to atomization. Sella et al. [16] described a FI on-line preconcentration system coupled to an electrothermal atomic absorption spectrometer for the determination of trace metals. Cabon et al. [271] studied the capabilities and limitations of the continuous FI hydride generation technique, coupled to atomic absorption spectrometry, for the speciation of major Sb species in seawater. Rodriguez et al. [272] developed a quartz tube atomic absorption spectrometric method for the determination of antimony by FI-HG in which stibine (SbH_3) is generated from the reaction between antimony in the injected solution and tetrahydroborate immobilized on a strong anion-exchange resin (Amberlite IRA-400). Zheng et al. [273] described a method for prevention of the oxidation of Sb(III) during sample pretreatment, preconcentration of Sb(III) and Sb(V) with nanometer size TiO_2 (rutile) and speciation analysis of Sb.

Inductively Coupled Plasma Mass Spectrometric Detection

Stroh et al. [43] used vapor generation FI-ICP-MS to determine As, Sb, and Hg. They demonstrate that this methodology enhances considerably the analytical capabilities of ICP-MS for volatile vapor forming elements.

Selenium

Electrochemical Detection

Ye et al. [274] fabricated a wall-jet gold electrode polymerized with 3,3'-diaminobenzidine (PDAB-Au) for the selective and sensitive determination of trace Se(IV) by using a FI system.

Atomic Fluorescence Detection

He et al. [275] developed a highly sensitive on-line pre-treatment methods for the direct determination of trace amounts of selenite and total inorganic selenium in seawater. For total inorganic selenium determination, acidified samples are reduced with sodium bromide and pre-reduced online by microwave irradiation. After pre-treatment, the sample zone is stored in a coil and then transported for reaction with $NaBH_4$. The hydride produced is separated and dried and Se is determined in a mini-hydrogen diffusion flame by non-dispersive atomic fluorescence spectrometry. The interface developed for this purpose makes possible the integration, within the same manifold, of two steps involving very different reaction times.

Spectrophotometric Detection

Gong et al. [276] proposed a simple, accurate, sensitive and selective FI catalytic kinetic spectrophotometric method for rapid determination of trace amounts of selenium. The method is based on the accelerating effect of Se(IV) on the reaction of EDTA and $NaNO_3$ with ammonium Fe(II) sulfate hexahydrate in acidic media. The absorbance intensity is registered in this reaction solution at 440 nm.

Atomic Absorption Spectrometric Detection

Cobo Fernandez et al. [277] proposed the on-line reduction of Se(VI) to Se(IV) and subsequent determination of Se(IV) by FI-hydride generation atomic absorption spectrometry (HGAAS).Hydride generation is performed in an ice-bath to prevent decomposition of $NaBH_4$. Tao et al. [278] described a FI procedure for the determination of ultra-trace amounts of Se(IV), which combines HGAAS with one-line preconcentration of the analyte by coprecipitation with $La(OH)_3$ generated in-situ. Cabon et al [279] determined Se(IV) in seawater by FI hydride generation, trapping within the graphite furnace, followed by ETAAS. Because this technique is specific to Se(IV), they investigated various physico-chemical treatments leading to the conversion of other chemical forms into Se(IV). Thus, thermal and

UV irradiation treatments of seawater samples in acidic or basic medium permitted the conversion of Se-II-selenomethionine and Se(VI) to Se(IV).

Yan et al. [280] developed a FI manifold to determine ultra-trace selenite and selenate with an air-segmented and air-transported operational sequence involving an on-line coupling of micro-column separation and preconcentration to ETAA. Selenite is determined by selective reaction with pyrrolidine dithiocarbamate (PDC), sorption of the resultant Se-PDC compound onto a conical micro-column packed with RP C18 sorbent, elution with ethanol, and ETAAS detection. Selenate was determined as the difference between selenite concentrations after and before pre-reduction of selenate to selenite.

Inductively Coupled Plasma Mass Spectrometric Detection

Olivas et al. [281] compared parameters for selenium determination in ICP-MS by pneumatic nebulization and FI-HG-ICP-MS. They evaluate, with both sample introduction modes, the effect of organic solvents (methanol, ethanol, propanol, acetone and acetonitrile) on the Se response. As a result, it is proven that the addition of alcohols leads to an important reduction of polyatomic interferences ($^{40}Ar^{37}Cl$, $^{40}Ar^{38}Ar$, $^{40}Ar_2H_2$) and to a 10-fold enhancement of the selenium signal. Lam et al. [47] determined arsenic and selenium (IV) following hydride generation (HG) by reaction with $NaBH_4$ in an automated FI system. The hydrides are trapped on reduced palladium in a graphite furnace and subsequently vaporized into an ICP for MS detection. Pozebon et al. [57] proposed a procedure for the determination of Cu, Cd, Pb, Bi and Se(IV) by electrothermal vaporization ICP-MS, after on-line separation using a FI system. Matrix separation and analyte preconcentration is accomplished by retention of the analytes complexed with the ammonium salt of O,O-diethyldithiophosphoric acid on C18 immobilized on silica in a minicolumn coupled directly to the autosampler arm of the vaporizer. Mesquita da Silva et al. [18] suggested a preconcentration method for low Ag, As, Au, Cd, Cu, Pb, and Se concentrations in water using cloud point extraction previous their determination by ICP-MS using ultrasonic nebulization and injecting the enriched phase with a FI system.

Tin

Bermejo-Barrera et al. [282] studied a method for the trace determination of tin in seawater samples by a HG technique using a FI system coupled with AAS. Tin hydride is generated using $NaBH_4$ in sodium hydroxide as the reductant. Woods et al. [283] combined a microcolumn technique with FI-ICP-MS for the determination of total organotin. Khoo et al. [284] fabricated an epoxy-C powder composite electrode bulk modified with 8-hydroxyquinoline for the continuous flow and FI anodic differential pulse voltammetric determination of Sn(II) after open circuit preconcentration

Titanium

Yan el al. [285] developed a matrix-analyte separation and preconcentration technique for ICP-MS. A miniature chelated ion-exchanger column of SO_3-CM quinolin-8-ol-cellulose is used to increase the sensitivity for multielement measurements by ICP-MS.

Thallium

Wei et al. [286] used a simple, inexpensive, laboratory built vapor generator with ICP-MS) to determine thallium in seawater. Applying vapor generation, ICP-MS alleviated non-spectroscopic interferences encountered when a conventional pneumatic nebulizer is used for sample introduction. The concentration of thallium is determined by the isotope dilution method. Vassileva et al. [287] used a FIA system incorporating a micro-column of ZrO2 for the development of an online multi-element method for the simultaneous preconcentration and determination of Al, Bi, Cd, Co, Cr, Cu, Fe, Ga, In, Mn, Mo, Ni, Pb, Tl, V, Sb, Sn, and Zn by inductively coupled plasma at. emission spectrometry (ICP-AES).

Vanadium

Shiller et al. [288] carried out a FI technique for the determination of dissolved vanadium in natural waters based on the V-catalyzed oxidation of Bindschedler's green leuco base by bromate with tiron and tartrate as reaction activators. The reaction product is quantified colorimetrically. A chelator column of immobilized 8-hydroxyquinoline reduces matrix effects but can be eliminated if samples with a constant matrix are analyzed.

Ebdon et al. [10] developed an online preconcentration matrix elimination procedure for the analysis of concentrated brines by ICP-MS. Elements of interest are concentrated and isolated from the sodium chloride matrix using a chelating ion-exchange column. Alves et al. [44] demonstrated the value of cryogenic desolvation with ICP-MS for the quantification of V, Ni, and As in matrices containing calcium and chloride. The addition of hydrogen to the central channel improves the analyte sensitivity but kept oxide ratios low. Willie et al. [91] described an automated low pressure FI method for the determination of Cu, Ni, Zn, Mn, Co, Pb, Cd, and V in seawater using ICP-MS. A column of commercially available iminodiacetate resin is used to sequester the trace elements from seawater samples. Lee et al. [93] proposed an on-line FI pre-treatment system using a disk-type chelating resin (iminodiacetate functional group) for the simultaneous multi-element determination of trace metals in sea-water samples by ICP-MS. The chelating resin is used for the collection of trace elements and the elimination of alkali and alkaline earth metals in highly concentrated salt solution. Field et al. [11] described a rapid, high-throughput method to determine trace elements (P, V, Mn, As, Mo, Ba, U) in seawater by ICP-MS with application to tracing ballast water exchange in ocean-going vessels. Maltez et al. [99] described an on-line pre-concentration system to simultaneously determine Cd, Cu, Ni, V, Zn, Co, and Pb by ICP-MS detection. The system is based on cationic retention of analytes in a mini-column filled with niobium(V) oxide-modified silica gel. Yan el al. [285] developed a matrix-analyte separation and preconcentration technique for ICP-MS determination of titanium and vanadium. For this, they used a miniature chelated ion-exchanger column of SO_3-CM quinolin-8-ol-cellulose to increase the sensitivity for multielement measurements by ICP-MS.

Zinc

Electrochemical Detection

Daih et al. [55] developed procedures based on the application of immobilized quinolin-8-ol silica gel and stripping voltammetry for the determination of Cu, Pb, Cd, Bi, In and Zn. Ang et al. [58] determined the levels of Cd, Cu, Pb, and Zn by an automated anodic stripping voltammetry (ASV) with a FI system. The ASV analysis is performed in the differential pulse mode at a glassy C-based Hg film electrode.

Fluorescence Detection

Nowicki et al. [289] proposed a sensitive technique for shipboard determination of zinc by FIA with fluorometric detection. A cation exchange column is used to preconcentrate and separate zinc from interfering alkali and alkaline earth ions. The organic indicator ligand, p-tosyl-8-aminoquinoline, is used to form a complex with Zn, the fluorescence of which is determined with a flow-through fluorometer. Worsfold et al. [60] described an integrated luminometer able to perform fluorescence, room temperature phosphorescence and chemiluminescence measurements on seawater samples. The method used to determine Zn is based on the fluorescence modification resulting from the binding of Zn to a polyazacrown covalently linked to a luminophore via one methylene group. The use of the Dowex 1X2-400 column not only resulted in a preconcentration of Zn on the column, but also allowed metal separation to prevent fluorescence quenching by other trace metals.

Spectrophotometric Detection

In the presence of borax-hydrochloric buffer solution of pH 8.3 with the surface agent (hexadecyl-trimethylammoniumbromide), zinc (II) reacted with salicylfluorone and forms a blue-purple ternary micelle complex. The ternary micelle complex has a sensitive absorption peak at 630 nm, so trace zinc could be determined by reverse FIA spectrophotometry [290].

Atomic Absorption Spectrometric Detection

Analytical methods proposed for zinc determination in seawater with AAS detection involve the application of minicolumn preconcentration by using commercially available resins such as Chelex-100 (iminodiacetate groups) [63-64, 291] or Chelite P (aminomethylphosphonic groups) [292], and solid materials as C18 bonded silica gel for adsorption of zinc complexes with 8-hydroxyquinoline [68] and 5,7-dichlorooxine [112].

Inductively-coupled Plasma Atomic Fluorescence Detection

Yuang et al. [293] presented an on-line preconcentration inductively-coupled plasma atomic fluorescence spectrometer (ICP-AFS) detection method for trace zinc controlled by FIA. A new type of chelating resin, carboxymethylated polyethylenimine-polymethylenepolyphenylene isocyanate, is used to concentrate trace Zn from seawater.

Inductively Coupled Plasma Atomic Emission Spectrometric Detection

Zinc an other metal ions are selectively preconcentrated by using a robust battery-powered field sampling unit on a commercially available iminodiacetate resin (Metpac CC-1) [85]. Zougagh et al. [294] described a simple and rapid method for the preconcentration and determination of zinc in water by ICP-AES. The method employs on-line preconcentration of zinc on a column packed with silica gel functionalized with 1,5-bis(di-2-pyridyl)methylene thiocarbohydrazide (DPTH-gel) and placed in the injection valve of a simple flow manifold.

Inductively Coupled Plasma Mass Spectrometric Detection

The determination of zinc by ICP-MS is developed performing a previous separation step mainly carried out for matrix elimination to minimize the interferents because highly saline solutions can cause both spectroscopic and non-spectroscopic interferences. Consequently, several chelating resins are used in FI systems [10]. The resins used are commercially available iminodiacetate-based resins (Chelex-100, Dionex MetPac CC-1, Toyopearl AF-Chelate 650 M, etc.) [89, 91-92, 98, 116], SO_3–quinolin-8-ol carboxymethylcellulose [295], 8-hydroxyquinoline immobilized onto silicone tubing (Sil-8-HQ) [94]or on fluorinated metal alkoxide glass (MAF-8HQ) [96] and niobium(V) oxide-modified silica gel [99]. Also, solid sorbents as C18 silica gel [101] are used to collect the diethyldithiocarbamate zinc complex [100].

ANIONIC SPECIES

Different features of FI methods for the determination of anionic species in sea and estuarine water are illustrated in Table 4.2. In the following paragraphs, some points observed in this table are highlighted due to their interest.

Alkalinity

Conductimetric Detection

Jardim et al. [296] carried out the determination of dissolved inorganic carbon (DIC) using FIA with a conductometric detector For comparison, DIC values are also determined by the classical approach via pH and alkalinity. The results obtained using FIA are systematically lower than those obtained via potentiometric titration. This difference is attributed to the presence of protolytes other than carbonates which are also titrated in the classical approach, but do not interfere when using FIA method. Hall et al. [219] also developed a conductimetric FI methodology to determine DIC. This method includes a gas-permeable membrane is remove CO_2 or NH_3 from acidic (DIC) or basic (NH_4^+) reagent streams into a receiving stream and conductivity detector.

Spectrophotometric Detection

Lu et al. [297] combined the technique of film diffusion with FIA. The optical determination system of DIC in seawater by FIA is based on the absorbance change of an indicator. Wei et al. [298] developed a gas diffusion unit constructed with double tubing. The inner tubing is a micro porous PTFE (polytetrafluoroethylene) tubing and the outer tubing is made of glass. This unit is coupled to a FIA system, which is used for the determination of total inorganic carbonate based on the absorbance change of an indicator. Bellerby et al. [299] described the use of a FI system with spectrophotometric detection for the measurement of seawater pH. For this, the acid-base absorption properties of phenol red injected into a seawater stream are measured. The performance of the technique in a shipboard environment is evaluated and shown to have a working precision of ±0.005 pH units. Furthermore, no standard is needed for calibration make. Bellerby et al. [300] studied for the spectrophotometric measurement of seawater pH selected sulfonephthalein indicators. For this, they describe a fully automated FI method, which uses the instantaneous spectral absorbance measurements afforded by a photodiode array spectrophotometer with charge coupled detection. The method has at sea a precision of ±0.002 pH units.

Table 4.2. Features of FI determinations of anionic species in sea and estuarine water

Analyte	Detection		DL (μM)	Linear range (μM)	SF (s/h)	RSD (%)	Ref
DIC	C		3	No data	60	No data	296
DIC	C		No data	100 - 20000	No data	0.4-1.4	219
DIC	SP		No data	No data	5	<1.2	297
DIC	SP		1	1 - 1000	No data	No data	298
pH	SP		-	-	25	-	299
pH	SP		-	-	30	-	300
Br^-	P		No data	No data	No data	No data	302
Br^-	P		0.05	No data	180	No data	304
Br^-	Ch		6.26	0 - 1250	40	<2.5	306
Br^-	SP		1.13	13 - 125	120	<1	307
Br^-	SP		1.25	0- 4	12	0.74	308
Cl^-	P		No data	No data	No data	No data	301
Cl^-	P		No data	10 - 10000	400	No data	303
Cl^-	P		0.01	No data	180	No data	304
Cl^-	A		0.05	No data	60	1.11	305
Cl^-	SP		No data	No data	300	1.4-1.6	309
Cl^-	SP		3.1	Up to 50	60	0.93	310
F^-	SP		No data	0 - 260000	40	0.4	311
F^-	SP		2.6	4.2 - 63	60	0.2	312
I^-	P		0.0004	No data	180	No data	304
IO_3^- I^-	SP		No data	0.0059 - 1.18 0.0043 - 0.86	3-7	1.3	313
I^-	SP		1.6	Up to 3150	50	2.0	314
I^-	SP		0.0005	No data	140	0.4-4	315
I^-	SP	RAMP RAML	0.0003	No data	140 220	0.4-9.4 0.4-1.5	316
I^-	FAAS		0.02	0.047 - 1.7	17	2.7	317
DO	F		0.49	2 -1200	18	<1	318
Analyte	Detection		DL (μM)	Linear range (μM)	SF (s/h)	RSD (%)	Ref
H_2O_2	F		0.005	0.005 - 0.7	No data	10	319
H_2O_2	Ch		0.005	0.005 - 0.5	120	1-2	320
H_2O_2	Ch		0.0106	No data	No data	1.8-2.5	321 322
H_2O_2	Ch		0.002	No data	No data	3	323
H_2O_2	Ch		No data	No data	No data	No data	324

Analyte	Detection	DL (µM)	Linear range (µM)	SF (s/h)	RSD (%)	Ref
H_2O_2	Ch	0.00035	No data	No data	15.0	325
H_2O_2	SP	0.012	No data	60	5	326
NO_2^-	F	0.0003	No data	No data	No data	327
NO_2^- NO_3^-	F	0.0046 0.0069	0.005 - 1.25	No data	1.03 1.12	220
NO_2^- /NO_3^-	SP	0.1	No data	75	<1	328
NO_2^-	SP	No data	No data	No data	No data	329
NO_2^-	SP	No data	No data	No data	No data	330
NO_3^-	SP	0.05	No data	10	No data	331
NO_3^-	SP	0.1	0.1 - 73	No data	No data	332
NO_3^-	SP	No data	No data	No data	No data	333
NO_2^- /NO_3^-	SP	0.45	0 - 100	45	5	334
NO_2^- /NO_3^-	SP	No data	No data	No data	No data	335
NO_2^- NO_3^-	SP	0.1-0.25 0.45	0.5 - 40 2-100	10	0.8-1.1 1.3	336
NO_2^- /NO_3^-	SP	0.1	No data	No data	No data	337
NO_2^- /NO_3^-	SP	No data	No data	No data	No data	338
NO_2^- NO_3^-	SP	0.02 0.16	0.07 - 40 0.48 - 30	20	1.10-1.60 1.14-1.64	339
NO_2^- NO_3^-	SP	0.46 0.51	1.1 - 20 1.3 - 30	10	3.9 9.2	231
NO_2^- /NO_3^-	SP	0.45	No data	40	1	340
NO_2^- NO_3^-	SP SP	0.1 0.45	0.05 - 40 2 - 100	10	0.8 1.3	341 341
NO_3^-	SP	0.023	0.023 - 12	No data	2.9-7.7	342
NO_3^-	SP	0.045	0.045 -1.6 1.6 - 32	No data	No data	343
PO_4^{3-}	Ch	0.002	0.005 - 0.194	No data	4.66	344
PO_4^{3-}	F	No data	0.01 - 3	15	1.2	345
PO_4^{3-}	F	0.3	No data	No data	No data	346
PO_4^{3-}	SP	0.05	Up to 4	90	1.5	347
P	SP	0.06	0 - 16	12	0.3-6	348
PO_4^{3-}	SP	50	No data	30	1.86	349
PO_4^{3-}	SP	No data	0.1 - 5	30	1 - 3	210
P	SP	0.002	Up to 0.58	No data	No data	350
PO_4^{3-}	SP	0.038	No data	No data	6.9	226
PO_4^{3-}	SP	0.062	No data	No data	4.2	351
PO_4^{3-}	SP	No data	No data	No data	No data	352
PO_4^{3-}	SP	0.24 0.22 0.17	5.2 - 210.5 2.1 - 105.2 1.05 - 42.1	50	3.8 3.4 2.7	353
PO_4^{3-}	SP	No data	No data	No data	No data	354
PO_4^{3-}	SP	0.00157	0.0032 - 0.0485	2	4.52	355
P	SP	0.06	No data	No data	< 2.5	356

Table 4.2. Continued

Analyte	Detection	DL (μM)	Linear range (μM)	SF (s/h)	RSD (%)	Ref
PO_4^{3-} PHP	SP	0.08 0.16	0.105 - 0.526 0.322 - 1.61	30-40	< 4	357
PO_4^{3-}	SP	0.02	0.02 - 1.05	No data	No data	358
PO_4^{3-}	SP	0.06	Up to 3.15	No data	1.4 - 6.6	359
PO_4^{3-}	SP	0.03	0.52 - 1.05	225	1.2	360
PO_4^{3-}	SP	0.2	0.0034 - 0.515	No data	4.45 - 6.75	361
P	ICP-MS	0.0016	No data	25	5.1	11
SO_4^{2-}	SP	No data	2000 - 30000	40	2	210
SO_4^{2-}	SP	No data	5.2 – 52.1	No data	1.6	362
SO_4^{2-}	SP	No data	No data	No data	No data	338
S^{-2}	SP	No data	1-90	120	1 - 3	210
S^{-2}	SP	No data	No data	No data	0.5	363
S^{-2}	SP	0.12 - 1.2	No data	No data	< 1	364
S^{-2}	SP	2.2	No data	No data	0.7	365
S^{-2}	SP	0.1 - 0.4	0 - 450	No data	< 1	366
S^{-2}	SP	a) 2.8 b) 4.7	a) 6.25 - 62.5 b) 15.63 - 156.25	a) 13 b) 27	a) 1.38 b) 1.44	367
S^{-2}	SP	0.09	0.625 – 6.25	2.5	0.7	368
S^{-2}	SP	0.04	0.625 – 15.625	5	2.2	369
SiO_3^{2-}	SP	No data	4 - 200	30	2 - 3	210
SiO_3^{2-}	SP	510	No data	No data	2.6	349
SiO_3^{2-}	SP	No data	No data	No data	0.6	363
SiO_3^{2-}	SP	0.5	No data	60 - 80	< 1	370
SiO_3^{2-}	SP	No data	0.33 – 16.64	40	No data	371
SiO_3^{2-}	SP	0.3	Up to 50	No data	1	372

[a] sandwich technique.

[b] multiple injection modality.

A: amperometry; C: conductimetry; Ch: chemiluminescence; DIC: dissolved inorganic carbon; DL: detection limit; DO: dissolved oxygen; F: fluorescence; FAAS: flame atomic absorption spectrometry; ICP-MS: inductively coupled plasma-mass spectrometry; P: potentiometric; PHP: phytase-hydrolysable phosphorus; RAML: reagent amylose; RAMP: reagent amylopectin; RSD: relative standard deviation; SF: sampling frequency; SP: spectrophotometry.

Halides: Bromide, Chloride, Fluoride and Iodide

Electrochemical Detection

Ilcheva et al. [301] studied the influence of the basic parameters in FI potentiometry (FIP) on the selectivity of the electrode membrane for the determination of chloride. Improvement in the selectivity in FIP appears only when the selectivity coefficient for conventional potentiometry is >1. A positive influence is achieved with the reduction of the time of contact between the sample and the electrode membrane and also with the presence of sensed ions in the carrier solution. Ilcheva et al. [302] applied direct FI potentiometry without pretreating the sample for bromide determination. The matrix effect of high chloride concentration is removed by calibrating the electrode with NaCl. Wang et al. [303] investigated a flow-through analytical system that utilizes a membrane-coated carbon rod ion-selective electrode. Electrodes for chloride are based on a bis(diphenylphosphino)propane-copper complex as ion exchanger. Dan et al. [304] developed a simple single channel ion-selective electrode (ISE)-FIA system, which consists of Mariot container, rotary sampling valve, PVC thin tube, and micro flow cell. It has been applied to the simultaneous determination of chloride, bromide and iodide in natural brines with citric acid-disodium citrate as the carrier stream. Lolic et al. [305] developed a rapid, indirect gas-diffusion FI method with amperometric detection for the selective determination of chloride. The method is based on the oxidation of chloride to chlorine by a saturated potassium permanganate solution. The chlorine diffuses through the micro-porous membrane in a gas-diffusion unit from the donor into the acceptor stream, and is amperometrically quantified on a platinum working electrode.

Chemiluminescence Detection

Borges et al. [306] proposed a simple flow-based procedure with chemiluminescence detection for bromide ion determination. The procedure is based on the oxidation of bromide to bromine by chloramine-T followed by the reaction of bromine with luminol resulting in chemiluminescence emission.

Spectrophotometric Detection

Anagnostopoulou et al. [307] automated the spectrophotometric phenol red method for the determination of bromide by using FI. Uraisin et al. [308] examined a sensitive and selective method for the determination of bromide in seawater by using a FI/stopped-flow detection technique. The detection system is developed for a new kinetic-spectrophotometric determination of bromide in the presence of chloride matrix without any separation step. The detection was based on the kinetic effect of bromide on the oxidation of methylene blue (MB) with H_2O_2 in a strongly acidic solution. The decolorization of the blue color of MB is used for the spectrophotometric detection of bromide at 746 nm. Ruzicka et al. [309] adapted to FIA the determination of chlorides based on spectrophotometric measurement of iron (III) thiocyanate. Moreno et al. [310] studied the determination of residual chloride with o-tolidine. The method proposed is based on the use of reverse FIA. Nishioka et al. [311] studied a FI method for the determination of fluoride in seawater with Alfusone as colorimetric reagent. León-Gonzalez et al. [312] proposed a stopped-flow reagent-injection method for the spectrophotometric determination of fluoride with lanthanum(III)-alizarine fluorine blue in the presence of sodium dodecyl sulfate at pH 4.6. Oguma et al. [313] developed a method for the determination of iodate and iodide with and without an anion-exchange column in the flow conduit, which involving spectrophotometric detection based on the catalytic, fading effect of either iodate or iodide on the indicator reaction of iron (III)thiocyanate and nitrite. Hakedal et al. [314] described a FI method for the determination of iodide based on the oxidation of iodide to iodine, which after permeation through a PTFE membrane is detected spectrophotometrically. The sample is injected into a carrier stream of water and merged with acidic dichromate reagent, oxidizing iodide to iodine. The iodine permeates through the membrane into a collector stream containing iodide. The iodine reacts with iodide forming triiodide, and is measured spectrophotometrically in a flow cell at 350 nm. Kuznetsov et al. studied under flow conditions complex formation between elemental iodine and polyvinyl alcohol [315], and elemental iodine and the two main starch polysaccharides, amylose and amylopectin [316], in order to determine the applicability of these compounds to the development of highly selective and highly sensitive procedures for the FIA determination of iodine.

Atomic Absorption Spectrometric Detection

Yebra et al. [317] proposed a continuous flow atomic absorption spectrometric system to develop an efficient on-line preconcentration-elution FI procedure for the indirect determination of iodide traces. Chromium (VI) is introduced into the flow system and is reduced to chromium (III) in acid medium proportionally to the iodide present in the sample. The Cr(III) reduced by iodide is retained on a minicolumn packed with a poly(aminophosphonic acid) chelating resin, which is only selective for this oxidation state, while unretained Cr(VI) is wasted. Cr(III) was preconcentrated by continuously passing the iodide containing sample through the system. It is then eluted with a small volume of hydrochloric acid into the FAAS spectrometer.

Hydrogen Peroxide and Dissolved Oxygen

Fluorescence Detection

Sakai et al. [318] developed a simple and highly sensitive FIA system for the spectrofluorometric determination of dissolved oxygen (DO). The decrease in fluorescence, which is based on the reaction of 2-thionaphthol with iodine liberated in Winkler's method, is used as the DO detection system. Amouroux et al. [319] developed the scopoletin-peroxidase fluorescence decay method using standard additions and a FI manifold to determine hydrogen peroxide.

Chemiluminescence Detection

Price et al [320] reported a rapid, FI procedure with chemiluminescence detection for the determination of hydrogen peroxide. It is based on the H_2O_2-induced oxidation of an alkaline solution of luminol in the presence of a catalyst (cobalt(II)). A portable, automated version of the monitor has been successfully deployed onboard ship to study a typical depth profile for hydrogen peroxide. Price et al. [321-322] presented further validation of the FI-chemiluminescence technique for shipboard deployment in marine environments. Shipboard results confirmed the feasibility of using this method for H_2O_2 determination in seawater. Yuan et al. [323] developed a reagent-injection chemiluminescent method for the aboard ship determination of H_2O_2. The method is based on the Co^{2+}-catalyzed oxidation of luminol by hydrogen peroxide in an alkaline solution. One mixed reagent is used for the analysis using the reagent injection method. Because Fe^{2+} oxidation is rapid compared to the rate of decay of H_2O_2, the Fe^{2+} interference can be eliminated by storing seawater samples for 1 h. Cooper et. al. [324] showed that 10-methyl-9-(p-formylphenyl)-acridinium carboxylate trifluoromethanesulfonate can be used as a chemiluminescent method to accurately quantify hydrogen peroxide. They adapted this method for FIA improving detection limits and sample throughput. King et al. [325] studied the mechanism of the reaction of H_2O_2 with 10-methyl-9-(p-formylphenyl)acridinium carboxylate trifluoromethanesulfonate in a FI system. For this, they evaluated the reaction rates of each step in the mechanism of the reaction and used a kinetic model to optimize the analysis of H_2O_2.

Spectrophotometric Detection

Johnson et al. [326] described a reagent-injection flow technique that allows automatic control of the volume of reagent to be injected. The sensitivity of the measurements can be adjusted over a wide range by changing the injection volume. This technique also eliminates problems in photometric determinations arising from refractive effects and from light scattering by suspended particles in the sample. The concentration of H_2O_2 is determined as the colored condensation product of N-ethyl-N-(sulfopropyl)aniline and 4-aminoantipyrene.

Nitrogen, Nitrite and Nitrate

Fluorescence Detection

Motomizu et al. [327] determined nitrite fluorometrically by FIA. In acidic medium, nitrite reacts with 3-aminonaphthalene-1,5-disulfonic acid (I) to form the diazonium salt, which is converted into the fluorescent azoic acid salt in an alkaline medium. The fluorescence intensity (excited at 365 nm) is measured at 470 nm. Masserini et al. [320] developed a fluorescence-based method to determine nitrite and nitrate (as excess nitrite following reduction of NO_3^- to NO_2^-). The technique capitalizes on the triple bond between the two nitrogen atoms within the diazonium ion formed via the well-known reaction between an acidified nitrite sample and an aromatic primary amine. Fluorescence of π-electrons within this bond allows this reaction to be probed with standard fluorescence spectroscopy. Reverse flow injection anal. (rFIA) is used to correct for background fluorescence from leachates and naturally occurring dissolved org. matter (DOM).

Spectrophotometric Detection

Johnson et al. [328] proposed an inexpensive and very reliable technique for the determination of nitrite and nitrate based on the conventional method of reducing NO_3^- to NO_2^- and colorimetric determination of nitrite as an azo dye, with a modification of the technique to permit FIA and to automate the processing of samples. Pai et al. [329] studied the kinetics of the reaction for the formation of the pink azo dye in the determination of nitrite at different acidities, temperatures and concentrations of N-1-naphthylethylenediamine (NED). It is found that the reaction is considerably faster in seawater than in fresh water, and that an increase in the acidity slightly increases the molar absorptivity. Ariza et al. [330] determined nitrite by photometric FIA by normal and reverse techniques. The methods are used to monitor the input and output streams of small tanks at fish breeding farms. Takeda et al. [331] described a rapid, sensitive, and maintenance-free FI method to determine nitrate based on the photo-induced conversion of nitrate to nitrite. McCormack et. al. [332] used for first time a single FI manifold capable of monitoring nitrate throughout an estuary, i.e., with salinities ranging from 0-3.5%, without any correction factor being applied for refractive index changes. The FI manifold includes a cadmium reductor column. The color reagent is a mixture of N-(1-naphthyl)ethylenediamine dihydrochloride and sulfamide reagents. The resultant pink-purple dye is detected on-line at 540 nm. Pai et al. [333] studied the analytical problems in the determination of nitrate using FIA coupled with an on-line cadmium column reductor. Thus, it is found difficult to prepare a nearly 100% efficient copperized Cd reductor, which maintains its efficiency over a lengthy period. Instead, the use of a narrow and lower efficiency Cd coil is recommended because it is more stable and therefore more suitable for FIA. Since the conversion of nitrate to nitrite is not quantitative, results for nitrate tend to be overestimated when nitrite is also present. This problem is solved using a simple correction scheme to compensate for the effect of nitrite, thus enabling the correct nitrate concentration to be evaluated. Daniel et al. [334] established a method with the aim of improve the precision and the sensitivity of nitrite and nitrate determination by FIA for in situ utilization. For this, two kinds of signal treatment are proposed in order to eliminate the refractive index provoked by the heterogeneous flow in FIA and the errors induced by temperature, salinity and pressure variations. McCormack et al. [335] described FI methods for the determination of nitrate and nitrite with the aim of application as in-situ monitors. Cerdà et al. [336] described an automated sequential injection system for monitoring nitrate and nitrite in water samples. The method enables the simultaneous determination of both parameters with a single injection of sample. For this, is adopted a sandwich arrangement, in which the sample is placed between two Griess reagent zones. Nitrite is determined in one end of the sample zone by diazotization-coupling reactions and spectrophotometric detection of the azo dye formed. In the other end, nitrate is similarly determined after its online reduction to nitrite using a copperized cadmium column. Ma et al. [337] investigated and applied a microFIA to minimize the ample size, reagent consumption and waste, to the simultaneous determination of nitrate and nitrite in water samples. Nitrate is reduced to nitrite on line with 99% reduction efficiency by passing through a mini column packed with 60-80-mesh powder Cd/Cu. The reagent solution contained sulfanilamide, N-(1-naphthyl)ethylenediamine, and hydrochloric acid. Teshima et al. [338] applied a FIA methodology to the simultaneous determination of nitrite and nitrate based on a diazotization/coupling reaction with a copperized cadmium (Cd/Cu) reducing column installed in the FIA system. Kazemzadeh et al. [339] developed a direct spectrophotometric method for the simultaneous determination of nitrite and nitrate by FIA The method is based on the catalytic effect of nitrite on the oxidation of pyrogallol red by bromate in acidic media and the decrease in absorbance of the system at 465 nm. The injected sample is split into 2 streams. One of the streams is directly treated with the above reagents and then is passed to the sample flow cell of the spectrophotometer. The decrease in absorbance at 465 nm is due to the nitrite. The other stream is passed through a reductor microcolumn containing copperized-cadmium, where nitrate reduces to nitrite and then the sample is treated with the mixed reagents and is passed through the same cell of the spectrophotometer. Tovar et al. [231] proposed a FIA method using spectrophotometric detection for the simultaneous determination of ammonium, nitrite and nitrate in marine waters. The method is based on absorbance measurement of azo dye obtained by the reaction of nitrite with N-(1-naphthyl)ethylenediamine dihydrochloride and sulfanilamide.

Several submersible FI analyzers for the in-situ determination of nitrate and nitrite have been proposed. Thus, Daniel et al. [340] developed a system based on the sample injection into a reagent stream. Three rotary valves (two injection and one stream selection) are developed to work under pressure. As a result, only a single valve is used to select six standards when four are needed in a system using pinch valves. The manifold was configured for nitrite and nitrate determination, but it may be adapted to other analyses (i.e. ammonium, silicate, phosphate, trace metals) without adding valves. As a result, the submersible analyzer described in this paper can be used for river as well as ocean waters without any modification and it is usable aboard large as well as small ships. David et al. [341] describes the design and construction of a remotely deployed submersible sensor for the determination of total oxidized nitrogen (nitrate plus nitrite) in seawater. It is based on the FI principle, with solid state spectrophotometric detection of the diazotization product from the reaction of nitrite with N-(1-naphthyl)ethylenediamine dihydrochloride and sulfanilamide. Nitrate is pre-reduced in-line with a copperized cadmium powder column. David et al. [342] depicted the design, construction, and performance of a remotely deployed submersible FI-based nutrient (total oxidized nitrogen) sensor. The sensor features a custom-built microcomputer and a solid-state, flow-through spectrophotometric detector, and the derivatization chemistry is based on in-line copper-cadmium reduction of nitrate to nitrite, and diazotization with N-(1-naphthyl)ethylenediamine and sulfanilamide. Gardolinski et al. [343] described a miniature, submersible FI analyzer, with solid-state spectrophotometric detection, for the in situ determination of nitrate. It uses the standard laboratory chemistry of cadmium reduction followed by diazotization.

Phosphorous and Phosphate

Chemiluminescence Detection

Liang et al. [344] established an on-line solid-phase extraction method coupled with FI and luminol chemiluminescence detection to determine ultratrace orthophosphate in seawater. This method applied the solid-phase extraction technique to extract molybdophosphoric heteropoly acid (MoP) paired with cetyltrimethylammonium bromide from seawater matrix using C18 sorbent. Chemiluminescence emission could be generated via MoP reaction with alkaline luminol.

Fluorescence Detection

Motomizu et al [345] proposed a sensitive method for the determination of trace amounts of phosphate by fluorescence-quenching detection/FIA. The fluorescence of Rhodamine B (RB) is quenched with the formation of the ion association of molybdophosphate with RB, λex and λem were 560 nm and 580 nm, respectively. Frank et al. [346] described a flow system that is based on sequential injection analysis (SIA) and is suitable for the fast determination of ammonia and phosphate. The determination of free reactive phosphate is based on the reaction of phosphate with acidic molybdate to phosphomolybdate, which forms nonfluorescent ion pairs with rhodamine 6G. The remaining rhodamine fluorescence is detected at 550 nm with an excitation at 470 nm.

Spectrophotometric Detection

Johnson et al. [347] described a high sensitivity variation of the technique of FIA. The variation consists of injecting small volumes of reagent into the flowing sample stream, rather than the conventional injection of sample into a flowing reagent (or inert carrier) stream. Injection of reagents into the sample or carrier stream in FIA has been used to minimize reagent consumption and to produce large pH gradients. The technique is tested by using it for the determination of dissolved phosphate with the colorimetric phosphomolybdate complex method. Aoyagi et al. [348] described a method for the determination of total phosphorous. For this, A coiled Teflon capillary digester, which contains a platinum wire as catalyst for oxidation with peroxodisulfate at 160°C, is directly connected to the color-development system based on ion pair formation between molybdophosphate and malachite green. Kang et al. [349] proposed an automatic reverse FIA method for continuous determination of phosphate. Shpigun et al. [210] reported some results on using FIA in marine chemical investigations. They described the new modifications of reversed FI manifolds for the determination of dissolved silicate, phosphate, sulfate, sulfide, and Mn(II) in seawater samples. The determination of phosphate is based on molybdenum blue formation after reaction with ammonium molybdate. Susanto et al. [350] proposed a sensitive method for the determination of microamounts of phosphorous existing as orthophosphate. The method is based on a filtration-dissolution preconcentration and FI with spectrophotometric detection. In an acidic medium orthophosphate reacts with molybdate to form molybdophosphate, which then reacts with a cationic dye, Malachite Green (MG) to form a colored ion assoc. The ion association is filtered through a membrane filter, then dissolved together with the membrane filter in methyl Cellosolve and their absorbance measured at 627 nm with a flow injection system equipped with an autosampler. Tovar et al. [226] developed a FIA method with spectrophotometric detection or the simultaneous determination of phosphate and ammonium. The method is based on the dual combination of a reversed FI system and a continuous flow system without injection. Tovar et al. [351] used the determination of phosphates to illustrate that by applying a very simple multivariate procedure, the results obtained are better than those obtained by using the univariate optimization. Worsfold et al. [352] described the FI determination of phosphorous with spectrophotometric detection to assess nutrient impact on water quality and provide decision support systems for catchment management. Karthikeyan et al. [353] describes a simple and rapid procedure for determination of traces of phosphate by molybdenum blue chemistry. These authors demonstrated that the use of a cost-effective home-made flow cell with a long path length in combination with a light emitting diode (LED) and a photodiode (PD) is as a simple absorbance detector for FIA. Thus, the color intensity of the resulting association complex, molybdenum blue, is measured photometrically (λmax 875 nm). Wang et al. [354] used the method of FIA-crystal violet molybdophosphate to determine phosphate in seawater, which can be applicable to the in-situ autonomous monitoring technology. Liang et al. [355] established an on-line solid phase extraction method coupled with FIA and colorimetric detection to determine nanomolar level orthophosphate in seawater. This method is based on the extraction of phosphomolybdenum blue (PMB) paired with cetyltrimethylammonium bromide (CTAB) using a solid phase extraction technique on C18 sorbent. A stopped flow technique is used to assure the complete formation of the PMB-CTAB compound, which is sequentially extracted on an in-line Sep-Pak C18 cartridge. The adsorbed PMB-CTAB can be rapidly eluted by sulfuric acid in ethanol, and determined with a spectrophotometer at 700 nm. Omaka et al. [356] proposed a four-channel FI method for the determination of filterable reactive phosphorus (FRP). The method is suitable for the determination of FRP in natural waters where arsenate and silicate are present in concentrations exceeding 0.300 mg/L As and 1.9 mg/L Si, respectively. Omaka et al. [357] established a FI method for the determination of phytase hydrolysable phosphorus (PHP) using immobilized phytase.

The Schlieren or refractive index (RI) effect is a major problem in the determination of dissolved reactive phosphorus in sea and estuarine waters using conventional FI manifolds with sample injection. This is because differences in RI between the injected sample zone and the carrier stream give rise to a lensing effect, which is superimposed on the blank response and causes significant error in quantitation. Auflitsch et al. [358] reported a simple reversed flow injection (rFI) manifold using spectrophotometric detection, which removes these quantitation errors. Acidic molybdate is injected into a sulfuric acid carrier stream of the same refractive index and sequentially merged with sample and reductant (ascorbic acid). Reduction of phosphomolybdate to phosphomolybdenum blue is carried out in a coil thermostated at 60 °C. Thus, dissolved reactive phosphorus has been successfully determined in estuarine waters with salinities ranging from 0 to 3.0%. McKelvie et al. [359] proposed conventional FI manifolds with sample injection for the determination of reactive phosphorus in estuarine waters. Hence, reported a simple FI manifold, which obviates the RI error in reactive phosphorus measurement. It involves the injection of acidic molybdophosphate reagent into a carrier stream of sodium chloride solution of similar refractive index, which is then sequentially merged with a sample (the salinity of which may vary widely from sample to sample) and a reductant. Despite the occurrence of sizeable blank signals, reactive phosphorus has been successfully measured in samples with salinities ranging from 0 to 3.4% using calibration standards prepared in deionized water. Ellis et al. [360] described a multi-reflection flow cell suitable for FIA Light from an LED is directed through an optical fiber into a silver coated capillary through a sidewall aperture, and emerges through a similar aperture along the capillary after undergoing an estimated 19 reflections. This process provides a sensitivity enhancement of approximately 2.5 compared with a conventional z-cell of the same nominal path length. This enhancement is due to both the increased optical path length achieved by multiple reflection of the light beam through the sample, and minimization of physical dispersion using a short, small internal diameter capillary as the flow cell. The optical design of this flow cell also minimizes the Schlieren effect. The flow cell is applied to the determination of reactive phosphorous in estuarine waters with wide variation in salinity and refractive index. Ma et al. [361] modified a method for the determination of nanomolar concentrations of orthophosphate in oligotrophic seawater [355] to make it fully feasible for shipboard application and for faster sample throughput with minimized sample volume. The technique is based on the FI method with solid phase extraction on a Sep-Pak C18 cartridge and calorimetric detector. The Schlieren effect is minimized by rinsing the cartridge sequentially with water and 95% ethanol solution. The modified method permits the analysis of samples over a wide range of concentrations.

Inductively Coupled Plasma Mass Spectrometric Detection

Field et al. [11] described a high-throughput method to determine trace elements (P, V, Mn, As, Mo, Ba, U) with application to tracing ballast water exchange in ocean-going vessels. Using direct flow injection and methane addition the method can perform >700 continuous determinations of ten-fold diluted seawater in less than 30 h.

Sulfate and sulfide

Sulfate is determined by a FIA procedure with reagent injection based on its turbidimetric reaction with barium chloride. The determination of dissolved sulfide by using a FIA manifold with reagent injection is based on the highly sensitive and specific chromogenic reaction of N,N-diethyl-p-phennylenediamine with sulfide in the presence of potassium dichromate as oxidant in acidic medium [210]. Sakuragawa et al. [362] described a rapid and simple FI method using a reaction column packed with barium chromate powder for the determination of micro amounts of sulfate in surface water and seawater samples. The absorbance of the chromate (which corresponds to the sulfate concentration) in the carrier solution from the reaction column is measured using a photometric detector at 370 nm. Teshima et al. [338] applied a FIA method using dimethylsulfonazo III-barium chelate for sulfate determination. Johnson et al. [363] developed a submersible chemical analyzer (SCANNER), which can perform analyses in situ in the ocean. The SCANNER is based on a modified FI system and can be used to automate virtually any spectrophotometric determination that can be done by FIA The SCANNER consists of a multichannel peristaltic pump, solid-state colorimeters, manifold tubing, valves, and an electronic module. All of the components are pressure-tolerant, except the electronic module, which is placed in a pressure housing. This analyzer has been applied to the determination of silicate and sulfide. Sakamoto-Arnold et al. [364] used FIA and flow analysis to automate the colorimetric determination of sulfide by the methylene blue method. This FIA method was tested onboard ship. Cassella et al. [365] proposed an automated FI methodology employing spectrophotometry and sodium nitroprussite to perform sulfide determination at 538 nm. Sarradin et al. [366] realized the laboratory adaptation of the methylene blue method for the analysis of sulfide to FIA before its integration on an in situ analyzer. Ferrer et al. [367] proposed a fully software-controlled multisyringe flow injection (MSFIA) spectrophotometric system for the determination of sulfide. The implementation of ancillary solenoid valves into the flow network allows a multitude of injection modalities to be explored, the selected modality being directly dependent on the aim of the assays. The multicommuted sandwich-type approach is introduced in this study as an efficient means to warrant high sensitivity for the particular assay with excellent repeatabilities and a considerable reagent saving. Also, a high injection frequency may be easily attained by carrying out a multiple injection modality during a single forward displacement of the piston driver bar. The interfacing of the robust and versatile multisyringe piston pump with an optical fiber plug-in spectrophotometer furnished with a light emitting diode enables the miniaturization of the flow analyzer, which is thus readily adaptable to in-situ and real-time monitoring schemes. The flow method is based on the coupling Fischer's reaction of sulfide with N,N-dimethyl-p-phenylenediamine in the presence of Fe(III) as oxidizing reagent in a hydrochloric acid medium. Ferrer et al. [368] developed a software-controlled flow-through optical fiber diffuse reflectance sensor capitalized on the implementation of disk-based solid-phase pre-concentration schemes in a MSFIA set-up for the trace determination of sulfide. The fully automated flowing methodology is based on Fischer's coupling reaction of sulfide with N,N-dimethyl-p-phenylenediamine in the presence of Fe(III) as oxidizing reagent in a hydrochloric acid medium. The online generated methylene blue dye is subsequently delivered downstream to a dedicated optode cell furnished with an octadecyl-chemically modified (C18) disk, while continuously recording the diffuse reflectance spectrum of the pre-concentrated compound. A double regeneration protocol is finally executed to warrant minimum background noise and negligible baseline. The interfacing of the robust and versatile multisyringe flow injection-based optode with a plug-in spectrophotometer furnished with a light emitting diode assures the miniaturization of the overall flow analyzer, which is, thus, readily adaptable to real-time monitoring schemes. Ferrer et al. [369] proposed a MSFIA system coupling a flow-through optical fiber diffuse reflectance sensor with in-line gas-diffusion (GD) separation for the isolation, preconcentration and determination of traces of volatile and gas-evolving compounds in samples containing suspended solids, with no need for any preliminary batch sample treatment. The proposed combination of techniques is applied satisfactorily to sulfide determination. The method based on multicommutation flow analysis involves the stripping of the analyte as hydrogen sulfide from the donor channel of the GD-module into an alkaline receiver segment, whereupon the enriched plug merges with well-defined zones of the chromogenic reagents (viz., N,N-dimethyl-p-phenylenediamine (DMPD) and Fe(III)). The in-line generated methylene blue dye is subsequently delivered downstream to the dedicated optrode cell furnished with a C18 disk, while recording continuously the diffuse reflectance spectrum of the pre-concentrated compound.

Silicate

Thomsen et al. [370] used FIA used to automate the determination of silicate by the molybdenum blue method avoiding the refractive index interference. Yokoyama et al. [371] developed a FIA system for the spectrophotometric determination of silicic acid based on the formation of a heteropoly blue complex. Ascorbic acid is used to reduce the yellow molybdosilicic acid to a heteropoly blue complex. Floch et al. [372] proposed a method to determine silicic acid in sea water with a submersible chemical analyzer. It is based on direct FIA, for fast and discrete measurements, and dual wavelength treatment of the signal to correct the refractive index interference, the main factor that affects in situ analysis. This colorimetric method is based on the formation of beta silicomolybdic acid reduced in intense colored molybdenum blue.

CONCLUSION

FI methodologies are shown to be an ideal tool for the automated continuous sampling, handling, pretreatment and determination of inorganic compounds in sea and estuarine water samples. They provide the possibility of developing in-situ measurements (shipboard determinations and submersible FI analyzers), in-situ preconcentration and matrix removal (microcolumn field sampling technique) or simply FI manifolds used at the laboratory. These FI manifolds usually included different on-line separation techniques to preconcentrate and/or to remove interferences produced by the high salt content of these samples (above all solid-phase extraction using a chelating resin or a solid sorbent to adsorb a complex formed between the analyte and a chelating ligand). Most of the reported FI methods are based on the adaptation of an existing usual method to the FI mode.

The accepted method for preserving water samples for trace metals determination involves the addition of a concentrated acid (hydrochloric or nitric acid) to reduce the pH below 2, which increase the possibility of analyte losses, sample contamination and a chemical modification of analyte species. However, when are utilized in-situ measurements or in-situ preconcentration manifolds, samples can be directly processed without the addition of substances for sample preservation, which make them an interesting approach for speciation studies. In addition, other advantages of FI methods are their simplicity, low cost, freedom from interferences, accuracy, precision, high sample throughput and low detection limits. These characteristics are very important because environmental legislations demand for these analytical properties. In this sense, it is important develop in-situ instrumentation for eliminating sample collection and transport. Thus, portable manifolds involving flow injection methodologies, with spectrophotometric and chemiluminescence detection, are a practical tool that can achieve these purposes and provide high quality analytical data to establish high temporal and spatial resolution measurements of waters from the marine environment.

REFERENCES

[1] Timerbaev, A.R. & Fukushi, K.(2003). *Mar.Chem.* 82, 221-238.

[2] Timerbaev, A.R. & Petrukhin, O.M. (2000). *Talanta* 52 1171-1173.

[3] Hanrahan, G., Ussher, S., Gledhill, M., Achterberg, E.P.& Worsfold, P.J. (2002). *Trends Anal.Chem.* 21, 233-239.

[4] David, A.R.J., McCormack, T. & Worsfold, P.J. (1999). *J. Autom. Methods Manag. Chem.* 21, 1-9.

[5] Coles, S., Nimmo M. & Worsfold, P. (2000). *Lab. Rob. Autom.* 12, 183-193.

[6] Ma R., Woods G. & McLeod C.W. (1999). Microcolumn field sampling and flow injection analysis: A strategy for enhanced trace analysis and element speciation. In: Sanz-Medel A., editor. Analytical Spectroscopy Library, Vol 9: F*low analysis with atomic spectrometric detectors*. Amsterdam: Elsevier.

[7] Lenehan, C.E., Barnett, N.W. & Lewis, S.W. (2002). *Analyst* 127, 997-1020.

[8] Thomas, O., Theraulaz. F., Cerdá, V., Constant, D. & Quevauviller, Ph. (1997). *Trends Anal. Chem.* 16, 419-424.

[9] Uehara, N., Aikawa, E., Sekine, S. & Shimizu, T. (2005). *Nippon Kaisui Gakkaishi* 59, 131-137.

[10] Ebdon, L., Fisher, A., Handley, H., Jones, P. (1993). *J. Anal. At. Spectrom.* 8, 977-981.

[11] Field, M.P., LaVigne, M., Murphy, K.R., Ruiz, G.M. & Sherrell, R.M. (2007). *J. Anal. At. Spectrom.* 22, 1145-1151.

[12] Yuan, Y., Wang, Y. & Qu, K. (1988). *Fenxi Huaxue* 16, 546-548.

[13] Fajardo, Y., Gomez, E., Garcias, F., Cerda, V. & Casas, M. (2007). *Talanta* 71, 1172-1179.

[14] Ye, R.D. & Khoo, S.B. (1997). *Electroanalysis* 9, 481-489.

[15] Shpigun, L.K. & Kopytova, N.E. (1997). *Zavodskaya Laboratoriya* 63, 5-7.

[16] Sella, S., Sturgeon, R.E., Willie, S.N., Campos, R.C. (1997). *J. Anal. At. Spectrom.* 12, 1281-1285.

[17] Dadfarnia, S., Haji Shabani, A.M. & Gohari, M. (2004). *Talanta* 64, 682-687.

[18] Mesquita da Silva, M.A., Azzolin Frescura, V. L.& Curtius, A. J. (2000). *Spectrochim. Acta* 55B, 803-813.

[19] Pozebon, D., Dressler, V. L. & Curtius, A.J. (2001). *Anal. Chim. Acta* 438, 215-225.

[20] Yang, L. & Sturgeon, R.E. (2002). *J. Anal. At. Spectrom.* 17, 88-93.

[21] Barriada, J.L., Truscott, J. B. & Achterberg, E.P. (2003). *J. Autom. Methods Manag. Chem.* 25, 93-100.

[22] Ndung'u, K., Ranville, M.A., Franks, R. P. & Flegal, A.R. (2006). *Marine Chem.* 98, 109-120.

[23] Resing, J.A. & Measures, C.I. (1994). *Anal. Chem.* 66, 4105-4111.

[24] de Jong, J.T.M., Boye, M., Schoemann, V.F., Nolting, R.F. & de Baar, H.J.W. (2000). *J. Environ. Monitor.* 2, 496-502.
[25] Obata, H., Nozaki, Y., Okamura, K., Maruo, M. & Nakayama, E. (2000). *Field Anal. Chem. Technol.* 4, 274-282.
[26] Vink, S., Boyle, E.A., Measures, C.I. & Yuan, J. (2000). *Deep-Sea Research, Part I: Oceanographic Research Papers* 47, 1141-1156.
[27] Kara, D., Fisher, A.& Hill, S.J.(2008). *Anal. Chim. Acta* 611, 62-67.
[28] Brown, M.T. & Bruland, K.W. (2008). *Limnol. Oceanography Methods* 6, 87-95.
[29] Yuan, D. & Shuttler, I.L. (1995). *Anal.Chim.Acta* 316, 313-322.
[30] Fujiwara, K., Kuramochi, A. & Tsubota, H. (1990). *Anal. Sci.* 6, 425-430.
[31] Yamamoto, M., Takada, K., Kumamaru, T., Yasuda, M., Yokoyama, S. & Yamamoto, Y. (1987). *Anal.Chem.* 59, 2446-2448.
[32] Lu, X., Li, J., Chen, S. & Dai, G. (1989). *Haiyang Xuebao (Zhongwenban)* 11, 444-449.
[33] Sperling, M., Yin, X. & Welz, B. (1991). *Spectrochim. Acta* 46B, 1789-1801.
[34] Willie, S.N. (1996). *Spectrochim. Acta* 51B, 1781-1790.
[35] Xu, G., Lu, X. & Yin, X. (1998). *Fenxi Huaxue* 26, 1037.
[36] Karthikeyan, S., Rao, T.P. & Iyer, C.S.P. (1999). *Talanta* 49, 523-530.
[37] Cabon, J.Y. & Cabon, N. (2000). *Fresenius' J. Anal. Chem.* 368, 484-489.
[38] Cabon, J.Y. & Cabon, N. (2000). *Anal.Chim. Acta* 418, 19-31.
[39] Tung-Ming, H. & Jing-Ming W. (2004). *J. Anal. At. Spectrom* 19, 923-928.
[40] Herbello-Hermelo, P., Barciela-Alonso, M.C.; Bermejo-Barrera, A. & Bermejo-Barrera, P. (2005). *J. Anal. At. Spectrom* 20, 662-664.
[41] Hou, J., Gong, Z.B., Guo, X.M. & Huang, B.L. (2000). *Fenxi Shiyanshi* 19, 13-16.
[42] Yan, X.P., Yin, X.B., He, X.W. & Jiang, Y. (2002). *Anal. Chem.* 74, 2162-2166.
[43] Stroh, A. & Voellkopf, U. (1993). *J. Anal. At. Spectrom.* 8, 35-40.
[44] Alves, L.C., Allen, L.A. & Houk, R.S. (1993). *Anal. Chem.* 65, 2468-2471.
[45] Huang, M.F., Jiang, S.J.& Hwang, C.J. (1995). *J. Anal. At. Spectrom* 10, 31-35.
[46] Chen, C.S. & Jiang, S.J. (1996). *Spectrochim. Acta* 51B, 1813-1821.
[47] Lam, J. W. & Sturgeon, R. E. (1999). *At. Spectrosc.* 20, 79-85.
[48] Lee, K.H., Oshima, M. & Motomizu, S. (2002). *J. Flow Injection Anal.* 19, 39-44.
[49] Holliday, A. E. & Beauchemin, D. (2003). *J. Anal. At. Spectrom* 18, 1109-1112.
[50] Ye, R. & Khoo, S.B. (1999). *Analyst* 124, 353-360.
[51] Falkner, K.K. & Edmond, J.M. (1990). *Earth Planetary Sci. Lett.* 98, 208-221.
[52] Falkner, K.K. & Edmond, J.M. (1990). *Anal. Chem.* 62, 1477-1481.
[53] Nose, K. & Zenki, M. (1991). *Analyst* 116, 711-714.
[54] Nishioka, H., Sai, N., Kumagai, T., Nagahiro, T. & Uesugi, K. (1992). *Nippon Kaisui Gakkaishi* 46, 391-396.
[55] Daih, B.J., Huang, H. J. (1992). *Anal. Chim. Acta* 258, 245-252.
[56] Moscoso-Perez, C., Moreda-Pineiro, J., Lopez-Mahia, P., Muniategui-Lorenzo, S., Fernandez-Fernandez, E. & Prada-Rodriguez, D. (2003). *Talanta* 61, 633-642.
[57] Pozebon, D., Dressler, V.L. & Curtius, Adilson J. (1998). *J. Anal. At. Spectrom.* 13, 363-369.
[58] Ang, K. P., Tay, B.T., Gunasingham, H., Khoo, S.B. & Koh, C.H. (1989). *Int. J. Environ. Studies* 32, 261-268.
[59] Tay, E.B.T., Khoo, S.B. & Ang, S.G. (1989). *Analyst* 114, 1271-1273.

[60] Worsfold, P.J., Achterberg, E.P., Bowie, A.R., Cannizzaro, V., C.S.; Costa, J.M., Dubois, F., Pereiro, R., San Vicente, B., Sanz-Medel, A., Vandeloise, R., Vander Donckt, E., Wollast, P. & Yunus, S. (2002). *J. Automat Meth Manag Chem*, 24, 41-47.

[61] Wu, H., Fang, C., Du, B. & Zhao, C. (2008). *Microchim. Acta* 160, 173-178.

[62] Ma, L., Zhang, X., Miao, P. & Li, H. (2006). *Pige Kexue Yu Gongcheng* 16, 33-36.

[63] Olsen, S., Pessenda, L.C.R., Ruzicka, J. & Hansen, E.H. (1983). *Analyst* 108, 905-917.

[64] Fang, Z., Ruzicka, J. & Hansen, E.H. (1984). *Anal. Chim.Acta* 164, 23-39.

[65] Ellis, L.A. & Roberts, D.J. (1998). *J. Anal. At. Spectrom.* 13, 631-634.

[66] Soylak, M. & Narin, I. (2005). *Chemia Analityczna* 50, 705-715.

[67] Fang, Z. & Welz, B. (1989). *J. Anal.At. Spectrom.* 4, 543-546.

[68] Zhuang, Z. & Hong, H. (1992). *Haiyang Yu Huzhao*, 23, 264-269.

[69] Sperling, M., Yin, X. & Weiz, B. (1991). *J. Anal.At. Spectrom.* 6, 295-300.

[70] Fang, Z., Guo, T. & Welz, B. (1991). *Talanta*, 38, 613-619.

[71] Welz, B., Yin, X. & Sperling, M. (1992). *Anal. Chim. Acta* 261, 477-487.

[72] Wang, M., Yuzefovsky, A.I. & Michel, R.G.(1993). *Microchem. J.* 48, 326-342.

[73] Ma, R., Van Mol, W. & Adams, F. (1996). *At. Spectrosc.* 17, 176-181.

[74] Su, P.G. & Huang, S.D. (1998). *Anal. Chim.Acta* 376, 305-311.

[75] Colbert, D., Johnson, K. S. & Coale, K.H. (1998). *Anal. Chim. Acta* 377, 255-262.

[76] Ali, A., Yin, X., Shen, H.,Ye, Y. & Gu, X. (1999). Anal. Chim.Acta 392, 283-289.

[77] Ceccarini, A., Cecchini, I.& Fuoco, R. (2005). *Microchem. J.* 79, 21-24.

[78] Anthemidis, A.N. & Karapatouchas, C.P.P. (2008). *Microchim. Acta* 160, 455-460.

[79] Hosten, E. & Welz, B. (1999). *Anal. Chim. Acta* 392, 55-65.

[80] Yebra, M. C., Garcia, A., Carro, N., Moreno-Cid, A.& Puig, L. (2002). *Talanta* 56, 777-785.

[81] Yebra, M. C., Salgado, J., Puig, L. & Moreno-Cid, A. (2002). *Anal. Bioanal. Chem.* 374, 530-534

[82] Yebra, M. C., Moreno-Cid, A., Cancela, S., Pedreira, S. & Puig, L. (2003). *Afinidad* 60, 529-533.

[83] Yebra-Biurrun, M. C.; Moreno-Cid, A. & Puig, L. (2004). *Anal. Chim.Acta* 524, 73-77.

[84] Bermejo-Barrera, P., Moreda-Pineiro, J., Moreda-Pineiro, A. & Bermejo-Barrera, A. (1996). *J. Anal.At. Spectrom.* 11, 1081-1086

[85] Nickson, R.A., Hill, S. J. & Worsfold, P.J. (1999). *Int. J. Environ.Anal. Chem.* 75, 57-69.

[86] Zougagh, M., Garcia Torres, A. & Cano Pavon, J. M. (2002). *Talanta* 56, 753-761.

[87] Kikuchi, M. & Kato, T. (1995). *Sendai-shi Eisei Kenkyushoho* 25, 223-228.

[88] Rosland, E. & Lund, W. (1998). *J. Anal. At. Spectrom.* 13, 1239-1244.

[89] Willie, S. N. & Sturgeon, R. E. (2001). Spectrochim. Acta 56B, 1707-1716.

[90] Huang, K.S. & Jiang, S.J. (1993). *Fresenius' J. Anal. Chem.* 347, 238-242.

[91] Willie, S. N., Iida, Y.& McLaren, J. W. (1998). *At. Spectrosc.* 19, 67-72.

[92] Willie, S. N., Lam, J. W. H., Yang, L. & Tao, G.(2001). *Anal.Chim.Acta* 447, 143-152.

[93] Lee, K.H., Oshima, M.& Motomizu, S. (2002). *Analyst* 127, 769-774.

[94] Willie, S. N., Tekgul, H.& Sturgeon, R. E. (1998). *Talanta* 47, 439-445.

[95] Howard, M., Jurbergs, H.A.& Holcombe, J.A. (1999). *J. Anal. At. Spectrom.* 14, 1209-1214.

[96] Nakatsuka, S., Okamura, K., Norisuye, K. & Sohrin, Y. (2007). *Anal. Chim. Acta* 594, 52-60.

[97] Song, M., Zhang, H., Wang, Z., Lee, F.S.C. & Wang, X. (2007). *Huaxue Yanjiu Yu Yingyong* 19, 979-986.
[98] Beck, N. G., Franks, R. P. & Bruland, K. W. (2002). *Anal. Chim.Acta* 455, 11-22.
[99] Maltez, H. F., Vieira, M.A., Ribeiro, A. S., Curtius, A.J. & Carasek, E. (2008). *Talanta* 74, 586-592.
[100] Bortoli, A., Gerotto, M., Marchiori, M., Mariconti, F., Palonta, M. & Troncon, A. (1996). *Microchem. J.* 54, 402-411.
[101] Yang, C., Wu, X., Gu, S., Zhuang, Z. & Wang, X. (2001). *Fenxi Huaxue* 29, 283-286.
[102] Ndung'u, K., Franks, R.P., Bruland, K. W. & Flegal, A. R. (2003). *Anal. Chim.Acta* 481, 127-138.
[103] Sakamoto-Arnold, C.M. & Johnson, K.S. (1987). *Anal. Chem.* 59, 1789-94.
[104] Hirata, S., Hashimoto, Y., Aihara, M.& Mallika, G.V. (1996). *Fresenius' J. Anal. Chem.* 355, 676-679.
[105] Cannizzaro, V., Bowie, A.R., Sax, A., Achterberg, E.P. & Worsfold, P.J. (2000). *Analyst* 125, 51-57.
[106] Tortajada-Genaro, L.A., Campins-Falco, P.& Bosch-Reig, F. (2003). *Anal. Chim. Acta* 488, 243-254.
[107] Yuzefovsky, A., Lonardo, R.F., Wang, M. & Michel, R.G.(1994). *J. Anal. At. Spectrom.* 9, 1195-202.
[108] Kawashima, T., Minami, T., Ata, M., Kamada, M. & Nakano, S. (1985). *J. Flow Injection Anal.* 2, 40-49.
[109] Malahoff, A., Kolotyrkina, I.Y. & Shpigun, L. K. (1996). *Analyst* 121, 1037-1041.
[110] Sperling, M., Yin, X. & Welz, B. (1991). *J. Anal. At. Spectrom.* 6, 615-622.
[111] Ma, R. & Adams, F. (1995). *Anal. Chim. Acta* 317, 215-222.
[112] Tony, K.A., Kartikeyan, S., Vijayalakshmy, B., Rao, T.P., Padmanabha I. & Chonatumatom S. (1999). *Analyst* 124, 191-195.
[113] Trojanowicz, M. & Pyrzynska, K. (1994). *Anal. Chim. Acta* 287, 247-252.
[114] Cassella, R.J., Salim, V.A., Jesuino, L.S., Santelli, R.E., Ferreira, S. L. C. & de Carvalho, M.S. (2001). T*alanta* 54, 61-67.
[115] Etoh, T., Tsuchiya, K. & Matsubara, M. (1994). *Kankyo Kagaku* 4, 390-391.
[116] Bloxham, M.J., Hill, S.J. & Worsfold, P.J. (1994). *J. Anal. At. Spectrom.* 9, 935-938.
[117] Chen, H.H. & Beauchemin, D. (2001). *J. Anal. At. Spectrom.* 16, 1356-1363.
[118] Ohura, H., Imato, T., Tamasaki, S. & Ishibashi, N. (1994). *Bunseki Kagaku* 43, 31-37.
[119] Yang, Y.J. & Huang, H.J. (2001). *Anal. Chem.* 73, 1377-1381.
[120] Paleologos, E.K., Stalikas, C.D., Tzouwara-Karayanni, S.M. & Karayannis, M.I. (2001). *Anal. Chim. Acta* 436, 49-57.
[121] Lynch, T. P., Kernoghan, N. J.& Wilson, J. N. (1984). *Analyst* 109, 839-842.
[122] Wang, Z., Xiao, X., Luo, Y. & Zhang, X. (2008). *Fenxi Shiyanshi* 27, 57-59.
[123] Shah, A. & Devi, S. (1990). *Anal. Chim. Acta* 236, 469-473.
[124] Prasada Rao, T., Karthikeyan, S., Vijayalekshmy, B. & Iyer, C.S.P. (1998). *Anal. Chim. Acta* 369, 69-77.
[125] Long, X., Miro, M. & Hansen, E.H. (2005). *Anal. Chem.* 77, 6032-6040.
[126] Anthemidis, A.N. & Koussoroplis, S.J.V. (2007). *Talanta* 71, 1728-1733.
[127] Hirata, S., Honda, K., Shikino, O., Maekawa, N. & Aihara, M. (2000). *Spectrochim. Acta* 55B, 1089-1099.

[128] Sun, Y.C., Lin, C.Y., Wu, S.F. & Chung, Y. T. (2006). *Spectrochim. Acta* 61B, 230-234.

[129] Zolotov, Y A., Shpigun, L.K., Kolotyrkina, I.Y., Novikov, E. A. & Bazanova, O.V. (1987). *Anal. Chim. Acta* 200, 21-33.

[130] Yamada, M. & Suzuki, S. (1987). *Anal. Chim. Acta* 193, 337-341.

[131] Sunda, W.G.& Huntsman, S. A. (1991). *Mar. Chem.* 36, 137-163.

[132] Coale, K.H., Johnson, K.S., Stout, P. M. & Sakamoto, C. M. (1992). *Anal. Chim. Acta* 266, 345-51.

[133] Zamzow, H., Coale, K. H., Johnson, K.S. & Sakamoto, C. M. (1998). *Anal. Chim. Acta* 377, 133-144.

[134] Achterberg, E.P., Braungardt, C.B., Sandford, R.C. & Worsfold, P. J. (2001). A*nal. Chim. Acta* 440, 27-36.

[135] Yuan, Y., Wang, P. & Wang, Y. (1991). *Fenxi Huaxue* 19, 468-470.

[136] Ma, L., Zhang, X., Miao, P. & Li, H. (2006). *Ziran Kexueban* 43, 638-641.

[137] Sung, Y.H., Liu, Z.S. & Huang, S.D. (1997). *J. Anal. At. Spectrom.* 12, 841-847.

[138] Ali, A.,Ye, Y., Xu, G. & Yin, X. (1999). *Fresenius' J. Anal. Chem.*, 365, 642-646.

[139] Gladis, J. M.; Biju, V. M.; Rao, T. P. (2002). *At. Spectrosc.* 23, 143-147.

[140] Cassella, R.J., Magalhaes, O.I.B., Couto, M.T, Lima, E.L.S., Neves, M.A.F.S. & Coutinho, F.M. B. (2005). *Talanta* 67, 121-128.

[141] Wu, D., Wang, A. & Guo, L. (2006). *Ziran Kexueban* 38, 80-83.

[142] Yuan, D., Wang, X., Yang, P., Huang, X. & Huang, B. (1991). *Ziran Kexueban* 30, 299-301.

[143] Yebra, M.C., Carro, N., Enriquez, M.F., Moreno-Cid, A. & Garcia, A. (2001). *Analyst* 126, 933-937.

[144] Yebra, M.C., Carro, N. & Moreno-Cid, A. (2002). *Spectrochim. Acta* 57B, 85-93.

[145] Zhuang, Z., Wang, X., Yang, P., Yang, C. & Huang, B. (1994). *Canadian J. Appl. Spectrosc.* 39, 101-107.

[146] Ivanova, E., Benkhedda, K. & Adams, F. (1998). *J. Anal. At. Spectrom.* 13, 527-531.

[147] Madrid, Y., Wu, M., Jin, Q. & Hieftje, G.M. (1993). *Anal. Chim. Acta* 277, 1-8.

[148] Lohan, M.C., Aguilar-Islas, A.M., Franks, R. P. & Bruland, K.W. (2005). *Anal. Chim. Acta* 530, 121-129.

[149] Chen, H.H. & Beauchemin, D. (2000). Canadian *J. Anal. Sci. Spectrosc.* 45, 84-89.

[150] O'Sullivan, D.W., Hanson, A.K. & Kester, D.R. (1995). *Mar. Chem.* 49, 65-77.

[151] Powell, R.T., King, D. W. & Landing, W.M. (1995). *Mar. Chem.* 50, 13-20.

[152] Bowie, A.R., Achterberg, E.P., Mantoura, R.F.C. & Worsfold, P.J. (1998). *Anal. Chim. Acta* 361, 189-200.

[153] de Jong, J.T.M., den Das, J., Bathmann, U., Stoll, M.H.C., Kattner, G., Nolting, R. F. & de Baar, H.J.W. (1998). *Anal. Chim. Acta* 377, 113-124.

[154] Croot, P.L. & Laan, P. (2002). *Anal. Chim. Acta* 466, 261-273.

[155] Bowie, A.R., Achterberg, E.P., Sedwick, P. N., Ussher, S. & Worsfold, P.J. (2002). *Environ. Sci. Technol.* 36, 4600-4607.

[156] Yaqoob, M., Ussher, S.J., Nabi, A., Achterberg, E.P. & Worsfold, P.J. (2003). *J. Flow Inject. Anal.* 20, 183-186.

[157] Bowie, A.R., Achterberg, E.P., Ussher, S. & Worsfold, P.J. (2005). *J. Autom. Methods Manag. Chem.* 2, 37-43.

[158] Ussher, S.J., Yaqoob, M., Achterberg, E.P., Nabi, A. & Worsfold, P. J. (2005). *Anal. Chem.* 77, 1971-1978.
[159] Lannuzel, D., de Jong, J., Schoemann, V., Trevena, A., Tison, J.L. & Chou, L. (2006). *Anal. Chim. Acta* 556, 476-483.
[160] Al-Gailani, B.R.M., Greenway, G. M. & Mccreedy, T. (2007). *Int. J. Environ. Anal. Chem.* 87, 637-646.
[161] Roy, E.G., Wells, M.L. & King, D. W. (2008). *Limnol. Oceanogr.* 53, 89-98.
[162] Elrod, V.A., Johnson, K.S. & Coale, K.H. (1991). *Anal.Chem.* 63, 893-898.
[163] Hirata, S., Yoshihara, H. & Aihara, M. (1999). *Talanta*, 49, 1059-1067.
[164] Kolotyrkina, I.Y., Shpigun, L. K., Zolotov, Y.A. & Malahoff, A. (1995). *Analyst* 120, 201-206.
[165] Measures, C. I., Yuan, J. & Resing, J. A. (1995). *Mar. Chem.* 50, 3-12.
[166] Weeks, D. A. & Bruland, K.W. (2002). *Anal. Chim.Acta* 453, 21-32.
[167] Laes, A., Vuillemin, R., Leilde, B., Sarthou, G., Bournot-Marec, C. & Blain, S. (2005). *Mar. Chem.* 97, 347-356.
[168] Lohan, M.C., Aguilar-Islas, A. M. & Bruland, K.W. (2006). *Limnol. Oceanogr. Methods* 4, 164-171.
[169] Planquette, H., Statham, P. J., Fones, G.R., Charette, M. A., Moore, C. M., Salter, I., Nedelec, F. H., Taylor, S. L., French, M., Baker, A. R., Mahowald, N. & Jickells, T. D. (2007). *Deep-Sea Research,* 54, 1999-2019.
[170] Carneiro, J.M.T., Dias, A.C.B., Zagatto, E.A.G. & Honorato, R. S. (2002). *Anal. Chim.Acta* 455, 327-333.
[171] Lopes da Conceicao, A. C., Tena, M. T., Correia dos Santos, M. M., Simoes Goncalves, M. L. & Luque de Castro, M. D. (1997). *Anal. Chim. Acta* 343, 191-197.
[172] Pons, C., Forteza, R. & Cerda, V. (2005). *Anal. Chim. Acta* 528, 197-203.
[173] Adams, M. L.& Powell, K. J. (2001). *Anal. Chim. Acta* 433, 289-297.
[174] Asan, A., Andac, M. & Isildak, I. (2003). *Anal. Sci.* 19, 1033-1036.
[175] Sarradin, P. M., Le Bris, N., Le Gall, C. & Rodier, P. (2005). *Talanta*, 66, 1131-1138.
[176] Asan, A., Andac, M., Isildak, I. & Tinkilic, N. (2008). *Chemical Papers* 62, 345-349.
[177] Andersen, J.E.T. (2005). *Analyst* 130, 385-390.
[178] Bowie, A. R., Ussher, S. J., Landing, W. M. & Worsfold, P. J. (2007). *Environ. Chem.* 4, 1-4.
[179] Richter, E.M., Augelli, M.A., Kume, G.H., Mioshi, R.N. & Angnes, L. (2000). *Fresenius J. Anal. Chem.* 366, 444-448.
[180] Bloxham, M.J., Hill, S.J. & Worsfold, P.J. (1996). *J. Anal. At. Spectrom.* 11, 511-514.
[181] Tseng, C.M., Balcom, P.H., Lamborg, C.H. & Fitzgerald, W.F. (2003). *Environ. Sci. Technol.* 37, 1183-1188.
[182] Leopold K., Harwardt, L., Schuster, M. & Schlemmer, G. (2008). *Talanta* 76, 382-388.
[183] Bermejo-Barrera, P., Moreda-Pineiro, J., Moreda-Pineiro, A. & Bermejo-Barrera, A. (1997). *J. Anal. At. Spectrom.* 12, 317-321.
[184] Ma, H.B., Fang, Z.L., Wu, J.F. & Liu, S.S. (1999). *Talanta* 49, 125-133.
[185] Rio-Segade, S. & Bendicho, C. (1999). *Spectrochim. Acta* 54B, 1129-1139.
[186] Wann, J.K. & Hung, J.J. (2000). *Huanjing Baohu* 23, 125-138.
[187] Capelo, J. L., Lavilla, I. & Bendicho, C. (2000). *Anal. Chem.* 72, 4979-4984.
[188] Bravo-Sanchez, L.R., San Vicente de la Riva, B., Costa-Fernandez, J.M., Pereiro, R. & Sanz-Medel, A. (2001). *Talanta* 55, 1071-1078.

[189] Ivanova, P. & Havezov, I. (2003). *Bulgarian Chemical Communications* 35, 100-103.
[190] Fernandez Garcia, M., Pereiro Garcia, R., Bordel Garcia, N. & Sanz-Medel, A. (1994). *Talanta* 41, 1833-1839.
[191] Menéndez-García, A., Fernández-Sánchez, M.L., Sánchez-Uría, J.E. & Sanz-Medel, A. (1996). *Mikrochim. Acta* 122, 157-166.
[192] Qvarnstrom, J., Tu, Q., Frech, W. & Ludke, C. (2000). *Analyst* 125, 1193-1197.
[193] Sanz, J., Raposo, J.C., Larreta, J., Martinez-Arkarazo, I., de Diego, A. & Madariaga, J. M. (2004). *J. Sep. Sci.* 27, 1202-1210.
[194] Cañada-Rudner, P., Cano-Pavon, J.M.; Sánchez-Rojas, F. & Garcia de Torres, A. (1998). *J. Anal. At. Spectrom.* 13, 1167-1171.
[195] Zara, L.F.; Rocha, J.C., Milani, M. R., Rosa, A. H., dos Santos, A. & Serudo, R. L. (2003). *Anal. Lett.* 36, 781-795.
[196] Cañada-Rudner, P., Garcia de Torres, A., Cano Pavon, J. M. & Rodriguez Castellon, E. (1998). *J. Anal. At. Spectrom.* 13, 243-248.
[197] Wei, M.T. & Jiang, S.J.. (1999). *J. Chinese Chem. Soc.* 46, 871-878.
[198] Bravo-Sanchez, L.R., Ruiz Encinar, J., Fidalgo Martinez, J.I. & Sanz-Medel, A. (2004). *Spectrochim. Acta* 59B, 59-66.
[199] Alibo, D.S., Amakawa, H. & Nozaki, Y. (1998). *Indian Academy of Sciences, Earth and Planetary Sciences* 107, 359-366.
[200] Murakami, M. & Furuta, N. (2006). *Anal. Chim. Acta* 556, 423-429.
[201] Chapin, T.P., Johnson, K.S. & Coale, K.H. (1991). *Anal. Chim. Acta* 249, 469-478.
[202] von Langen, P. J., Johnson, K.S., Coale, K.H. & Elrod, V.A. (1997). *Geochim. Cosmochim. Acta* 61, 4945-4954.
[203] Olafsson, J. (1986). *Sci.Total Environ.* 49, 101-113.
[204] Resing, J.A. & Mottl, M.J. (1992). *Anal. Chem.* 64, 2682-2687.
[205] Aguilar-Islas, A.M., Resing, J.A. & Bruland, K.W. (2006). *Limnol. Oceanogr. Methods* 4, 105-113.
[206] Nakano, S., Tanaka, K., Oki, R. & Kawashima, T. (1999). *Talanta* 49, 1077-1082.
[207] Su, L., Li, J., Ma, H. & Tao, G. (2004). *Anal. Chim. Acta* 522, 281-288.
[208] Kolotyrkina, I. Y., Shpigun, L.K. & Zolotov, Y.A. (1988). *Zhurnal Analiticheskoi Khimii* 43, 284-288.
[209] Kolotyrkina, I.Y., Shpigun, L.K., Zolotov, Y.A. & Tsysin, G.I. (1991). *Analyst* 116, 707-710.
[210] Shpigun, L.K., Kolotyrkina, I.Y. & Zolotov, Y.A. (1992). *Anal. Chim. Acta* 261, 307-314.
[211] Mallini, L.J. & Shiller, A.M. (1993). *Limnol. Oceanogr.* 38, 1290-1295.
[212] Chin, C.S., Johnson, K.S. & Coale, K.H. (1992). *Mar. Chem.* 37, 65-82.
[213] Sarzanini, C., Abollino, O. & Mentasti, E. (2001). *Anal. Chim.Acta* 435, 343-350.
[214] Iyer, C.S.P., Rao, T.P., Kartikeyan, T. & Damodaran, A.D. (1994). *At. Spectrosc.* 15, 234-236.
[215]Yin, X.F. & Liu, M. (1994). *Gaodeng Xuexiao Huaxue Xuebao* 15, 1766-1769.
[216] Chen, H., Xu, S. & Fang, Z. (1995). *J. Anal. At. Spectrom.* 10, 533-537.
[217] Furuta, N., Brushwyler, K.R. & Hieftje, G.M. (1989). *Spectrochim. Acta* 44, 349-358.
[218] Oms, M.T., Cerda, A. & Cerda, V. (1996). *Electroanalysis* 8, 387-390.
[219] Hall, P.O. & Aller, R.C. (1992). *Limnol. Oceanogr.* 37, 1113-1119.
[220] Masserini, R.T. & Fanning, K.A. (2000). *Mar. Chem.* 68, 323-333.

[221] Aminot, A., Kerouel, R. & Birot, D. (2001). *Water Res.* 35, 1777-1785
[222] Watson, R. J., Butler, E.C.V., Clementson, L.A. & Berry, K. M. (2005). *J. Environ. Monitor.* 7, 37-42.
[223] Amornthammarong, N. & Zhang, J.Z. (2008). *Anal. Chem.* 80, 1019-1026.
[224] Yamane, T. & Saito, M. (1995). *Nippon Kaisui Gakkaishi* 49, 318-320.
[225] Muraki, H., Higuchi, K., Sasaki, M., Korenaga, T. & Toei, K. (1992). *Anal. Chim. Acta* 261, 345-349.
[226] Tovar, A., Moreno, C., Manuel-Vez, M.P. & Garcia-Vargas, M. (2001). *Quim. Anal.* 20, 37-45.
[227] Willason, S.W. & Johnson, K.S. (1986). *Mar. Biol.* 91, 285-290.
[228] Zhang, Y.S., Zou, X.C., Zhang, M.P. & Ren, L. (2002). *Haiyang Huanjing Kexue* 21, 57-59.
[229] Fukui, K., Ohno, S., Higuchi, K., Teshima, N. & Sakai, T. (2007). *Bunseki Kagaku* 56, 757-763.
[230] Lu, X., Kang, D. & Hong, W. (1991). *Haiyang Yu Huzhao* 22, 389-395.
[231] Tovar, A., Moreno, C., Manuel-Vez, M.P. & Garcia-Vargas, M. (2002). *Anal. Chim. Acta* 469, 235-242.
[232] Zhang, W., Zhang, X., Miao, P. & Yuan, D. (2006). *Pige Kexue Yu Gongcheng* 16, 29-31.
[233] Ali, A., Ye, Y., Xu, G., Yin, X. & Zhang, T. (1999). *Microchem. J.* 63, 365-373.
[234] Aldstadt, J., King, D.F. & Dewald, H.D. (1994). *Analyst* 119, 1813-1818.
[235] San Vicente de la Riva, B., Costa-Fernandez, J.M., Pereiro, R. & Sanz-Medel, A. (1999). *Anal. Chim. Acta* 395, 1-9.
[236] Cheng, X.S., Yang, Y., Qin, X.G., Zhang, Y. & Zhang, L.X. (2005). *Guangpu Shiyanshi* 22, 1245-1248.
[237] Novikov, E.A., Shpigun, L.K. & Zolotov, Y.A. (1990). *Anal. Chim. Acta* 230, 157-162.
[238] Zou, Y., Zhang, X., Li, H., Tu, J. & Yuan, D. (2006). *Fenxi Huaxue* 34, 1471-1474.
[239] Sun, X. (1997). *Fenxi Kexue Xuebao* 13, 34-37.
[240] Hernandez Tilves, M.C., de la Fe Napoles, C. & Nodarse, I C. (2003). *Afinidad* 60, 525-528.
[241] Yebra, M.C., Carro, N., Moreno-Cid, A. & Puig, L. (2001). *Field Anal. Chem. Technol.* 5, 239-243.
[242] Yebra, M.C., Rodriguez, L., Puig, L. & Moreno-Cid, A. (2002). *Mikrochim. Acta* 140, 219-225.
[243] Di Pietro Roux, K.C., Maltez, H. F., Carletto, J. S., Martendal, E. & Carasek, E. (2008). *Anal. Sci.* 24, 365-370.
[244] Fang, Z., Sperling, M. & Welz, B. (1990). *J. Anal. At. Spectrom.* 5, 639-646.
[245] Nakajima, J., Hirano, Y. & Oguma, K. (2003). *Anal. Sci.* 19, 585-588.
[246] Madrid, Y., Chakraborti, D. & Camara, C. (1995). *Mikrochim. Acta* 120, 63-72.
[247] Ren, J., Wang, H., Zhu, L. & Zhang, G. (1996). *Haihuyan Yu Huagong* 25, 14-15.
[248] Wang, S.Z., Liu, Y.C. & Song, Y.R. (2003). *Guangpu Shiyanshi* 20, 513-516.
[249] Ke, S.H., Huang, L.S., Huang, J.S. & Lin, KC. (2001). *Appl. Spectrosc.* 55, 604-610.
[250] Seki, T., Hirano, Y. & Oguma, K. (2002). *Anal. Sci.* 18, 351-354.
[251] Huang, Z.Y., Chen, F.R., ZhuZ.Z., Wang, X.R. & Leeb, F.S.C. (2003). *Fenxi Kexue Xuebao* 19, 301-304.

[252] Huang, Z.Y., Chen, F.R., Zhuang, Z.X., Wang, X.R. & Lee, F.S.C. (2004). *Anal. Chim. Acta* 508, 239-245.

[253] Wang, Z., Jing, M., Lee, F S.C. & Wang, X. (2006). F*enxi Huaxue* 34, 459-463.

[254] Ke, C.B. & Lin, K.C. (1999). *Anal. Chem.* 71, 1561-1567.

[255] Vincente, O., Padro, A., Martinez, L., Olsina, R. & Marchevsky, E. (1998). *Spectrochim. Acta* 53B, 1281-1287.

[256] Eroglu, A.E., McLeod, C.W., Leonard, K.S. & McCubbin, D. (1998). *Spectrochim. Acta* 53B, 1221-1233.

[257] Eroglu, A.E., McLeod, D., Leonard, K.S. & McCubbin, D. (1997). *Plasma Source Mass Spectrometry* 202, 166-174.

[258] Zhu, Z. & Lu, J. (1995). *Ziran Kexueban* 23, 59-62.

[259] Yang, L.J., Du, J.X., Li, J.J. & Lu, J.R. (2007). *Huaxue Fence* 43, 348-353.

[260] Economou, A., Fielden, P.R. & Packham, A. J. (1994). Analyst 119, 279-285.

[261] Dadfarnia, S. & McLeod, C.W. (1994). *Appl. Spectrosc.* 48, 1331-1336.

[262] Karpas, Z., Lorber, A., Halicz, L. & Gavrieli, I. (1998). Determination of uranium in clinical and environmental samples by FIAS-ICPMS. In R.W. Morrow & J.S. Crain (Eds.), *Applications of inductively coupled plasma-mass spectrometry to radionuclide determinations* (89-98). West Conshohocken, PA, USA: ASTM International.

[263] Halicz, L., Segal, I., Gavrieli, I., Lorber, A. & Karpas, Z. (2000). *Anal. Chim.Acta* 422, 203-208.

[264] Oshita, K., Oshima, M., Gao, Y., Lee, K.H. & Motomizu, S. (2002). *J. Flow Injection Anal.* 19, 143-146.

[265] Oshita, K., Takayanagi, T., Oshima, M. & Motomizu, S. (2008). *J. Flow Injection Anal.* 25, 73-76.

[266] Jimenez, M.S., Velarte, R. & Castillo, J.R. (2000). *At. Spectrosc.* 21, 156-164.

[267] Wu, H., Jin, Y., Tian, Y., Tao, X. & Bi, S. (2007). *At. Spectrosc.* 28, 220-226.

[268] De la Calle Guntinas, M.B., Madrid, Y. & Camara, C. (1991). *Anal. Chim. Acta* 252, 161-166.

[269] Yan, X.P., Van Mol, W. & Adams, F. (1996). *Analyst* 121, 1061-1067.

[270] Ding, W.W. & Sturgeon, R. E. (1996). *J. Anal. At. Spectrom.* 11, 225-230.

[271] Cabon, J.Y. & Madec, C.L. (2004). *Anal. Chim. Acta* 504, 209-215.

[272] Rodriguez, Y. & Tyson, J.F. (2006). *J. Anal. At. Spectrom.* 21, 757-762.

[273] Zheng, F.Y., Qian, S.H., Li, S.X., Huang, X.Q. & Lin, L.X. (2006). *Anal. Sci.* 22, 1319-1322.

[274] Ye, R. & Khoo, S.B. (2001). *Anal. Sci.* 17, a113-a116.

[275] He, Y., El Azouzi, H., Cervera, M.L. & de la Guardia, M. (1998). *J. Anal. At. Spectrom.* 13, 1291-1296.

[276] Gong, Z., Zhang, X., Chen, G. & Xiao, X. (2005). *Talanta* 66, 1012-1017.

[277] Cobo Fernandez, M.G., Palacios, M.A. & Camara, C. (1993). *Anal. Chim. Acta* 283, 386-392.

[278] Tao, G. & Hansen, E.H. (1994). *Analyst* 119, 333-337.

[279] Cabon, J. Y. & Erler, W. (1998). *Analyst* 123, 1565-1569.

[280] Yan, X.P., Sperling, M. & Welz, B. (1999). *Anal. Chem.* 71, 4353-4360.

[281] Olivas, R.M., Quetel, C.R. & Donard, O.F.X. (1995). *J. Anal. At. Spectrom.* 10, 865-870.

[282] Bermejo-Barrera, P., Ferron-Novais, M., Gonzalez-Campos, G. & Bermejo-Barrera, A. (1999). *At. Spectrosc.* 202, 159-165.
[283] Woods, G.D., McLeod, C.W., Kawabata, K., Nagaoka, S., Okochi, H. & Otsuki, A. (1997). *Plasma Source Mass Spectrometry* 202, 166-174.
[284] Khoo, S.B. & Ye, R. (2000). *Analyst* 125, 895-902.
[285] Yang, K.L., Jiang, S.J. & Hwang, T.J. (1996). *J. Anal. At. Spectrom.* 11, 139-143.
[286] Wei, M.T. & Jiang, S.J. (1990). *J. Anal. At. Spectrom* 14, 1177-1181.
[287] Vassileva, E. & Furuta, N. (2001). *Fresenius' J. Anal. Chem.* 370, 52-59.
[288] Shiller, A.M., Mao, L. & Cramer, J. (1998). *Limnol. Oceanogr.* 43, 526-529.
[289] Nowicki, J.L., Johnson, K.S., Coale, K.H., Elrod, V.A. & Lieberman, S.H. (1994). *Anal. Chem.* 66, 2732-2738.
[290] Long, F., Zhang, X., Zou, Y., Xia, X., Jiang, R. & Zhang, W. (2006). *Pige Kexue Yu Gongcheng* 16, 44-46.
[291] Hirata, S. & Honda, K. (1987). *Bunseki Kagaku* 36, 213-216.
[292] Yebra-Biurrun, M.C. & Cespón-Romero, R. M. (2006). *Anal. Sci.* 22, 51-55.
[293] Yuan, D., Wang, X., Yang, P. & Huang, B. (1992). *Fenxi Huaxue* 20, 162-164.
[294] Zougagh, M., Cañada-Rudner, P., Garcia de Torres, A. & Cano Pavon, J.M. (2000). *J. Anal. At. Spectrom.* 15, 1589-1594.
[295] Hwang, T.J. & Jiang, S.J. (1997). *Analyst* 122, 233-237.
[296] Jardim, W.F., Guimaraes, J.R. & Allen, H.E. (1991). *Ciencia e Cultura* 43, 454-456.
[297] Lu, M., Zhang, L., Wang, B. & Zhang, J. (2002). *Qingdao Haiyang Daxue Xuebao* 32, 94-100.
[298] Wei, Y. & Lin, J. (2005). *Chemical Journal on Internet* 7. http://www.chemistrymag.org/cji/2005/07c085nc.htm
[299] Bellerby, R.G.J., Turner, D.R., Millward, G.E. & Worsfold, P.J.(1995). *Anal. Chim. Acta* 309, 259-270.
[300] Bellerby, R.G.J., Goyet, C.& Eischeid, G. (1996). High frequency shipboard determination of sea water pH using flow injection analysis with photodiode array and charge coupled detection. *Book of Abstracts, 211th ACS National Meeting*. Washington, D.C.: American Chemical Society.
[301] Ilcheva, L. & Yanakiev, R. (1990). *Izvestiya po Khimiya* 23, 167-171.
[302] Ilcheva, L., Yanakiev, R. & Cammann, K. (1991). *J. Indian Chem.Soc.* 68, 154-157.
[303] Wang, E. & Kamata, S. (1992). *Anal. Chim. Acta* 261, 399-404.
[304] Dan, D., Wan, Y. & Du, G. (1991). *Fenxi Shiyanshi* 10, 59-61.
[305] Lolic, A., Nikolic-Mandic, S. & Polic, P. (2001). *J. Serbian Chem. Soc.* 66, 637-646.
[306] Anagnostopoulou, P.I. & Koupparis, M.A. (1986). *Anal. Chem.* 58, 322-326.
[307] Borges, E.P., Lavorante, A.F & Freire dos Reis, B. (2005). *Anal. Chim. Acta* 528, 115-119.
[308] Uraisin, K., Nacapricha, D., Lapanantnoppakhun, S., Grudpan, K. & Motomizu, S. (2005). *Talanta* 68, 274-280
[309] Ruzicka, J., Stewart, J.W.B. & Zagatto, E.A. (1976). *Anal.Chim. Acta* 81, 387-396.
[310] Moreno, C., Galindo-Riano, M.D., Roman, M.P. & Garcia-Vargas, M. (2006). Eurasian J. Anal. Chem. 1. http://www.eurasianjournals.org/ejac/Volume1(1)/V1_(1)1.pdf.
[311] Nishioka, H., Sai, N., Kumagai, T., Nagahiro, T. & Uesugi, K.. (1994). *Nippon Kaisui* Gakkaishi 48, 10-14.

[312] Leon-Gonzalez, M.E., Santos-Delgado, M.J. & Polo-Diez, L.M. (1989). *Anal. Chim. Acta* 219, 329-333.
[313] Oguma, K., Kitada, K. & Kuroda, R. (1993). *Mikrochim. Acta* 110, 71-77.
[314] Hakedal, J.T. & Egeberg, P.K. (1997). *Analyst* 122, 1235-1237.
[315] Kuznetsov, V.V., Ermolenko, Y.V. & Seffar, L. (2004). *J. Anal. Chem.* 59, 688-693.
[316] Kuznetsov, V.V., Ermolenko, Y.V. & Seffar, L. (2007). *J. Anal. Chem.* 62, 479-485.
[317] Yebra, M.C. & Cespón, R.M. (2000). *Fresenius' J. Anal. Chem.* 367, 24-28.
[318] Sakai, T., Takio, H.,Teshima, N. & Nishikawa, H. (2001). *Anal. Chim. Acta* 438, 117-121
[319] Amouroux, D. & Donard, O.F.X. (1995). *Oceanol. Acta* 18, 353-361.
[320] Price, D., Worsfold, P.J. & Mantoura, R.F.C. (1994). *Anal. Chim. Acta* 298, 121-128.
[321] Price, D., Mantoura, R.F.C. & Worsfold, P.J. (1998). *Anal. Chim. Acta* 371, 205-215.
[322] Price, D., Mantoura, R.F.C. & Worsfold, P.J. (1998). *Anal. Chim. Acta* 377, 145-155.
[323] Yuan, Jinchun; Shiller, Alan M. (1999). Anal. Chem. 71, 1975-1980.
[324] Morgan, C.R., Burr, A., Theberge, S.M. & King, D.W. (2001). Determination of hydrogen peroxide in ocean water using a chemiluminescent flow-injection analysis system. *Book of Abstracts, 211th ACS National Meeting*. Washington, D.C.: American Chemical Society.
[325] King, D.W., Cooper, W.J., Rusak, S.A., Peake, B.M., Kiddle, J.J., O'Sullivan, D.W., Melamed, M.L., Morgan, C.R. & Theberge, S.M. (2007). *Anal. Chem.* 79, 4169-4176.
[326] Johnson, K.S., Sakamoto-Arnold, C.M., Willason, S.W. & Beehler, C.L. (1987). *Anal. Chim. Acta* 201, 83-94.
[327] Motomizu, S., Mikasa, H. & Toei, K. (1986). *Talanta* 33, 729-732.
[328] Johnson, K.S. & Petty, R.L. (1983). *Limnol. Oceanogr.* 28, 1260-1266.
[329] Pai, S.C.,Yang, C.C. & Riley, J.P. (1990). *Anal. Chim.Acta* 232, 345-349.
[330] Ariza, A.C., Linares, P., Luque de Castro, M.D. & Valcarcel, M. (1992). *J. Autom. Chem.*14, 181-183.
[331] Takeda, K. & Fujiwara, K. (1993). *Anal. Chim. Acta* 276, 25-32.
[332] McCormack, T., David, A.R.J., Worsfold, P.J. & Howland, R. (1994). *Anal. Proc.* 31, 81-83.
[333] Pai. S.C. & Riley, J.P. (1994). *Int. J. Environ. Anal. Chem.* 57, 263-277.
[334] Daniel, A., Birot, D., Lehaitre, M. & Poncin, J. (1995). *Anal.Chim. Acta* 308, 413-424.
[335] McCormack, T., David, A.R.J. & Worsfold, P.J. (1995). Flow injection procedures with spectrophotometric detection for the determination of nitrate and nitrite in riverine, estuarine, and coastal waters. In Andrews, D.L. & Davies, A.M.C. (Eds.), *Frontiers in Analytical Spectroscopy* (pp. 129-135). Cambridge, U.K.: Royal Society of Chemistry.
[336] Cerdà, A., Oms, M.T., Forteza, R. & Cerdà, V. (1998). *Anal.Chim. Acta* 371, 63-71.
[337] Ma, L., Oshima, M., Motomizu, S. & Hattori, T. (1998). *Bunseki Kagaku* 47, 375-380.
[338] Teshima, N., Sakai, T., Yasuda, S. & Higuchi, K. (2000).*Bunseki Kagaku* 49, 455-462
[339] Kazemzadeh, A. & Ensafi, A.A. (2001). *Microchem. J.* 69, 159-166.
[340] Daniel, A., Birot, D., Blain, S., Treguer, P., Leielde, B. & Menut, E. (1995). *Mar. Chem.* 51, 67-77.
[341] David, A.R.J., McCormack, T., Morris, A.W. & Worsfold, P.J. (1998). *Anal. Chim. Acta* 361, 63-72.
[342] David, A.R.J., McCormack, T. & Worsfold, P.J. (1999). *J. Autom. Methods Manag. Chem.* 21, 1-9.

[343] Gardolinski, P.C.F.C., David, A.R.J. & Worsfold, P.J. (2002). *Talanta* 58, 1015-1027.
[344] Liang, Y., Yuan, D., Li, Q. & Lin, Q. (2006). *Anal. Chim.Acta* 571, 184-190.
[345] Motomizu, S., Oshima, M. & Katsumura, N. (1995). *Anal. Sci. Technol.* 8, 843-848.
[346] Frank, C., Schroeder, F., Ebinghaus, R. & Ruck, W. (2006). *Microchim. Acta* 154, 31-38.
[347] Johnson, K.S. & Petty, R.L. (1982). *Anal. Chem.* 54, 1185-1187.
[348] Aoyagi, M., Yasumasa, Y. & Nishida, A. (1988). *Anal. Chim. Acta* 214, 229-237.
[349] Kang, D., Chen, L. & Lu, X. (1991). *Haiyang Xuebao* 13, 60-67.
[350] Susanto, J. P., Oshima, M., Motomizu, S., Mikasa, H. & Hori, Y. (1995). *Analyst* 120, 187-191.
[351] Tovar, A., Moreno, C., Manuel-Vez, M.P. & Garcia-Vargas, M. (2002). *Spectrosc. Lett.* 35, 715-728.
[352] Worsfold, P, Achterberg, E., Bowie, A., Gardolinski, P., Gledhill, M., Hanrahan, G., McKelvie, I. & Ussher, S. (2003). *Chiang Mai Journal of Science* 30, 153-164.
[353] Karthikeyan, S., Hashigaya, S., Kajiya, T. & Hirata, S. (2004). *Anal. Bioanal. Chem.* 378, 1842-1846.
[354] Wang, P., Zhang, X.S. & Wu, J.M. (2005). *Pige Kexue Yu Gongcheng* 15, 18-21
[355] Liang, Y., Yuan, D., Li, Q. & Lin, Q. (2007). *Mar. Chem.* 103, 122-130.
[356] Omaka, O.N., Keith-Roach, M.J. & Worsfold, P. J. (2007). *J. Chem. Soc. Nigeria* 32, 143-149.
[357] Omaka, O.N., Keith-Roach, M., McKelvie, I.D. & Worsfold, P.J. (2008). *Int. J. Environ. Anal. Chem.* 88, 91-101.
[358] Auflitsch, S., Peat, D.M.W., Worsfold, P.J., Auflitsch, S. & McKelvie, I.D. (1997). *Analyst* 122, 1477-1480.
[359] McKelvie, I.D., Peat, D.M.W., Matthews, G.P. & Worsfold, P.J. (1997). *Anal. Chim. Acta* 351, 265-271.
[360] Ellis, P.S., Lyddy-Meaney, A.J., Worsfold, P.J. & McKelvie, I.D. (2003). *Anal. Chim. Acta* 499, 81-89.
[361] Ma, J., Yuan, D., Liang, Y. & Dai, M. (2008). *J. Oceanogr.* 64, 443-449.
[362] Sakuragawa, A., Nakayama, S. & Okutani, T. (1994). *Anal. Sci.* 10, 77-81.
[363] Johnson, K.S., Beehler, C.L. & Sakamoto-Arnold, C.M. (1986). *Anal. Chim. Acta* 179 245-257.
[364] Sakamoto-Arnold, C.M.,Johnson, K.S. & Beehler, C.L. (1986). *Limnol. Oceanogr.* 31, 894-900.
[365] Cassella, R.J. & Santelli, R. E. (1995). *Quimica Nova*, 18, 536-540.
[366] Sarradin, P.M., Le Bris, N., Caprais, J.C. & Birot, D. (1999). *Anal. Commun.* 36, 157-160.
[367] Ferrer, L., de Armas, G., Miró, M., Estela, J.M. & Cerdà, V. (2004). *Talanta* 64, 1119-1126.
[368] Ferrer, L., de Armas, G., Miró, M., Estela, J.M. & Cerdà, V. (2005). *Analyst* 130, 644-651.
[369] Ferrer, L., de Armas, G., Miró, M., Estela, J.M. & Cerdà, V. (2005). *Talanta* 68, 343-350.
[370] Thomsen, J., Johnson, K. S. & Petty, R.L. (1983). *Anal. Chem.* 55, 2378-2382.

[371] Yokoyama, T., Hirai, Y., Yoza, N., Tarutani, T. & Ohashi, S. (1982). *Bull. Chem. Soc. Japan* 55, 3477-3481.
[372] Floch, J., Blain, S., Birot, D. & Treguer, P. (1998). Anal. Chim. Acta 377, 157-166.

Chapter 5

MARINE AND ESTUARINE SEDIMENTS

ABSTRACT

In this chapter, the state-of-the-art of flow injection (FI) methodologies proposed for the determination of organic and inorganic analytes (cationic and anionic species) in sea and estuarine sediment samples is presented and discussed. Thus, are described FI methods applied to the determination of bio and molecular markers, dissolved organic carbon (in sediment pore waters), polycyclic aromatic hydrocarbons, silver, aluminium, arsenic, bismuth, cadmium, cobalt, chromium, copper, iron, germanium, mercury, iridium, magnesium, manganese, nickel, lead, platinum, rare earths, plutonium, thorium, uranium, rhenium, antimony, selenium, tin, tellurium, zinc, carbonate, sulfide and silicate. Analytical figures of merit, characteristics, features and interferences are also discussed for each analyte.

INTRODUCTION

Recently, a great progress has been made with analytical methods of solid samples, which do not involve dissolving the samples using instruments for analysis such as emission spectroscopic analysis and fluorescent X-ray analysis. Nevertheless, common analytical instrumentation such as electrochemical instrumentation, gas or liquid chromatography, UV-VIS spectrophotometry, atomic absorption spectrophotometry (AAS), inductively coupled plasma optical emission spectrometers (ICP-OES), inductively coupled plasma mass spectrometers (ICP-MS), etc. usually require that the analytes are found in a liquid phase to facilitate their introduction [1]. Although elemental analysis of a solid material by AAS, ICP-OES and ICP-MS can be carried out directly by direct introduction of slurries or through laser ablation. However, in these cases the fast sample treatment is compensated for by an increase of calibration requirements, which eventually sacrifices some of the advantages of these direct solid introduction techniques. For this, a common step including in the preparation of a solid sample is the dissolution of the entire sample or the extraction of the target analyte(s).

The sample preparation for the determination of organic substances in sediment samples includes analyte extraction. Thus, sediment samples are freeze dried before analyte(s) extraction, and extraction is performed with an organic solvent or a mixture of organic solvents manually by using a separating funnel, with an automated solvent extractor or with the aid of ultrasound energy. After phase separation, the organic layer is filtered through a

Wathman No 4 filter-paper. To remove salt, the extracts are washed with double-distilled water, then the volume of the organic phase is reduced by rotary evaporation and the concentrated extract treated with anhydrous sodium sulfate to remove any water. The residue is dissolved in a suitable organic solvent. These analytical operations are carried out in off-line mode, and FI is applied for automatic sample introduction into the analytical instrument and/or to perform on-line reactions involving enzyme reactors (Figure 5.1). Supercritical fluid extraction (SFE) is an effective alternative to conventional liquid-liquid extraction for the isolation of analytes from a variety of matrices. The combined liquid-liquid solvating capabilities and gas-like transport properties of supercritical fluids afford efficient, rapid extraction of analytes, thus considerably simplifying the analytical process. This separation technique is the unique integrated within a FI system for the extraction of organic substances (polycyclic aromatic hydrocarbons) from sediment samples.

Most of the existing procedures for preparing siliceous samples as sediments for analysis of cationic species involve dry ashing or wet oxidation, followed by an acid digestion of the remaining ash. Wet digestions with nitric acid and perchloric acid, or the use of oxygen bombs are examples of such procedures. Thus, wet digestion methods have been increasingly succeeded by microwave assisted digestion methods using closed vessels. Nevertheless, these sample preparation procedures have numerous drawbacks because they are: very time-consuming (requiring nearly three hours), labor-intensive (multiple interventions are required for reagent additions to the individual samples), and if the preparation occurs in an open environment, this gives greater opportunity for sample contamination and/or analyte losses. These drawbacks are even more severe when very low amount of a volatile element is to be determined as for instance mercury. A means of overcoming these drawbacks in sample preparation of sediment samples is to use on-line microwave digestion. This sample preparation methodology can greatly reduce the time required for a digestion by superheating the sample and digestion reagents, thus requiring less than a minute for a sample digestion and reduces analyst intervention and analytical errors. Due to the on-line nature of the digestion, additional reagents can be added in an automated manner, reducing the effort required of the analyst within the preparation procedure. Furthermore, on-line microwave digestion coupled to a FIA manifold or to an automatic analyzer provides a completely automated analytical system. Ultrasound-assisted leaching is an alternative and effective way of extracting cationic species from sediment samples. The influence of extremely high effective temperature and pressure due to the collapse of gas or vapor bubbles (which result in increased solubility and diffusivity that favor penetration and transport at the interface of an aqueous or organic phase subjected to ultrasonic energy and a solid matrix) combined with the oxidative energy of radicals created during sonolysis of the solvent (hydroxyl and hydrogen peroxide of water) may result in high extraction power using ultrasound. Nevertheless, this sample preparation method only has been proposed in an off-line mode for sediments samples.

Flow injection (FI) on-line matrix removal and analyte preconcentration has been the most active area in the field of flow injection analysis, especially for atomic absorption spectrometry determinations. Because this analytical technique is insensitive (FAAS) or nonselective due to pronounced matrix interference (ETAAS). Hence, preconcentration coupled to AAS is often advocated, as it offers higher sensitivity, selectivity, better precision and accuracy. Conventional off-line preconcentration procedures, although effective, are

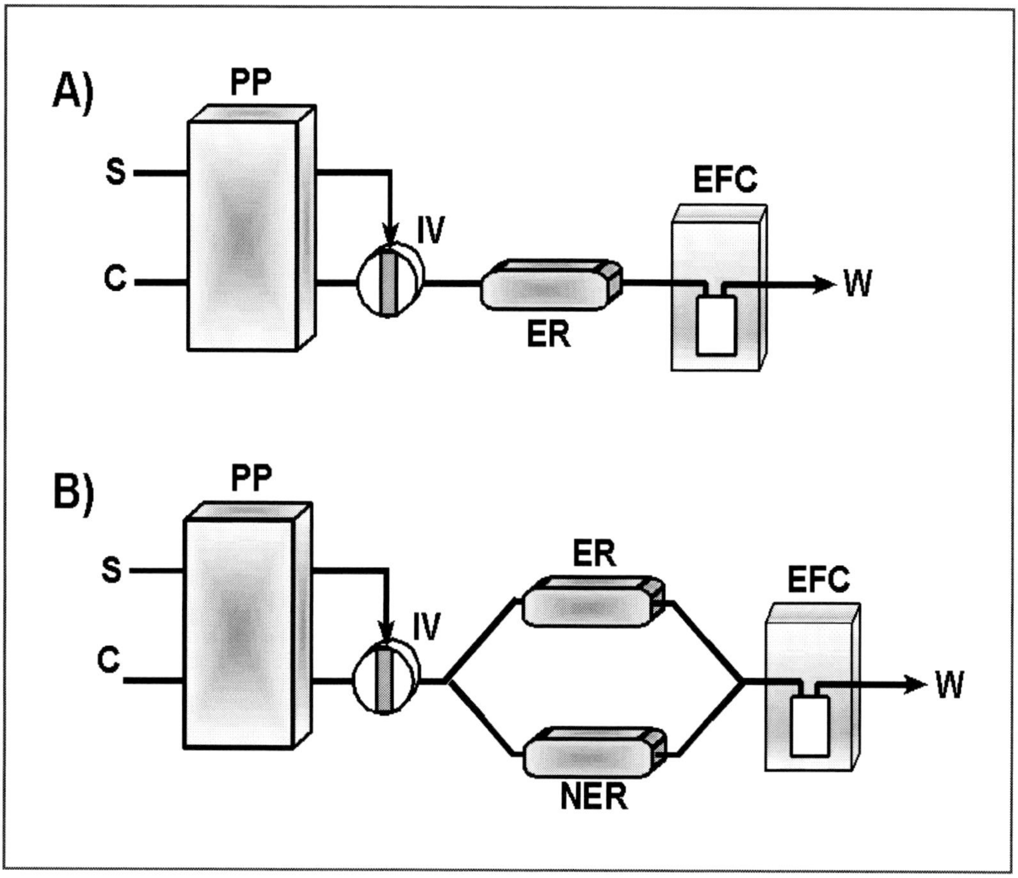

Figure 5.1. Schematic manifolds proposed for the determination of organic substances in sediment samples involving an enzyme reactor. A) manifold only with an enzyme reactor. B) manifold also including a non-enzyme reactor for blank measurements. C: carrier; ER: enzyme reactor; IV: injection valve; NER: non enzyme reactor; PP: peristaltic pump; S: sample; SFC: spectrophotometric flow cell; W: waste.

usually time consuming and tedious, require large quantities of sample and reagents and are vulnerable to contamination and analyte losses. On the other hand, FI methods provide an opportunity to avoid contamination and large reagent consumption by working with closed systems. In addition, FI methods offer higher sample throughput and much better precision and accuracy compared to off-line methods. The use of on-line solid phase extraction (SPE) preconcentration techniques coupled with FAAS (FI-FAAS) have been shown to be a promising alternative to ETAAS thus, became increasingly important for trace metal determinations. These techniques have been based predominantly on incorporation of a minicolumn, packed with various polar or non-polar sorbent materials. Therefore, in most on-line FI–FAAS methods, the carrier and reagent solutions are continuously fed to the flow system, mixed on-line and passed through the mini or microcolumns (packed with a material that retains the analyte) prior to detection. One such methodology is to use chelate sorbed, functionalized sorbents and chelating resins as column materials in on-line FI-FAAS. Typical configurations are shown in the Figure 5.2. Figure 5.2A illustrates a FI manifold where the minicolumn is placed between the injection valve and the detector. With the valve in load position the sample stream flows through the column to waste and analyte is retained onto the minicolumn. The valve is then switched, and eluent is injected in the FI manifold flowing through the column and eluting analyte. Figure 5.2B shows a FI manifold where the minicolumn is installed across opposite ports of a six-port injection valve. With the valve in one position, the sample stream flows through the column to waste, and analyte is retained onto the column. The valve is then switched, and eluent then flows through the minicolumn, eluting analyte.

For analysis of anionic species, sediment samples are dried in an oven and analyte extraction is performed by shaking the sediment sample with diluted hydrochloric acid in a horizontal shaker or in a closed system. Extracts are separated from the sediment by centrifugation, followed by filtration with 0.45 mm cellulose acetate membranes.

ORGANIC SPECIES

The organic analytes that have been determined in sediment by using FI methodologies were the following: bio and molecular markers, dissolved organic carbon (in sediment pore waters) and polycyclic aromatic hydrocarbons. Different features of FI methods for the determination of organic species in sea and estuarine sediments are illustrated in Table 5.1. In the following paragraphs, some points observed in this table are highlighted due to their interest.

Table 5.1. Features of FI determinations of organic species in sea and estuarine sediments

Analyte	Detection	DL (μM)	Linear range (μM)	SF (s/h)	Recovery (%)	RSD (%)	Ref
Biomarkers	APCI-MS	No data	No data	No data	No data	No data	2
Cholestanol	SP	2×10^{-6}	2.2×10^{-5} - 1.8×10^{-4}	No data	95-118	No data	3
Coprostanol	SP	7×10^{-6}	2.2×10^{-4} - 9.0×10^{-4}	No data	96-112	No data	4
DOC	SP	No data	No data	30	No data	No data	5
PAHs	F	No data	No data	120	No data	No data	6
PAHs	F	0.009 μg/mL	0.0125-1.25 μg/mL	No data	89-110	3.9-5.5	7

APCI-MS: atmospheric pressure chemical ionization mass spectrometry; DL: detection limit; DOC: dissolved organic carbon; F: fluorescence; PAHs: polycyclic aromatic hydrocarbons; RSD: relative standard deviation; SF: sampling frequency; SP: spectrophotometry.

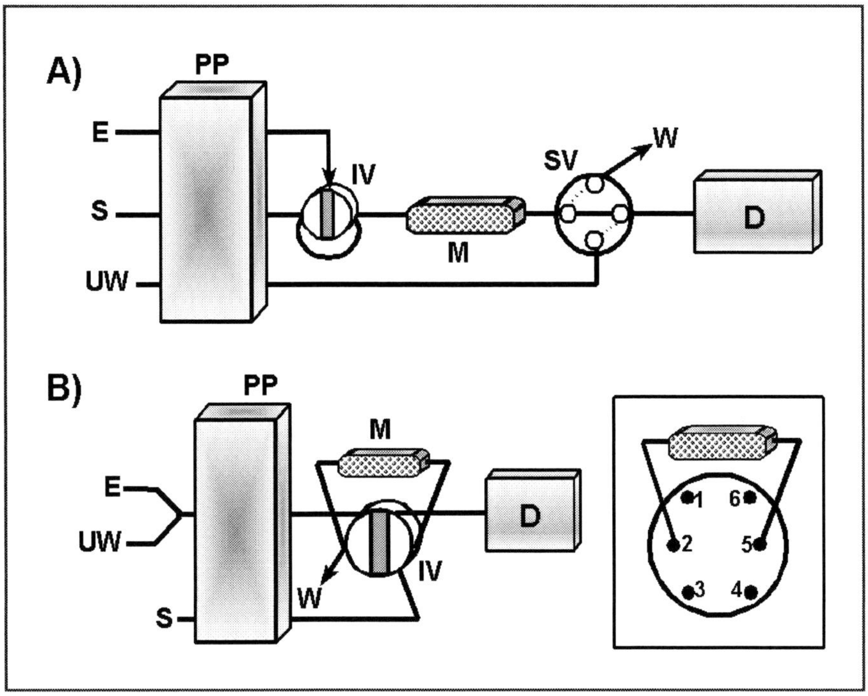

Figure 5.2. Flow injection manifolds proposed for solid-phase extraction. A) minicolumn placed between the injection valve and the detector. B) minicolumn installed across opposite ports of a six-port injection valve (positions 5-6). 1.-ultrapure water or eluent in; 3.- sample out (waste); 4.- sample in; 6: to the detector. D: detector; E: eluent; IV: injection valve; M: minicolumn; PP: peristaltic pump; S: sample; SV, switching valve; UW: ultrapure water; W: waste.

Bio and Molecular Markers

Smittenberg et al. [2] performed repeated semi-preparative normal-phase high performance liquid chromatography (HPLC) to isolate selected biomarkers from sediment extracts for radiocarbon analysis. FIA-mass spectrometry is used for rapid analysis of collected fractions to evaluate the separation procedure, taking only 1 min per fraction, while GC or HPLC-MS analyses take more than 1 h per isolated fraction. In this way, 100-1000 μg of glycerol dialkyl glycerol tetraethers, sterol fractions and chlorophyll-derived phytol are isolated from typically 100 g of marine sediment, i.e., in sufficient quantities for radiocarbon analysis, without significant carbon isotopic fractionation or contamination. Piñeiro-Avila et al. [3] developed a procedure for the spectrophotometric determination of cholestanol in sediments based on its extraction with chloroform-methanol, dissolution of the extracts, after preconcentration, in pH 7.0 buffer-saturated toluene containing p-anisidine and enzymatic determination in non-aqueous media using a bienzymic reactor consisting of cholesterol oxidase and of horseradish peroxidase non-covalently co-immobilized on controlled pore glass beads carried out in order to determine accurately both compounds in the same sample. A linear relationship is obtained between cholestanol dissolved or extracted in toluene and the absorbance at 458 nm, which corresponds to the oxidized form of p-anisidine, used as a test molecule to monitor the enzymatic reaction spectrophotometrically. This method has been applied by the same authors [4] for the determination of coprostanol in sediments.

Dissolved Organic Carbon (in Sediment Pore Waters)

Deflandre et al [5] developed and tested a simple analytical system for estimating dissolved organic carbon (DOC) concentrations in nanoliter samples of sediment pore waters. The system consists of a loop injector that introduces two hundred nanoliters of sample directly into the capillary tubing connected to a scanning UV-VIS detector equipped with a flow-cell.

Polycyclic Aromatic Hydrocarbons

Utsumi et al. [6] determined polycyclic aromatic hydrocarbons (PHAs) in spilled oil, sea sand, and water using HPLC equipped with a fluorescence detector. Although the HPLC method is accurate and quantitative, it takes too much time for rapid analysis of many oil-polluted environmental samples. Thus, a FIA method is used as a rapid screening test of C-fuel oil pollution. Zougagh et al. [7] reported a fast, simple method that allows the direct screening of sediment samples for PAHs using a flow manifold coupled online to a fluorometric detector. The method avoids the need to use expensive instruments such as a liquid chromatograph equipped with a scanning fluorescence detector. The proposed experimental set-up allows PAHs in supercritical fluid extracts to be selectively retained on a column packed with polytetrafluoroethylene (PTFE) accommodated in the loop of a high-pressure injection valve. Retained PAHs are subsequently eluted with acetonitrile and directly driven to the fluorometric detector. Samples testing positive for PAHs are subjected to liquid chromatography in order to separate, identify, and quantify the analytes. This allows anthracene, pyrene, benzo(a)anthracene, 1,2-benzodiphenylene sulfide and benzo(k) fluoranthene to be quantified.

Cationic Species

The cationic species that have been determined in sediment by using FI methodologies were the following: silver, aluminium, arsenic, bismuth, cadmium, cobalt, chromium, copper, iron, germanium, mercury, iridium, magnesium, manganese, nickel, lead, platinum, rare earths, plutonium, thorium, uranium, rhenium, antimony, selenium, tin, tellurium and zinc.

Different features of FI methods for the determination of cationic species in sea and estuarine sediments are illustrated in Table 5.2. In the following paragraphs, some points observed in this table are highlighted due to their interest.

Silver

Elmahadi et al. [8] compared the properties of two covalently immobilized algae, Chlamydomonus reinhartii and Selenestrum capricornutum, for the preconcentration of Cu(II), Ag(I), Cr(III) and Cr(VI). The preconcentration method is proposed for the determination of these trace metals in estuarine sediment (using both reagents) by FAAS. Yang et al. [9] described an online FI method for the direct determination of silver in seawater using isotope dilution inductively coupled plasma mass spectrometry (ID-ICP-MS). A mini-column packed with Dowex 1-X8 anion exchange resin is used to separate and concentrate silver.

Table 5.2. Features of FI determinations of cationic species in sea and estuarine sediments

Analyte	Detection	Separation Technique	DL (μg/L)	SF (s/h)	RSD (%)	Ref
Ag	FAAS	On-line preconcentration on immobilized algae	2.0	No data	1.2-1.5	8
Ag	ICP-MS	On-line anion-exchange Dowex 1-X8	6×10^{-5}	10	0.5-1.2	9
Ag	SP	No data	No data	90-100	No data	10
Al	SP	No data	No data	80-90	< 0.9	11
Al	ETAAS	No data	No data	No data	No data	12
As^{5+}	P		299.7	40	0.1	13
As	AF	On-line gas-liquid separation	9	No data	2.1-8.9	14
As	FIAS-200[a] AAS	On-line gas-liquid separation	No data	No data	No data	15
As	FIAS-200[a] FAAS	On-line gas-liquid separation	0.15	No data	No data	16
As	FIAS-400[a] AAS	On-line gas-liquid separation	No data	17-28	No data	17
As	FIAS-200[a] AAS	On-line gas-liquid separation	0.15	No data	No data	18
As	AAS	On-line gas-liquid separation	0.19-2.8[c]	No data	No data	19
As	AAS	On-line gas-liquid separation	0.14	No data	4.1	20
As(III) As	AAS	On-line gas-liquid separation	0.07 0.06	No data	0.84 0.23	21
As	AAS	On-line gas-liquid separation	1.4	No data	6.4	22
As	AAS	On-line gas-liquid separation	1.9	No data	<5	23
As(III)	ICP-MS	On-line gas-liquid separation	0.6	No data	No data	24
^{75}As	ICP-MS	On-line gas-liquid separation	0.04-0.19	No data	2.5-9.1	25
^{75}As	ICP-MS	On-line gas-liquid separation	5-6[b]	No data	<2	26
Bi	FIAS-200[a] AAS	On-line gas-liquid separation	No data	No data	No data	15
Bi	FIAS-400[a] AAS	On-line gas-liquid separation	No data	17-28	No data	17
Bi	AAS	On-line gas-liquid separation	0.7	No data	2.7	27
Bi(III)	ICP-MS	On-line gas-liquid separation	1.2	No data	No data	24
209Bi	ICP-MS	On-line gas-liquid separation	0.08-0.13	No data	2.6-9.0	25
Cd	AF	On-line 717-strong alkaline anion exchange resin On-line gas-liquid separation	0.058	No data	1.09	28

Table 5.2. Continued

Analyte	Detection	Separation Technique	DL (µg/L)	SF (s/h)	RSD (%)	Ref
Cd	FAAS	On-line chelating resin (Muromac A-1)	0.14	13	0.7-1.7	29
Cd	FAAS	On-line SC covalently immobilized on CPG	0.2	No data	1.3	30
Cd	FAAS	On-line adsorption: analyte complexes with DDDC or DDPA on SC18	0.8	No data	No data	31
Cd	FAAS	On-line adsorption: analyte complexes with DADTP on SC18	0.18	No data	<2	32
Cd	ETAAS	On-line gas-liquid separation	0.01	No data	12	33
Cd	FAAS	On-line adsorption: analyte complexes with DTZ on SC18	0.6	30	3	34
Cd	FAAS	On-line chelating resin (Chelex-100)	0.017	No data	No data	35
Cd	ICP-MS	On-line gas-liquid separation	0.026	No data	1.9-3.5	36
Cd	ICP-MS	On-line gas-liquid separation	No data	No data	0.2	37
Cd	FIAS-400[a] ICP-MS	On-line gas-liquid separation	0.15[c]	60	2.3	38
Co	Ch		0.0006	120	1.5	39
Co	Ch		0.00006	120	3.9	40
Co	Ch		0.3	No data	1.4	41
Co	FAAS	On-line solid phase extraction (polymeric materials)	10-80	30	2.1-2.6	42
Co	ICP-MS	On-line chelating resin (Chelite S)	No data	30	<3	43
Cr(III) Cr(VI)	F		20 50	30	0.1-2.2	44
Cr(III) Cr(VI)	FAAS	On-line preconcentration on immobilized algae	20-25 30-40	No data	1.5-1.6 1.3-1.5	8
Cr	FAAS	On-line chelating resin (Muromac A-1)	0.31	13	0.7-1.7	29
Cr(VI)	FAAS	On-line adsorption: analyte complexes with APDC on PTFE	0.8	18	3.2	45
Cr(III) Cr(VI)	FAAS	On-line separation on alumina	1.2-2.9[c] 0.38-0.45[c]	No data	8.7 7.1	46
Cu	Ch		0.06	120	3.1	47
Cu	SP		50	100	3.3-5	48
Cu	FAAS	On-line preconcentration on immobilized algae	0.05-1	No data	1.2-1.5	8
Cu	FAAS	On-line chelating resin (Muromac A-1)	0.72	13	0.7-1.7	29

Analyte	Detection	Separation Technique	DL (μg/L)	SF (s/h)	RSD (%)	Ref
Cu	FAAS	On-line SC covalently immobilized on CPG	0.7	No data	1.4	30
Cu	FAAS	On-line adsorption: analyte complexes with DDDC or DDPA on SC18	1.4	No data	No data	31
Cu	FAAS	On-line adsorption: analyte complexes with DADTP on SC18	1.4	No data	<2	32
Cu	FAAS	On-line adsorption: analyte complexes with DTZ on SC18	1.2	30	5	34
Cu	FAAS	On-line oxine-loaded on activated C column	4.4	20	2.0	49
Cu	FAAS	On-line PTFE fiber sorbent	0.20	55	1.2	50
Fe^{2+}	Ch		0.3	No data	5.2	41
Fe	SP		20	No data	2.0	51
Fe	FAAS	On-line chelating resin (Muromac A-1)	0.59	13	0.7-1.7	29
Fe	FAAS	On-line SC covalently immobilized on CPG	0.6	No data	1.2	30
^{74}Ge	ICP-MS	On-line gas-liquid separation	0.02-0.25	No data	6.7-7.7	25
Hg	A	On-line gas-liquid separation	0.9	12	5.29	52
Hg	AF		0.4[b]	No data	No data	53
MHg	AF	On-line gas-liquid separation	0.01[b]	No data	<5	54
MHg	AF	Capillary gas chromatography	0.005[b]	No data	1.3-2.5	55
Hg(II) MHg EtHg PhHg	AF	On-line gas-liquid separation	0.010 0.020 0.030 0.070	No data	<3	56
Hg	AF	On-line gas-liquid separation	0.06	No data	No data	57
Hg	FIAS-200[a] FAAS	On-line gas-liquid separation	0.15	No data	No data	16
Hg	AAS	On-line gas-liquid separation	No data	No data	No data	58
Hg	AAS	On-line gas-liquid separation	0.035	32	1.1	59
Hg	AAS	On-line gas-liquid separation	0.2[b]	20-30	<10	60
Hg	AAS	On-line gas-liquid separation	No data	No data	No data	61
Hg	AAS	On-line gas-liquid separation	0.34	30	0.95	62
Hg	AAS	Gold-coated silica column and a precolumn containing copper(II) oxide	0.065	30	7	63
Hg	AAS	On-line gas-liquid separation	0.01	No data	<3	64
Hg	ETAAS	Preconcentration in a KR	0.0062	22	1.1	65
Hg	AAS	On-line gas-liquid separation	0.2	No data	<5	66
Hg	AAS	On-line gas-liquid separation	No data	No data	No data	67
MHg	ICP-AES	SFE and GC	0.1[b]	No data	No data	68

Table 5.2. Continued

Analyte	Detection	Separation Technique	DL (μg/L)	SF (s/h)	RSD (%)	Ref
Hg	ICP-MS	On-line chelating resin (Chelite S)	No data	30	<3	43
^{202}Hg	ICP-MS	On-line gas-liquid separation	18[b]	No data	<5	26
Hg	FIAS-400[a] ICP-MS	On-line gas-liquid separation	0.09[b]	60	0.9	38
Hg	ICP-MS		1[b]	No data	No data	69
Ir	ICP-MS	Off-line anion exchange resin (Biorad AGl-x8, Cl^- form)	0.006[b]	No data	20	70
Mg	FAAS		60	10	0.82	71
Mn	FAAS	On-line chelating resin (Muromac A-1)	0.81	13	0.7-1.7	29
Mn	SP		300	60	<1	72
Ni	FAAS	On-line PTFE fiber sorbent	0.25	55	1.6	50
Pb	Ch		No data	No data	No data	73
Pb	AF	On-line gas-liquid separation	0.31	No data	<7.8	74
Pb	AF	On-line gas-liquid separation	0.26	No data	<6.5	75
Pb	FAAS	On-line chelating resin (Muromac A-1)	2.1	13	0.7-1.7	29
Pb	FAAS	On-line SC covalently immobilized on CPG	8	No data	1.1	30
Pb	FAAS	On-line adsorption: analyte complexes with DDDC or DDPA on SC18	10	No data	No data	31
Pb	FAAS	On-line adsorption: analyte complexes with DTZ on SC18	3	30	6	34
Pb	FAAS	On-line macrocycle immobilized on silica gel	5	63	1.9	76
Pb	FAAS	On-line adsorption: analyte complexes with APDC on PTFE	0.8	15	2.6	77
Pb	FAAS	On-line resins immobilized DDDC MCR	3 1.3	20	1.44 4.36	78
Pb	ICP-AES	On-line gas-liquid separation	1-2	No data	1.2-1.3	79
Pb	FIAS-400[a] ICP-MS	On-line gas-liquid separation	6[c]	60	1.1	38
Pb	ICP-MS		No data	No data	<1.1	80
Pb	ICP-MS		7.1[c]	No data	No data	81
Pb	ICP-MS	On-line gas-liquid separation	0.0007	No data	0.21	82
Pt	ICP-MS	Off-line anion exchange resin (Biorad AGl-x8, Cl^- form)	0.014[b]	No data	5	70
La	ICP-MS		2.7[c]	No data	No data	81

Analyte	Detection	Separation Technique	DL (µg/L)	SF (s/h)	RSD (%)	Ref
REEs	ICP-MS	On-line sorption (MA-PTFE)	$(1\text{-}20)\times10^{-6}$	22	0.5-1.1	83
Th	ICP-MS		0.09[b]	No data	1.5	84
^{230}Th	ICP-MS	On-line resin (Uteva resin)	0.35 mBq/g	No data	No data	85
U	ICP-MS		0.07[b]	No data	0.5	84
^{234}U	ICP-MS	On-line resin (Uteva resin)	0.05 mBq/g	No data	No data	85
^{239}Pu ^{240}Pu ^{242}Pu	ICP-MS	On-line resin (Teva resin)	0.0034[b] 0.00031[b] 0.00029[b]	2	8.8-10.5	86
Re	ICP-MS	Off-line anion exchange resin (Biorad AGl-x8, Cl^- form)	0.005[b]	No data	15	70
Sb	FIAS-200[a] AAS	On-line gas-liquid separation	No data	No data	No data	15
Sb(V)	FIAS-400[a] AAS	On-line gas-liquid separation	No data	17-28	No data	17
Sb(III) Sb(V)	AAS	On-line gas-liquid separation	0.007 ng	No data	0.4	87
Sb	AAS	On-line gas-liquid separation	No data	No data	No data	88
Sb	AAS	On-line resin (Chelex-100) On-line gas-liquid separation	4.3	2	4.1	89
Sb	ICP-AES	On-line gas-liquid separation	No data	No data	No data	90
Sb(III)	ICP-MS	On-line gas-liquid separation	0.8	No data	No data	24
^{121}Sb ^{123}Sb	ICP-MS	On-line gas-liquid separation	0.06-0.09 0.07-0.09	No data No data	1.7-12.7 2-16.1	25
^{121}Sb	ICP-MS	On-line gas-liquid separation	23[b]	No data	<2	26
Se(IV) Se(VI)	F		1 10	28 22	0.1-2 0.1-2	91
Se	AF	On-line gas-liquid separation	8.7	No data	0.4-5.8	14
Se	FIAS-200[a] AAS	On-line gas-liquid separation	No data	No data	No data	15
Se	FIAS-200[a] FAAS	On-line gas-liquid separation	0.17	No data	No data	16
Se(IV)	FIAS-400[a] AAS	On-line gas-liquid separation	No data	17-28	No data	17
Se	FIAS-200[a] AAS	On-line gas-liquid separation	0.20	No data	No data	18
Se	AAS	On-line gas-liquid separation	1.8	No data	<10	27
Se	AAS	On-line gas-liquid separation	0.3	No data	2.6	92
^{77}Se ^{78}Se	ICP-MS	On-line gas-liquid separation	0.44 0.25	 No data	23.3	25
^{77}Se	ICP-MS	On-line gas-liquid separation	60-110[b]	No data	<2	26
Se	ICP-MS	On-line gas-liquid separation	0.7	No data	2.7	38

Table 5.2. Continued

Analyte	Detection	Separation Technique	DL (µg/L)	SF (s/h)	RSD (%)	Ref
Se	ICP-MS	On-line gas-liquid separation	No data	No data	No data	93
Sn	Ch		2.5	No data	No data	94
Sn	FIAS-400[a] AAS	On-line gas-liquid separation	No data	17-28	No data	17
Sn	AAS	On-line gas-liquid separation	0.17	No data	5.3	67
TBT	ETAAS	On-line gas-liquid separation	0.33 66[b]	No data	3.3	95
Sn	ICP-MS	On-line gas-liquid separation	0.8	No data	No data	24
^{118}Sn ^{120}Sn	ICP-MS	On-line gas-liquid separation	0.04-0.41 0.15-0.54	No data	5.2-14.8 4.9-17.4	25
^{120}Sn	ICP-MS	On-line gas-liquid separation	70[b]	No data	<2	26
TBT	ISMSIMS		0.2[c]	50	No data	96
Sn org.	Capillary gas chromatography with atomic emission detection		No data	No data	No data	97
Te	FIAS-400[a] AAS	On-line gas-liquid separation	No data	17-28	No data	17
Te	AAS	On-line gas-liquid separation	3 0.10[c]	No data	6-16	98
Zn	FAAS	On-line chelating resin (Muromac A-1)	0.04	13	0.7-1.7	29
Zn	FAAS	On-line SC covalently immobilized on CPG	0.1	No data	2.6	30
Zn	SP	On-line immobilized BCA	No data	No data	4.9	99

[a] Perkin-Elmer; [b] ng/g; [c] µg/g

A: amperometry; AAS: atomic absorption spectrometry; AF: atomic fluorescence; APDC: ammonium pyrrolidine dithiocarbamate; BCA: bovine carbonic anhydrase; Ch: chemiluminescence; CPG: controlled pore glass; DADTP: dialkyldithiophosphates; DDDC: diethylammonium-N,N-diethyldithiocarbamate; DDPA: ammonium diethyldithiophosphate; DL: detection limit; DTZ: dithizone; ETAAS: electrothermal atomic absorption spectrometry; EtHg: ethylmercury; F: fluorescence; FAAS: flame atomic absorption spectrometry; GC: gas chromatography; ICP-AES: induced plasma atomic emission spectrometry; ICP-MS: inductively coupled plasma-mass spectrometry; ISMSIMS: ion spray mass spectrometry/mass spectrometry; KR: knotted reactor; MA-PTFE: maleic acid grafted polytetrafluoroethylene fiber; MCR: sorbed Merrifield chloromethylated resin; MeHg: methylmercury; P: potentiometry; PhHg: phenylmercury; PTFE: polytetrafluoroethylene; REEs: rare earth elements; RSD: relative standard deviation; SC: Saccharomyces cerevisiae; SC18: silica C18; SF: sampling frequency; SFE: supercritical fluid extraction; SP: spectrophotometry; TBT: tributyltin

Aluminium

Zhao [10] described a new FI method based on sum and difference technique. The sensitivity of the new method is 2.7 fold higher than the original method. The method is used for the determination of aluminium in sediments with spectrophotometric detection. The same author [11] carried out the FI determination of microamounts of aluminum in sediments by spectrophotometry using a simple dual-beam set-up. The sensitivity of the method is increased by approximately one fold compared to that of the conventional method. Zhao et al. [12] developed a new FI analyzer, applied to the atomic absorption spectrophotometer, relying on it without flame in place of visible spectrophotometer, and studied the appropriate condition for the determination of aluminum in sediments.

Arsenic

Electrochemical Detection

Barrado et al. [13] described the construction and evaluation of a tubular potentiometric detector sensitive to As(V) ions. This electrode, with no inner reference solution, is comprised of a FeOOH-SiO_2-graphite composite agglutinated with an epoxy resin. The sensor is used as the selective electrode in a FIA system along with a commercial Ag/AgCl reference electrode.

Atomic Fluorescence Detection

Wei et al. [14] used the hydride vapor generator-coupled atomic fluorescence spectroscopic system to replace the atomic absorption spectroscopic detector, which was used in the ordinary hydride generation technique. For better sensitivity, membrane dryers were used that use a hygroscopic, ion-exchange membrane in a continuous drying process between hydride generator (separator) and atomic fluorescence detector to selectively remove water vapor from mixed hydride gas streams.

Atomic Absorption Spectrometric Detection

Guo et al. [15] carried out the determination of hydride forming elements with the FIAS-200 FI mercury hydride system. Saraswati et al. [16] developed a FIA-AAS method for the determination of trace amounts of arsenic, selenium and mercury. The samples were prepared in two manners: a) a wet digestion procedure with nitric acid, sulfuric acid, and perchloric acid using a reflux column and b) a microwave-oven digestion procedure utilizing nitric acid, sulfuric acid and hydrochloric acid. Microwave-oven digestion provides results comparable to those found by reflux column digestion and reduces the sample preparation time by a factor of 10. Tsalev et al. [17] studied the in-situ collection of volatile hydrides in an electrothermal atomizer with an integrated platform pre-treated with zirconium or wolframium and iridium for permanent modification. For this, was elaborated an optimization study of the performance characteristics of an automated FI-HG-ETAAS system based on an FI hydride generator interfaced with a transverse-heated graphite atomizer and longitudinal Zeeman-effect background correction. Thus, it was proven that Ir-Zr platforms are more suitable than those treated with Ir-W for trapping hydrides of As in an automated FI-HG-ETAAS system, owing to lower atomization temperatures, longer lifetime of the atomizer, less critical hydride trapping parameters and absence of double peaks. Zhou et al. [18] compared five closed-vessel microwave digestion methods to accurately determine As and Se. Digestion methods use five different acid mixtures (HNO_3/ H_2SO_4, $HNO_3/HClO_4$, HNO_3/HCl, $HNO_3/HCl/HF$, $HNO_3/H_2SO_4/HClO_4$), being all reliable to determine the analytes in sediment samples.

Considering the safety and suitability for the analysis, methods using HNO_3/HCl and HNO_3/HCl/HF are recommended for closed vessel microwave digestion with pressure control. Capelo et al. [19] developed a new sample pretreatment method based on ultrasonic extraction in hydrochloric acid medium and subsequent oxidation of the extracts by sonozone (i.e., sonolysis-ozonolysis) for determination of reactive arsenic toward sodium tetrahydroborate [mainly As(III) + As(V)] by FI-HG-AAS. This method avoids the use of intensive treatments with concentrated and corrosive acids, high pressures, and temperatures that are inherent with traditional wet or dry ashing procedures and entails reduced waste production and reagent consumption. Furthermore, a sonozone process at room temperature is optimized to break the bond of As to proteins and macromolecules constituents, which are an essential requirement for effective reduction by L-cysteine prior to arsine generation. Sakamoto et al [20] described a method for the determination of total As by FI-HG-AAS using a mixed acid as a pretreatment. Hydride generation is done by the FI method. The authors investigated in detail the temperature and time of decomposition using inorganic, organic arsenic and environmental standard samples, pretreated with nitric-perchloric-sulfuric mixed acid. By using a mixed acid as a pretreatment agent at 220°C, the decomposition time could be shortened and the blank value of As from the reagents used is reduced. Gonzalez et al. [21] studied non-chromatographic speciation approaches for determination of water-soluble and phosphate-exchangeable As(III) and As(V) by FI-HGAAS. Determination of both oxidation states of As in the extracts could be accomplished following arsine generation under different reaction conditions, namely, (a) selective determination of As(III) in citric acid medium or using soft generation conditions (i.e. low HCl and $NaBH_4$ concentrations); (b) determination of total As in each extract using thioglycolic acid as reaction medium or after pre-reduction of As(V) to As(III) with a potassium iodide plus ascorbic acid mixture. The As(V) content is estimated by difference between both measurements. Quevedo et al. [22] mounted and optimized a method of FI-HG generation for the arsenic determination in sediments. The digestion of the sample is carried out through acid digestion with HNO_3-H_2SO_4-$HClO_4$. Bentlin et al. [23] proposed an analytical methodology for the determination of As in marine sediment using ultrasound for sample preparation. Thus, arsenic can be quantitatively extracted from marine sediments using 20% (V/V) hydrochloric acid and sonication. The slurry is centrifuged and the analyte is determined in the supernatant by HG-AAS. A FI system is employed for hydride generation, with $NaBH_4$ used as reductant and a HCl used as sample carrier.

Inductively Coupled Plasma Atomic Emission Spectrometric Detection

Feng et al. [24] established a hydride generation system using a small concentric hydride generator combined with inductively coupled plasma atomic emission spectrometry (ICP-AES) to determine tin, arsenic, bismuth and antimony in a marine sediment material with L-cysteine as a pre-reductant. The interferences from transition ions are found to be insignificant for determination of the four elements in presence of L-cysteine.

Inductively Coupled Plasma Mass Spectrometric Detection

Abranko et al. [25] performed a simultaneous multi-elemental measurement of As, Bi, Ge, Sb, Se, and Sn by flow injection-hydride generation-inductively coupled plasma-time-of-flight mass spectrometry (FI-HG-ICP-TOFMS). An off-line pre-reduction treatment using a potassium iodide and ascorbic acid is described by presenting its advantages and disadvantages and compared with the results achieved without pre-reduction. So, it has been shown that the applied pre-reduction treatment enables to carry out accurate measurements in spite of the fact that the abundance of different species (with different oxidation states) in the sample and in the standard solution is not necessarily the same. Because of the additional uncertainty sources provided by the required treatment, the limits of detection became poorer, nevertheless it was only measuring with prior treatment that enabled us to achieve reliable results for both digested and not digested samples. Consequently, the use of the described pre-reduction treatment is recommendable when applying the presented method, but in this case selenium should be eliminated from the set of analytes. Ribeiro et al [26] proposed a method for the determination of As, Hg, Sb, Se and Sn as acidified slurries, by FI coupled to a HG system and detection by ICP-MS. The HG unit has a gas-liquid separator and a drying unit for the generated vapor. The slurries are prepared by two procedures. The sample ground to a particle size $\leq$50 μm, is mixed with acid solutions in an ultrasonic bath. In Procedure A, the medium is a solution of hydrochloric acid and in Procedure B, the medium is aqua regia plus a hydrochloric acid solution. Harsh conditions are used in the slurry preparation to reduce the hydride forming analytes to their lower oxidation states, As (III), Se(IV), Sb(III) and Sn(II), before reacting with sodium tetrahydroborate.

Bismuth

Atomic Absorption Spectrometric Detection

FI-hydride generation-atomic absorption spectrometry has been used to determine bismuth in sediment samples [15,17]. Klassen et al. [27] also employed a FI hydride generation system with a metal furnace atomizer for Bi and Se determination.

Inductively Coupled Plasma Mass Spectrometric Detection

A simultaneous multi-elemental measurement of Bi and other elements: tin, arsenic and antimony [24] or germanium, tin, arsenic and antimony is performed by flow injection-hydride generation-inductively coupled plasma-spectrometry [25].

Cadmium

Atomic Fluorescence Detection

Wang et al. [28] developed a method for the determination of ultra-trace cadmium in sediment by atomic fluorescence spectrometry. The method comprises a FI on-line separation and preconcentration technique coupled with an intermittent injection vapor generation technique. It was showed that Cd(II) can be preconcentrated effectively and interferences can be completely eliminated by this improved method. The cadmium vapor generation efficiency could be greatly enhanced in the presence of Co(II) and 1,10-phenanthroline.

Atomic Absorption Spectrometric Detection

Hirata et al. [29] proposed an on-line column preconcentration technique for FI atomic absorption spectrometry. Diverse metal ions (Cd^{2+}, Zn^{2+}, Cu^{2+}, Mn^{2+}, Pb^{2+}, Fe^{3+} and Cr^{3}+) in solution are concentrated quantitatively by a microcolumn packed with Muromac A-1 (iminodiacetate chelate resin), in a FI system. Maquieira et al. [30] packed in a minicolumn the yeast Saccharomyces cerevisiae covalently immobilized on controlled pore glass using a modified method of enzyme immobilization. This minicolumn is incorporated in a FI manifold system for trace determination of some metals prior to quantification with atomic absorption spectrometry. A number of investigations on the ability of the microorganism to remove metal ions from solutions have shown that different metal ions bind to the cell wall of yeast. The authors have observed that the pH-dependence behavior is different for Pb(II) and Fe(III) compared to the other metal ions. Between pH 6.0 and 7.5 a variety of metal ions bind strongly to the cell surface. All the metal ions tested are deabsorbed by lowering the pH to <2.0. Ma et al. [31] studied a FI on line sorbent extraction system in conjunction with flame atomic absorption spectrometry (FAAS) to determine cadmium, copper and lead in digest solutions of solid environmental samples using octadecyl functional groups (C18) bonded silica gel as sorbent with diethylammonium-N,N-diethyldithiocarbamate or ammonium diethyldithiophosphate (DDPA) as complexing agent and methanol as eluent. The same authors [32] used octadecyl functional groups (C18) bonded to silica gel as sorbent and methanol as eluent. This FI sorbent extraction system involving the use of dialkyldithiophosphates as the chelating agent for Cd, Cu and lead determination by FAAS. Matusiewicz et al. [33] generated a volatile cadmium species (presumed to be the hydride) from aqueous solutions by merging sample and tetrahydroborate reductant in a continuous-flow system. The gaseous analyte was transferred onto the inner wall of a graphite tube furnace for in situ preconcentration at 200 °C. Kartikeyan et al. [34] described a sensitive atomic absorption spectroscopy (AAS) method is for the determination of copper, cadmium and lead based upon on-line preconcentration on a microcolumn packed with C18 material. These heavy metals are complexed with dithizone from ammonia solutions in the FI system and adsorbed on the column. The preconcentrated species are eluted with acidified methanol and injected directly into the nebulizer for atomization in an air-acetylene flame for measurement. Soylak et al. [35] proposed a simple, environmentally friendly, economical and sensitive method for the determination of trace cadmium. The method used FI on-line solid phase extraction coupled with FAAS. Determination is based on the adsorption of cadmium(II) ions on a mini-column packed with Chelex-100.

Inductively Coupled Plasma Mass Spectrometric Detection

Hwang et al. [36] used a simple and inexpensive laboratory built vapor generator with isotope dilution ICP-MS for the determination of cadmium. The application of vapor generation-ICP-MS alleviated the spectroscopic interferences in cadmium determination encountered when a conventional pneumatic nebulizer is used for sample introduction. Valles Mota et al. [37] characterized experimental parameters governing the instrumental precision and accuracy for isotope ratio measurements of cadmium in ICP-MS, including sampling time, mass bias, detector dead-time and spectroscopic interferences. Two alternative flow approaches for the determination of ultratrace concentrations of cadmium by isotope dilution (ID) were explored and compared with the more conventional ID methodology: 1) on-line mixing of the sample solution with the spike solution just before the ICP-MS nebulizer using a peristaltic pump, and 2) the generation of volatile cadmium species using sodium tetraethylborate by merging zones FI-ICP-MS (Figure 5.3). The online ID method proved to be the most convenient for the determination of cadmium in sediment samples because it is fast, provides similar results to those of conventional ID and requires less sample preparation. Vieira et al. [38] proposed a method for the determination of Cd, Hg, Pb and Se in sediments by slurry sampling chemical vapor generation (CVG) using isotopic dilution (ID) calibration and detection by ICP-MS. Two different systems were used for the study: an on-line FI system (FI-CVG-ICP-MS) and an off-line system with in situ trapping electrothermal vaporization (CVG-ETV-ICP-MS). About 100 mg of the sample, ground to a particle size ≤50 μm, is mixed with acid solutions (aqua regia, HF and HCl) in an ultrasonic bath. The enriched isotopes ^{111}Cd, ^{198}Hg, ^{206}Pb and ^{77}Se are then added to the slurry to produce an altered isotopic ratio close to 1. For the on-line system, a standing time for the slurry of 12 h before measurement is required, while for the batch system, no standing time is needed to obtain accurate results.

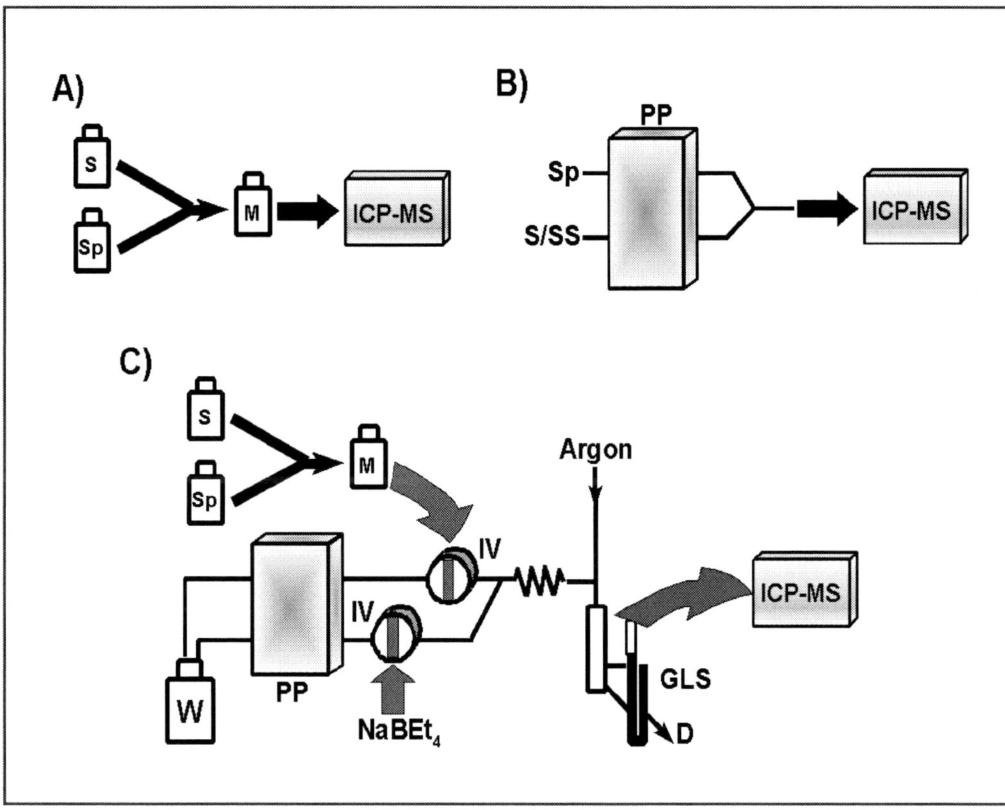

Figure 5.3. Diagrams of the three isotope dilution inductively coupled plasma mass spectrometry (ID-ICP-MS) methods. A) conventional ID; B) on-line ID; C) ID-FI-vapor generation using NaBEt4. D: drain; GLS: gas-liquid separator; IV: injection valve; M: mixture; PP: peristaltic pump; S: sample; SS: standard solution; Sp: spike; W: water.

Cobalt

Chemiluminescence Detection

Yamada et al. [39] used a FIA manifold with chemiluminescence detection based on a gallic acid/H_2O_2/NaOH/MeOH system, for the sensitive and rapid determination of Co(II).

Komatsu et al. [40] proposed a novel method for the determination of ultratraces of Co(II) based on the measurement of sonic chemiluminescence enhanced by the presence of Co(II). It provides analytical characteristics comparable to those for the conventional luminol chemiluminescence method using H_2O_2 as an oxidizing agent. Sonic chemiluminescence is the intense light emission produced by ultrasonic irradiation of alkaline luminol solution as a result of the reaction of luminol with reactive species, which are generated ultrasonically from water molecules. Makita et al. [41] utilized the catalytic decomposition of peroxomonosulfate, which is found to be light-emissive, for the chemiluminescent determination of Co(II) and Fe(II) by a FI method. By use of sensitizer, brilliant sulfoflavine in an octadecyltrimethylammonium chloride micellar solution, the chemiluminescent signal is enhanced by 100-400 times and its linear dynamic range is extended over three orders of magnitude.

Atomic Absorption Spectrometric Detection

Praveen et al. [42] prepared a dichloroquinoline-8-ol-embedded styrene-ethylene-glycol dimethacrylate polymer materials via the bulk, precipitation, and suspension polymerization methods using similar compositions. The polymerization is carried out by thermal means in the presence of 2,2'-azobisisobutyronitrile as initiator and 2-methoxy ethanol as porogen. These polymeric materials are packed into homemade micro columns and used for on-line solid phase extraction preconcentration of trace amounts of cobalt. The preconcentrated cobalt species are eluted with diluted nitric acid and injected directly into the nebulizer of a flame atomic absorption spectrometer (FAAS) for quantification.

Inductively Coupled Plasma Mass Spectrometric Detection

Zara et al. [43] described a FI system for multielemental analysis with a mercury(II) preconcentration step using a resin Chelite-S (sulfhydryl groups) packed minicolumn by ICP-MS.

Chromium

Fluorescence Detection

Paleologos [44] developed a relatively simple, sensitive, selective, automatic fluorometric method for the simultaneous determination of Cr(III) and Cr(VI) by FIA. The method is based on the selective oxidation of the nonfluorescing reagent 2-(α-pyridyl)thioquinaldinamide (PTQA), which with Cr(VI) yields an intensely fluorescent product ($\lambda ex = 360$ nm; $\lambda em = 500$ nm). Cr(III) is oxidized on-line to Cr(VI) with sodium metaperiodate and the Cr(VI) is subsequently treated with PTQA. Fluorescence due to the sum of Cr(III) and Cr(VI) is measured and Cr(III) is determined from the difference in fluorescence values.

Atomic Absorption Spectrometric Detection

Elmahadi et al. [8] compared the properties of two covalently immobilized algae, Chlamydomonus reinhartii and Selenestrum capricornutum, for the preconcentration of Cu^{2+}, Ag^{+}, Cr^{3+} and Cr^{6+}. These reagents were found to have different properties, with

Chlamydomonus reinhartii having a higher capacity for Cr^{6+}. The effects of sodium chloride, sodium hydrogencarbonate and humic acid are studied and it was found that interfere by competing for the metal ions to be complexed. Hirata et al. [29] developed an on-line column preconcentration technique for FI-FAAS involving the chelating resin Muromac A-1. Anthemidis et al. [45] elaborated a new, sensitive and robust time-based FI method for on-line preconcentration and determination of ultra trace amounts of Cr(VI) by FAAS). The sample is initially mixed on-line with ammonium pyrrolidine dithiocarbamate (APDC) and the Cr(VI)-PDC chelate is absorbed on a mini-column packed with polytetrafluoroethylene (PTFE). The complex is subsequently eluted with isobuthylmethyl ketone and introduced directly into the nebulizer-burner system. Pazos-Capeans et al. [46] developed an analytical FI methodology with the purpose of realize speciation studies for chromium. The research builds on previous publications using an alumina column and eluent solutions of ammonia for Cr(VI) and nitric acid for Cr(III). However, this approach facilitates either the determination of Cr(VI) or Cr(III), but not both from the same sample injection.

Copper

Chemiluminescence Detection

Yamada et al. [47] used FIA with chemiluminescence detection to determine traces of Cu(II) by means of the FMN-hydrogen peroxide-phosphate buffer system. This permits the determination of Cu(II) more selectively than any other chemiluminescence system.

Spectrophotometric Detection

Yuan [48] determined copper by FI spectrophotometry with a carrier stream containing Tritron X-100, bathocuproine, and hydroxylamine as reducing agent in acetate buffer at pH 4.5. The detection wavelength is 479 nm.

Atomic Absorption Spectrometric Detection

Elmahadi et al. [8] compared the properties of two covalently immobilized algae, Chlamydomonus reinhartii and Selenestrum capricornutum for the preconcentration of Cu^{2+}, Ag^{+}, Cr^{3+} and Cr^{6+} and with Chlamydomonus reinhartii having a lower capacity for Cu^{2+}. Hirata et al. [29] developed an on-line column preconcentration technique based on the utilization of a microcolumn packed with Muromac A-1 chelating resin for the determination by FI-AAS of diverse metal ions (Cd^{2+}, Zn^{2+}, Cu^{2+}, Mn^{2+}, Pb^{2+}, Fe^{3+} and Cr^{3+}). Maquieira et al. [30] used the yeast Saccharomyces cerevisiae covalently immobilized on controlled pore glass for on-line preconcentration of trace metals prior to their quantification by FAAS. Ma et al. studied a FI on-line sorbent extraction system in conjunction with FAAS to determine Cd, Cu and Pb using octadecyl functional groups (C18) bonded silica gel as sorbent with diethylammonium-N,N-diethyldithiocarbamate (DDDC) or ammonium diethyldithiophosphate (DDPA) [31], or dialkyldithiophosphates as complexing agent [32]. Kartikeyan et al. [34] using C18 material as sorbent with dithizone. Murakami et al. [49] determined trace amounts of copper by preconcentration with oxine-loaded activated carbon column, and elution with diluted nitric acid, followed by FAAS detection. Wang et al. [50] extended to FI on-line microcolumn preconcentration-FAAS determination of trace copper and nickel by using an acrylic acid-grafted PTFE fiber sorbent. Thus, on-line preconcentration of trace analytes is achieved on a microcolumn packed with acrylic acid-grafted PTFE fibers, and the retained analytes are eluted with diluted hydrochloric acid for on-line FAAS determination.

Iron

Chemiluminescence Detection

Makita et al. [41] utilized the catalytic decomposition of peroxomonosulfate for the chemiluminescent determination of Co(II) and Fe(II) by a FI method.

Spectrophotometric Detection

Yuan [51] developed a FIA system for the spectrophotometric determination of Fe(II) based on the Fe(II)-bathophenanthroline-Triton X-100 system.

Atomic Absorption Spectrometric Detection

Hirata et al. [21] developed an on-line column preconcentration technique for FI-FAAS by using Muromac A-1 resin. Maquieira et al. [30] used the yeast Saccharomyces cerevisiae covalently immobilized on controlled pore glass for the on-line preconcentration of some metals prior to quantification by FAAS.

Germanium

Abranko et al. [25] performed the simultaneous multi-elemental measurement of As, Bi, Ge, Sb, Se, and Sn by flow injection-hydride generation-inductively coupled plasma-time-of-flight mass spectrometry (FI-HG-ICP-TOFMS).

Mercury

Electrochemical Detection

Amini et al. [52] described a gas-diffusion FI system for the determination of Hg(II) in at trace levels. The analytical procedure involves injection of Hg(II) samples and standards into a stream of a diluted solution of sulfuric acid, which subsequently merges with a reagent stream of $SnCl_2$ in acid medium to reduce Hg(II) to metallic Hg. The gas-diffusion cell is thermostated at 85° to enhance the vaporization of Hg. The Hg vapor diffused across a Teflon membrane into an acceptor stream where it is detected amperometrically at a Au electrode set at +600 mV vs. Ag/AgCl reference electrode.

Atomic Fluorescence Detection

Wang et al. [53] determined directly without preconcentration trace amounts of Hg in different marine sediments by FI cold vapor generation-nondispersive-atomic fluorescence spectrometry (FI-CV-NDAFS). KBH_4 is used as reducing agent. The method was suitable for marine sediment samples having Hg content 0.006-0.172 μg/g. Tseng et al. [54] developed a simple and robust mercury speciation analyzer (MSA) for measurement of monomethylmercury (MMHg) in environmental matrixes with the aim of improve the precision by directly coupling and automating the numerous steps involved with analysis of this toxic mercury species. This on-line hyphenated system couples the main analytical steps, including sample introduction, aqueous-phase ethylation, Tenax preconcentration and gas chromatographic separation, to cold vapor atomic fluorescence detection and data acquisition. Shi et al. [55] described a comprehensive method for the determination of methylmercury by capillary gas chromatography online coupled with atomic fluorescence spectrometry. Yin et al. [56] reported a method for direct vapor generation of mercury species on nano TiO_2 under UV irradiation in the presence of a formic acid and sodium formate mixture as a hole scavenger. A novel designed UV/TiO_2 photocatalysis reaction device is used for the first time as an effective sample introduction unit and an interface for mercury species determination by atomic fluorescence spectrometry (AFS) and speciation by HPLC-AFS. Liang et al. [57] proposed the method of slurry sampling FI-microwave digestion with cold vapor generation-atomic fluorescence detection for the determination of mercury. Samples are dispersed in 50% (V/V) aqua regia and agitated magnetically to obtain well-proportioned and stable sample solutions.

Atomic Absorption Spectrometric Detection

Saraswati et al. [16] developed a FIA atomic absorption spectrometric method for the determination of trace amounts of arsenic, selenium and mercury. The samples are prepared using a wet digestion procedure with HNO_3, H_2SO_4, and $HClO_4$ using a reflux column and a microwave-oven digestion procedure utilizing HNO_3 for Hg. Microwave-oven digestion provides results comparable to those found by reflux column digestion and reduces the sample preparation time by a factor of 10. Fostier et al. [58] digested lyophilized sediment samples in a microwave oven. Samples are analyzed by FI cold vapor at. absorption spectrometry. Hanna et al. [59] described a system that fully automates the digestion and determination of total Hg in environmental samples. Once appropriate digestion reagents are added to the sample, system operation occurs in a stand-alone manner. Time required for sample digestion is reduced from 0.5-2 h to <1 min. Murphy et al. [60] developed a simple and accurate method for the determination of total Hg. The method uses an off-line microwave digestion state followed by analysis using a FI system with detection by cold vapor atomic absorption spectrometry. Zhou et al. [61] compared four closed-vessel microwave digestions methods for the accurate determination of Hg in sediment by FI cold-vapor atomic absorption spectrometry. Several acid mixtures (HNO_3/H_2SO_4, $HNO_3/HClO_4$, HCl/HNO_3 and $HCl/HNO_3/HF$) completely digested the soil matrix for the determination of Hg. The method using aqua regia is preferred because it is time saving, less dangerous and suitable for other trace metal analyses. The merits of pressure-feedback microwave digestion are that it simplifies soil sample digestion, and there is no loss of Hg. Bauza de Mirabo et al. [62] proposed a sequential injection analysis system is for the determination of mercury by cold vapor at absorption spectrometry. Both the sample and the reagent are sequentially aspirated using a Crison automatic Compact Titrator and impelled into a gas-liquid separation cell. Once there, a N_2 flow sweeps the reduced mercury into a measuring cell of an atomic absorption spectrometer. The system proposed allows the detection of mercury in addition to data acquisition and treatment in an automatic way. Magalhaes et al. [63] described a pyrolysis chamber for the direct determination of Hg in sediments by atomic absorption spectrometry. A finely ground, dried sample is weighed, transferred to a quartz cup, inserted in a pyrolysis chamber, and heated at 1100° for 30 s by 3 focused IR lamps in the presence of an air stream as the carrier gas. Interfering substances are destroyed by a pre-column containing Al_2O_3, SiO_2, and Cu(II) oxide; the released Hg is amalgamated on to a Ag-coated collector. Hg vapor is later measured by atomic absorption spectroscopy following thermal desorption. Martinez-Garcia et al. [64] developed a FI-cold vapor atomic absorption spectrometric method for the determination of mercury at trace level in estuarine sediments using sodium tetrahydroborate (III) as reductant. The mercury is solubilized with nitric acid in closed vessels and microwave oven heating. No interferences are recorded excepting for copper and nickel, which caused a serious depressing effect. Li et al. [65] carried out a novel methodology for the determination of trace mercury by on-line coupling of FI displacement sorption preconcentration in a knotted reactor (KR) to electrothermal at. absorption spectrometry (ETAAS). The developed methodology involved the on-line formation of copper pyrrolidine dithiocarbamate (Cu-PDC), presorption of the resulting Cu-PDC onto the inner walls of the KR, and selective retention of the analyte Hg(II) onto the inner walls of the KR through on-line displacement reaction between Hg(II) and the presorbed Cu-PDC. The retained analyte is subsequently eluted with ethanol and on-line detected by ETAAS. Interferences from coexisting heavy metal ions with lower stability of their APDC complexes relative to Cu-PDC are minimized without the need of any masking reagents. Collasiol et al. [66] studied a method for Hg determination without sample digestion. Hg determination is performed by cold vapor atomic absorption spectrometry (CV-AAS) using a FI system. Hg quantitatively leaches out from the studied marine sediment into 30% (V/V) nitric acid assisted by ultrasonic irradiation (during 90-120 s) when sample particles size are $\leq$120 µm. Navarro et al. [67] optimized the closed vessel microwave assisted digestion of sediments for the determination of Sn and Hg in the same extract. The optimization of the digestion conditions was made with FI-cold vapor-quartz furnace AAS measurements of Hg. The optimum conditions were: 2.1 bar of pressure during 10 min of irradiation and two local optima composition of the acid mixtures: 80% HCl-20% HNO_3 and 60% H_2O-20% HCl-20% HNO_3.

Inductively Coupled Plasma Atomic Emission Spectrometric Detection

Zara et al. [43] described a FI system for multielemental analysis with a mercury(II) preconcentration step using a resin Chelite-S packed minicolumn by inductively coupled plasma atomic emission spectroscopy. A mercury reductive elution procedure with a mixture of $SnCl_2$/HCl is used, which allows use of diluted hydrochloric acid solution instead of concentrated hydrochloric acid.

Emteborg et al [68] presented a method employing supercritical fluid extraction (SFE) and gas chromatography (GC) coupled with microwave-induced plasma atomic emission spectrometry (MIP-AES) for the determination of methylmercury in sediments. Butylmagnesium chloride is used to derivatize the target compound to butylmethylmercury, which is amenable to GC.

Inductively Coupled Plasma Mass Spectrometric Detection

Ribeiro et al. [26] proposed a method for the determination of As, Hg, Sb, Se and Sn in environmental and in geological materials, as acidified slurries, by FI coupled to a hydride generation system and detection by inductively coupled plasma mass spectrometry. Vieira et al. [38] developed a method for the determination of Cd, Hg, Pb and Se in sediments by slurry sampling chemical vapor generation using isotopic dilution calibration and detection by inductively coupled plasma mass spectrometry. Woller et al. [69] described a method for the determination of total Hg in sediment samples. Extraction of Hg from a sample matrix is carried out in an open vessel microwave digestion system while maintaining mild conditions during digestion to avoid any loss of Hg due to volatilization. A complexing agent (EDTA) and a surfactant (Triton X-100) are added to the samples to eliminate memory effects associated with Hg determinations and to obtain reproducible linear calibration curves. Standard additions and internal standardization are used for calibration and correction in an FI-ICP-MS detection system.

Iridium

Colodner et al. [70] developed methods to measure Re, Ir, and Pt in sediments by isotope dilution inductively coupled plasma mass spectrometry (ID-ICPMS). In each case, a stable isotope-enriched spike is added to the sample before processing. Sediments are dissolved in all-Teflon digestion vessels using a modified standard kitchen microwave oven. Anion exchange of the chloro complexes of iridium is used to preconcentrate the elements and to separate them from concomitants, which produce molecular ions in the argon plasma resulting in isobaric interferences. Samples are then introduced into the ICPMS in a small volume using FI.

Magnesium

de Almeida et al [71] tested a FI system based on online microwave-assisted digestion as a tool to perform silicate rock dissolution in acid medium (HF + HNO_3). A volume of the digested sample slurry is introduced into an aqueous carrier stream and mixed with a reagent stream containing boric acid before introduction into the AAS system.

Manganese

Hirata et al. [29]described an on-line column preconcentration technique using a chelating resin (Muromac-A1) for FI atomic absorption spectrometry to determine diverse metal ions. Mesquita et al. [72] proposed a FI procedure based on permanganate formation for spectrophotometric determination of manganese. The oxidation by periodate is carried out under acidic conditions at about 95°C. The rate of this reaction is accelerated by exploiting the autocatalytic effect, which is enhanced with the addition of a permanganate confluent stream.

Nickel

Wang et al. [50] achieved the preconcentration of trace metals on a minicolumn packed with acrylic acid-grafted PTFE fibers, and the retained analytes were eluted with diluted hydrochloric acid for on-line FAAS determination.

Lead

Chemiluminescence Detection

Gong et al. [73] determined Pb(II) by catalysis on the oxidation of lucigenin by hydrogen peroxide and FI chemiluminescence measurement. The chemiluminescence intensity is enhanced by ethanol.

Atomic Fluorescence Detection

Cheng et al. [74] studied the determination of trace lead in marine sediment by FI hydride generation atomic fluorescence spectrometry. This method is based on a $K_3Fe(CN)_6$-HCl-KBH_4 system with oxalic acid as a masking agent. The fluorescence value is linear with lead concentrations at 0-160 μg/L. Cheng et al. [75] proposed a method for the determination of lead in sediments, which combines the slurry technique with hydride-generation atomic fluorescence spectrometry. The sample is ground to 0.088 mm and suspended uniformly in solution by an electromagnetic stirrer. In this method, potassium ferricyanide is used as an oxidant.

Atomic Absorption Spectrometric Detection

Hirata et al. [29] developed an on-line column preconcentration technique for FI atomic absorption spectrometry for diverse metal ions using a microcolumn packed with Muromac A-1. Ma et al. [31] studied a FI on-line sorbent extraction system in conjunction with FAAS to determine Cd, Cu and Pb in digest solutions using octadecyl functional groups (C18) bonded silica gel as sorbent with diethylammonium-N,N-diethyldithiocarbamate (DDDC) or ammonium diethyldithiophosphate (DDPA) as complexing agent. Kartikeyan et al. [34] described a rapid and sensitive AAS method for the determination of Cu, Cd and Pb based upon on-line preconcentration on a microcolumn packed with C18 material. These heavy metals are complexed with dithizone in the FI system and adsorbed on the column. Yan et al. [76] developed a simple and highly selective FI on-line preconcentration and separation FAAS method for routine analysis of trace amounts of lead. The selective preconcentration of lead is achieved in a wide range of sample acidity on a microcolumn packed with a macrocycle immobilized on silica gel. The lead retained on the column was effectively eluted with an EDTA solution. The results showed that a combination of increase in column capacity, quantitation based on peak area, and use of diluted nitric acid for column wash before elution efficiently avoid potential interferences. Zachariadis et al. [77] reported a rapid and sensitive time-based FI method for on-line preconcentration and determination of lead by FAAS, using polytetrafluoroethylene (PTFE) turnings as packing material in a micro-column. The sample was mixed on-line with ammonium pyrrolidine dithiocarbamate (APDC) and the noncharged Pb(II)-PDC complex is absorbed quantitatively on the hydrophobic PTFE material. The preconcentrated complex is effectively eluted with isobuthylmethylketone (IBMK) and introduced into the nebulizer-burner system. Praveen et al. [78] described the preparation of dithiocarbamate immobilized/functionalized and diethylammonium dithiocarbamate (DDTC) sorbed Merrifield chloromethylated resin (MCR) beads. These two materials enrich lead quantitatively over an identical optimal pH range (8.0-9.0), a preconcentration/loading time (up to 4 min) and elution with acidified methanol. Valdés-Hevia y Temprano et al. [79] investigated continuous flow generation of two different types of volatile lead species, PbH_4, ($NaBH_4$, as the reducing agent) or tetraethyllead ($NaBEt_4$, as the alkylating agent) as a means of gaseous sample introduction for inductively coupled plasma atomic emission spectrometry (ICP-AES). The advantages of both compounds for sample introduction are critically compared in terms of sensitivity, selectivity and accuracy for the determination of low levels of lead by ICP-AES.

Inductively Coupled Plasma Mass Spectrometric Detection

Vieira et al. [38] proposed a method for the determination of Cd, Hg, Pb and Se in sediments by slurry sampling chemical vapor generation (CVG) using isotopic dilution (ID) calibration and detection by ICP-MS. Lasztity et al. [80] developed an on-line isotope dilution technique that is coupling to commercial FI instruments with an inductively coupled plasma mass spectrometer. The approach simplifies sample preparation and analysis and matrix effects have no influence on the determinations. Persaud et al. [81] combined acid leaching (with diluted nitric acid) with slurry nebulization in ICP-MS, with mixed-gas plasmas and FI in an attempt to facilitate the quantitative analysis of heterogeneous materials such as sediments. Slurries are prepared by grinding the material to <3 μm in high-purity water and then diluting with 1 mol/L nitric acid, which served as both a dispersing and a leaching agent. The resulting slurries are injected into a deionized distilled water carrier. Li et al. [82] explored a method for the determination of lead by hydride generation inductively coupled plasma mass spectrometry using oxalic acid-ammonium cerium(IV) nitrate-sodium tetrahydroborate as the reaction matrix and a homemade hydride generator. The sensitivity achieved is approximately 7.5 times higher than that obtained using an ultrasonic nebulizer.

Platinum

Colodner et al. [70] developed analytical methods to measure Re, Ir, and Pt by isotope dilution inductively coupled plasma mass spectrometry (ID-ICPMS). Sediments are dissolved in all-Teflon digestion vessels using a modified standard kitchen microwave oven. Anion exchange of the chloro complexes of Ir and Pt and of the perrhenate ion (ReO_4^-) is used to preconcentrate the elements and to separate them from concomitants, which produce molecular ions in the argon plasma resulting in isobaric interferences. Samples are then introduced into the ICPMS in a small volume using FI.

Rare Earths, Plutonium, Thorium and Uranium

Persaud et al. [81] combined acid leaching with slurry nebulization in inductively coupled plasma mass spectrometry, with mixed-gas plasmas and FI to determine lanthanum in marine sediments. Wang et al. [83] prepared and evaluated a new sorbent, maleic acid grafted polytetrafluoroethylene fiber (MA-PTFE), for on-line solid-phase extraction coupled with inductively coupled plasma mass spectrometry (ICP-MS) for fast, selective, and sensitive determination of (ultra)trace rare earth elements (REEs) in environmental samples. The REEs are selectively retained onto a microcolumn packed with the MA-PTFE fiber, and the adsorbed REEs are subsequently eluted on-line with diluted nitric acid for ICP-MS determination. The new sorbent extraction system allows effective preconcentration and separation of the REEs from the major matrix constituents of alkali and alkali earth elements. Halicz et al. [84] determined the uranium and thorium concentrations using FI-ICP-MS with thallium as an internal standard. Godoy et al. [85] developed a determination method for ^{234}U and ^{230}Th based on an extraction chromatographic separation with uteva resin (diamyl amylphosphonate groups), on a FI system coupled to a quadruple ICP-MS. Epov et al. [86] described a simple, rapid on-line analytical method for the determination of plutonium isotopes. Microwave leaching is used to extract Pu from the major sample matrix. L-Ascorbic acid is found to be the most appropriate agent to convert all the Pu to Pu(IV) for chromatography. FI chromatography using TEVA resin (active component: aliphatic quaternary amine) is used to separate, preconcentrate and elute Pu from other constituents of the sample including ^{238}U. An APEX desolvation unit with a PEEK Mira Mist nebulizer is used to minimize clogging and hydride interferences.

Rhenium

Colodner et al. [70] developed methods to measure Re, Ir, and Pt by isotope dilution inductively coupled plasma mass spectrometry (ID-ICPMS).

Antimony

Atomic Absorption Spectrometric Detection

Guo et al. [15] determined hydride-forming elements with the FIAS-200 flow injection-mercury/hydride system. Tsalev et al. [17] studied the in-situ collection of volatile hydrides in an electrothermal atomizer with an integrated platform pre-treated with Zr or W and Ir for permanent modification. De la Calle Guntinas et al. [87] realized a comparative study of several methods to speciate Sb(III) and Sb(V) by AAS: (1) selective extraction of Sb(III) with lactic acid/malachite green and graphite furnace-AAS, (2) Sb(III) and total antimony determination by hydride generation-AAS coupled to FI, batch, and continuous flow systems. For soils an sediment samples different sample pretreatments are used: HNO_3-H_2SO_4-$HClO_4$, HF-HNO_3-H_2SO_4-$HClO_4$, cold aqua regia and slurry formation procedures in water and diluted hydrochloric acid. D'ulivo et al. [88] developed a pretreatment procedure for the reduction of Sb(V) to Sb(III) to remove the effect of HF, which strongly interferes with the reduction of pentavalent Sb to the trivalent state when antimony is determined by hydride generation techniques. It is based on the combined action of L-cysteine and boric acid at 80°. The pretreatment is effective in both nitric and HCl media. The method is applied to the determination of total Sb in sediments after pressurized microwave digestion with HNO_3-HCl-HF. Bolea et al. [89] proposed a method for the removal of the interference caused by iron on electrochemical generation of stibine. It consists of a chelating resin Chelex 100 column integrated into a FI system and coupled to the electrochemical hydride generator quartz tube atomic absorption spectrometer (EcHG-QT-AAS). Fe, as Fe(II), is retained in the column with high efficiency. No significant retention is observed for Sb(III) under same conditions and a 97 ± 5% signal recovery is achieved.

Inductively Coupled Plasma Atomic Emission Spectrometric Detection

Liu et al. [90] used an activated carbon microcolumn for dynamic adsorption and separation of antimony. The adsorbed Sb could be eluted from the activated C microcolumn with an aqueous solution of nitric acid. A new liquid-gas phase separator is designed and used for hydride generation of Sb in alkaline mode followed by ICP-AES determination.

Inductively Coupled Plasma Mass Spectrometric Detection

Feng et al. [24] established a hydride generation system using a small concentric hydride generator combined with inductively coupled plasma atomic emission spectrometry (ICP-AES) to determine tin, arsenic, bismuth and antimony in a marine sediment material with L-cysteine as a pre-reductant. Abranko et al. [25] developed a simultaneous multi-elemental measurement of As, Bi, Ge, Sb, Se, and Sn by flow injection-hydride generation-inductively coupled plasma-time-of-flight mass spectrometry (FI-HG-ICP-TOFMS). Pre-reduction is carried out off-line using a solution of KI and ascorbic acid for 15 min at 80°C. Ribeiro et al. [26] proposed a method for the determination of As, Hg, Sb, Se and Sn in solid environmental samples, as acidified slurries, by FI coupled to a hydride generation system and detection by inductively coupled plasma mass spectrometry.

Selenium

Fluorescence Detection

Ahmed et al. [91] developed a simple, sensitive, highly selective, automatic spectrofluorometric method for the simultaneous determination of Se(IV) and Se(VI) as selenite-selenate by FIA. The method is based on the selective oxidation of the nonfluorescent reagent 2-(α-pyridyl)thioquinaldinamide (PTQA) in acidic solution by Se(IV) to give an intensely fluorescent oxidation product ($\lambda ex = 350$ nm; $\lambda em = 500$ nm). Se(VI) is reduced on-line to Se(IV), in a reduction coil installed in a photo-reactor, which is then treated with PTQA and the fluorescence due to the sum of Se(IV) and Se(VI) is measured; Se(VI) is determined from the difference in fluorescence values.

Atomic Fluorescence Detection

Wei et al. [14] applied a hydride vapor generator-coupled atomic fluorescence spectroscopic system with the aim to replace the atomic absorption spectroscopic detector, which was used in the ordinary hydride generation technique.

Atomic Absorption Spectrometric Detection

Guo et al. [15] determined hydride-forming elements with the FIAS-200 flow injection-mercury/hydride system. Saraswati et al. [16] established a FIA atomic absorption spectrometric method for the determination of trace amounts of arsenic, selenium and mercury. The samples are prepared using a microwave-oven digestion procedure. Tsalev et a. [17] reported the in-situ collection of volatile hydrides in an electrothermal atomizer with an integrated platform pre-treated with Zr or W and Ir for permanent modification was studied. Zhou et al. [18] compared closed-vessel microwave digestion methods to determine As and Se by FI hydride-generation at. absorption spectrometry. Digestion methods with different acid mixtures (HNO_3/ H_2SO_4, $HNO_3/HClO_4$, HNO_3/HCl, $HNO_3/HCl/HF$, $HNO_3/H_2SO_4/HClO_4$) are all reliable to determine the analytes. Klassen et al. [27] employed a FI hydride generation system with a metal furnace atomizer for Bi and Se determination. Chan et al. [92] applied FIA to sample introduction in conjunction with automated hydride generation and AAS for the determination of As and Se in environmental samples. The samples are digested with a mixture of nitric, sulfuric, and perchloric acids. Se(VI) in the digested solutions is pre-reduced to Se(IV) by exothermic reaction in 6-8 M HCl. The analyte is then converted to hydride by $NaBH_4$ in an automated hydride generation system. The evolved hydride is carried through to a heated quartz tube by a stream of argon, and the atomic absorption of the analyte is measured. 1,10-Phenanthroline is used as masking agent to control interferences from Cu and Ni on Se.

Inductively Coupled Plasma Mass Spectrometric Detection

Abranko et al. [25] performed a simultaneous multi-elemental measurement of As, Bi, Ge, Sb, Se, and Sn by flow injection-hydride generation-inductively coupled plasma-time-of-flight mass spectrometry (FI-HG-ICP-TOFMS). Ribeiro et al. [26] proposed a method for the determination of As, Hg, Sb, Se and Sn in environmental solid samples, as acidified slurries, by FI coupled to a hydride generation system and detection by ICP-MS. Vieira et al. [38] applied a method for the determination of Cd, Hg, Pb and Se in sediments by slurry sampling chemical vapor generation using isotopic dilution calibration and detection by ICP-MS. Moor et al. [93] compared two approaches for the determination of selenium in marine sediments. For this, a special hydride technique is used to minimize chemical interferences. The technique is based on a double loop FI-hydride system with a short reaction, followed by immediate separation of the volatile products. To ensure a quantitative reduction of Se, a pre-reduction had to be done to convert all Se to Se(IV). The hydride system is coupled to a quadrupole ICP-MS.

Tin

Chemiluminescence Detection

Zhu et al. [94] studied the lucigenin-H_2O_2-Sn(IV) chemiluminescence system by using reversed FI. This system is applied to the determination of Sn(IV) in water and sediment samples.

Atomic Absorption Spectrometric Detection

Tsalev et al. [17] studied the in-situ collection of volatile hydrides in an electrothermal atomizer with an integrated platform pre-treated with Zr or W and 2 Ir for permanent modification. The best leveling-off effect on the integrated absorbance for different analyte species (isoformation) is observed for As and the worst for organotins, particularly for trialkylated species such as tributyltin, trimethyltin and trimethylselenonium. Relatively better isoformation is achieved for organotins on Ir-W- and for organoselenium on Ir-Zr-treated platforms. Navarro et al. [67] optimized the determination of tin by FI-hydride generation-quartz furnace atomic absorption spectrometry (FI-HG-QFAAS) following different experimental designs. Scriver et al. [95] described a procedure for the quantitation of tributyltin in aqueous samples and extracts based on its relatively high volatility in halide media, permitting vapor phase sampling from the headspace above such samples. Tributyltin chloride (TBT-Cl) is purged from various chloride containing aqueous matrixes and collected on the surface of an Ir treated graphite tube for subsequent quantitation by graphite furnace atomic absorption.

Inductively Coupled Plasma Atomic Emission Spectrometric Detection

Feng et al. [24] established a hydride generation system using a small concentric hydride generator combined with inductively coupled plasma atomic emission spectrometry (ICP-AES) to determine tin, arsenic, bismuth and antimony with L-cysteine as a pre-reductant.

Inductively Coupled Plasma Mass Spectrometric Detection

Abranko et al. [25] performed a simultaneous multi-elemental measurement of As, Bi, Ge, Sb, Se, and Sn by flow injection-hydride generation-inductively coupled plasma-time-of-flight mass spectrometry (FI-HG-ICP-TOFMS) with off-line pre-reduction using potassium iodide and ascorbic acid. Ribeiro et al. [26] proposed a method for the determination of As, Hg, Sb, Se and Sn in solid environmental samples, as acidified slurries, by FI coupled to a hydride generation system and detection by ICP-MS.

Ion Spray Mass Spectrometry/mass Spectrometry

Siu et al. [96] determined tributyltin (TBT) in a sediments by using ion spray mass spectrometry/mass spectrometry. TBT is extracted into isooctane or 1-butanol, diluted with methanol containing ammonium acetate, delivered to the ion spray tandem mass spectrometer by using FI, and quantitated by means of selected reaction monitoring of the daughter/parent pair of m/z 179/291.

Capillary Gas Chromatography with Atomic Emission Detection

Liu et al. [97] used capillary gas chromatography with atomic emission detection (GC-AED) for the simultaneous determination of organotin, organolead, and organomercury compounds in environmental samples. Pentylmagnesium bromide is used to pentylate ionic organotin, organolead, and organomercury compounds. The derivatives are then separated by GC and determined by AED. Their minimum detectable concentrations are approximately 1.0 to 2.5 ng/mL using a 0.5 μL on-column injection.

Tellurium

Tsalev et al. [17] carried out the in situ collection of volatile hydrides in an electrothermal atomizer with an integrated platform pre-treated with of Zr or W and Ir for permanent modification. Moya et al. [98] compared three sample preparation techniques for the determination of Te. Two acid digestion procedures with and without HF on a hot plate and one without HF in a microwave oven were applied. Tellurium is quantified by FI hydride generation atomic absorption spectrometry. The mineralization procedure using the microwave oven proved to be the most efficient sample preparation technique, with recovery values from (98 ± 2)% to (103 ± 4)%.

Zinc

Hirata et al. [29] developed an on-line column preconcentration technique for FI atomic absorption spectrometry for the determination of Cd^{2+}, Zn^{2+}, Cu^{2+}, Mn^{2+}, Pb^{2+}, Fe^{3+} and Cr^{3+}. Maquieira et al. [30] preconcentrated Cd^{2+}, Zn^{2+}, Cu^{2+}, Pb^{2+}, and Fe^{3+} by using a FI manifold involving a minicolumn packed with yeast Saccharomyces cerevisiae covalently immobilized on controlled pore glass and flame atomic absorption spectrometric detection. Kashiwabara et al. [99] used a small column packed with immobilized bovine carbonic anhydrase for determination of traces of Zn based on the measurement of recovered esterase activity of the metal-free apoenzyme after taking up Zn from the sample solution.

ANIONIC SPECIES

The anionic species that have been determined in sediment by using FI methodologies were the following: carbonate, phosphorous/phosphate, sulfur/sulfide and silicate.

Different features of FI methods for the determination of anionic species in sea and estuarine sediments are illustrated in Table 5.3. In the following paragraphs, some points observed in this table are highlighted due to their interest.

Carbonate

Perez-Ponce et al. [100] described a rapid, sensitive and direct procedure for the determination of carbonate in sediments based on vapor-generation FTIR spectrometry. A volume of diluted hydrochloric acid is injected into a vessel containing the sediment sample and the vessel is heated at 40°C. The CO_2 evolved under these conditions is swept by a stream of nitrogen to an IR gas cell. The FI recordings are registered as a function of time and the areas of the FI recording obtained in the wavenumber range 2500-2150 cm^{-1} are measured and interpolated in a calibration equation obtained from known amounts of $CaCO_3$ treated in the same way as the samples.

Phosphorus/phosphate

McKelvie et al. [101] reported a study of the factors affecting separation and detection of dissolved organic and inorganic phosphorus species. The system involved the use of gel filtration and FIA. Orthophosphate and myo-inositol hexakisphosphate (Phytic acid), as model solutes representative of low molecular weight P (LMWP) and high molecular weight P (HMWP), are separated on a Sephadex G25 column incorporated into a FI manifold, which utilized photo-oxidation and spectrophotometry for detection of dissolved reactive and dissolved organic phosphorus. Galhardo et al. [102] presented a spectrophotometric sequential injection (SI) determination of phosphate and silicate using the molybdenum blue reaction. The interference of silicate in the determination of phosphate is eliminated by using a reagent composed of ammonium molybdate in diluted nitric acid, containing oxalic acid to avoid the formation of molybdosilicic acid. To perform this method in the single line SI system, obtaining a total sample and reagent zones penetration, it is used as a combination of sandwiching the sample zone between reagent zones and flow reversal through an auxiliary reaction coil. Anderson et al. [103] streamlined the five-step SEDEX (sedimentary extraction) procedure for characterizing sedimentary phosphorus to a four-step procedure. The procedure combines extraction of adsorbed and oxide-associated pools into a single step, retaining steps for authigenic, detrital, and organic P. In addition, the method uses automated spectrophotometric FIA to determine phosphorous concentrations.

Sulfur/sulfide

Makishima et al. [104] developed a new, simple, and accurate method for the determination of total sulfur at microgram per g levels in milligram-sized silicate materials with isotope dilution high-resolution inductively coupled plasma mass spectrometry equipped with a FI system. In this method, sulfur can be quantitatively oxidized by bromine into sulfate with achievement of isotope equilibrium between the sample and spike. Xia et al. [105] established a FI spectrophotometric method for the determination of acid hydrolyzable sulfides. The sample is analyzed in an hydrochloric acid solution containing dimethylphenylenediamine and $FeCl_3$ medium, and methylene blue, being monitored at 745 nm.

Table 5.3. Features of FI determinations of anionic species in sea and estuarine sediments

Analyte	Detection	DL	Linear range	SF (s/h)	RSD (%)	Ref
CO_3^{2-}	FTIR	0.02 g/g	No data	15	0.7	100
PO_4^{3-} Phytic acid	SP	1.1-12 μg/L 27-44 μg/L	No data	12-15	No data	101
PO_4^{3-}	SP	0.1 mg/L	0.2-0.7 mg/L	75	No data	102
Oxide-associated P Authigenic P Detrital P Organic P	SP	21.1 μg/g 19.5 μg/g 21.7 μg/g 7.8 μg/g	No data	No data	10-62	103
^{32}S ^{34}S	ICP-MS	0.07 μg/g 0.3 μg/g	No data	No data	<9	104
S^{-2}	SP	0.04 mg/L	0.05-2 mg/L	50	No data	105
SiO_3^{2-}	SP	1 mg/L	5-50 mg/L	40	No data	102

DL: detection limit; FTIR: Fourier transform infrared spectrometry; ICP-MS: inductively coupled plasma mass spectrometry; RSD: relative standard deviation; SF: sampling frequency; SP: spectrophotometry.

Silicate

Galhardo et al. [102] reported a spectrophotometric sequential injection (SI) determination of phosphate and silicate using the molybdenum blue reaction. The interference of phosphate in the determination of silicate is avoided adding a oxalic acid solution to the reaction zone where the molybdophosphoric and molybdosilicic acids are previously formed, to destroy the molybdophosphoric acid.

Conclusion

FIA is shown to be an effective tool for the determination of analytes in sea and estuarine sediments. Thus, have been proposed analytical methodologies including FIA strategies for sample introduction into the analytical instrument, preconcentration and/or matrix removal (above all by using solid-phase extraction) and to perform on-line different chemical reactions. Nevertheless, only one method is established where the preparation of a solid sample, as are sediments, is accomplished fully on-line by microwave digestion coupled to a FIA manifold to provide a completely automated analytical system.

References

[1] Mitra, S. (2003). *Sample preparation techniques in analytical chemistry*. New York: Wiley-IEEE Press.

[2] Smittenberg, R.H., Hopmans, E.C., Schouten, S. & Sinninghe-Damste, J. S. (2002). *J. Chromatogr.* 978A, 129-140.
[3] Piñeiro-Avila, G., Salvador, A. & de la Guardia, M. (1998). *Analyst* 123, 2291-2295.
[4] Piñeiro-Avila, G., Salvador, A. & de la Guardia, M. (2000). *Ciencia* 8, 77-84.
[5] Deflandre, B. & Gagne, J.P. (2001). *Water Research* 35, 3057-3062.
[6] Utsumi, A., Nakashima, A., Ando, K., Kizu, R. & Hayakawa, K. (1998). Anal. Sci. 14, 845-847.
[7] Zougagh, M., Rios, A. & Valcarcel, M. (2004). *Anal. Chim. Acta* 524, 279-285.
[8] Elmahadi, H.A.M. & Greenway, G.M. (1994). *J. Anal. At. Spectrom.*, 9, 547-551.
[9] Yang, L. & Sturgeon, R.E. (2002). *J. Anal. At. Spectrom.*, 17, 88-93.
[10] Zhao, Z. (1996). *Huaxue Tongbao* 12, 25-27.
[11] Zhao, Z. (1996). *Fenxi Huaxue* 24, 456-458.
[12] Zhao, Z., Han, G.; Song, X. & Luo, Z. (2008). *Guangpuxue Yu Guangpu Fenxi* 28, 1413-1415.
[13] Barrado, E., Rodriguez, J.A., Quinaz, M.B. & Lima, J.L.F.C. (2006). *Int. J. Environ. Anal. Chem.* 86, 563-572.
[14] Wei, L., Gupta, P., Hernandez, R. & Farhat, F. (1999). *Microchem. J.* 62, 83-98.
[15] Guo, T., Erler, W., Schulze, H. & McIntosh, S. (1990). *At. Spectrosc.* 11, 24-28.
[16] Saraswati, R., Vetter, T.W. & Watters, R.L. (1995). *Mikrochim. Acta* 118, 163-175.
[17] Tsalev, D.L., D'Ulivo, A., Lampugnani, L., Di Marco, M. & Zamboni, R. (1996). *J. Anal. At. Spectrom.* 11, 979-988.
[18] Zhou, C.Y., Wong, M.K., Koh, L.L. & Wee, Y.C. (1997). *Mikrochim. Acta* 127, 77-83.
[19] Capelo, J.L., Lavilla, I. & Bendicho, C. (2001). *Anal. Chem.* 73, 3732-3736.
[20] Sakamoto, H., Susa, Y., Ishiyama, H., Tomiyasu, T. & Anazawa, K. (2001). *Anal. Sci.* 17, 1067-1071
[21] Gonzalez, J.C., Lavilla, I. & Bendicho, C. (2003). *Talanta* 59, 525-534.
[22] Quevedo, O., Villanueva, M., Luna, B., Garrido, M.L. & Rodriguez, A.C. (2005). *Ciencias Quimicas* 36, 3-5.
[23] Bentlin, F.R.S., Duarte, F.A., Dressler, V.L. & Pozebon, D. (2007). *Anal. Sci.* 23, 1097-1101.
[24] Feng, Y.L., Chen, H.Y., Chen, H.W. & Tian, L.C. (1998). *Fresenius J. Anal. Chem.*, 361, 155-157
[25] Abranko, L., Stefanka, Z. & Fodor, P. (2003). *Anal. Chim.Acta* 493, 13-21.
[26] Ribeiro, A. S., Vieira, M.A. & Curtius, A.J. (2004). *Spectrochim. Acta* 59B, 243-253.
[27] Klassen, A., Kim, M.L., Tudino, M.B., Baccan, N. & Arruda, M.A. Z. (2008). *Spectrochim. Acta* 63B, 850-855.
[28] Wang, Y., Qiu, H., Tang, Z., Gu, Y. & Zhang, S. (2007). *Guangpuxue Yu Guangpu Fenxi* 27, 2581-2584.
[29] Hirata, S., Honda, K. & Kumamaru, T. (1989). *Anal. Chim. Acta* 221, 65-76.
[30] Maquieira, A., Elmahadi, H.A.M. & Puchades, R. (1994). Anal. Chem. 66, 1462-1467.
[31] Ma, R., Van Mol, W. & Adams, F. (1994). *Anal. Chim. Acta* 285, 33-43.
[32] Ma, R. & Adams, F. (1996). *Spectrochim. Acta* 51B, 1917-1923.
[33] Matusiewicz, H., Kopras, M. & Sturgeon, R.E. (1997). *Analyst* 122, 331-336.
[34] Kartikeyan, S., Vijayalekshmy, B., Chandramouleeswaran, S., Prasada Rao, T. & Iyer, C.S.P. (1997). *Anal. Lett.* 30, 1037-1050.
[35] Soylak, M. & Narin, I. (2005). *Chemia Analityczna* 50, 705-715.

[36] Hwang, T.J. & Jiang, S.J. (1997). *J. Anal.At. Spectrom.* 12, 579-584.
[37] Valles Mota, J.P.; Fernandez de la Campa, M. R., Garcia Alonso, J.I. & Sanz-Medel, A. (1999). *J. Anal.At. Spectrom.* 14, 113-120.
[38] Vieira, M.A., Ribeiro, A.S., Dias, L.F. & Curtius, A.J. (2005). *Spectrochim. Acta* 60B, 643-652.
[39] Yamada, M., Komatsu, T., Nakahara, S. & Suzuki, S. (1983). *Anal. Chim. Acta* 155 259-262.
[40] Komatsu, T., Ohira, M., Yamada, M. & Suzuki, S. (1986). *Bull. Chem. Soc. Japan* 59, 1849-1855.
[41] Makita, Y., Suzuki, T., Yamada, M. & Hobo, T. (1994). *Nippon Kagaku Kaishi* 8, 701-706.
[42] Praveen, R.S., Daniel, S. & Rao, T.P. (2006). *At. Spectrosc.* 27, 35-43.
[43] Zara, L.F., Rocha, J.C., Milani, M.R., Rosa, A.H., dos Santos, A. & Serudo, R.L. (2003). *Anal. Lett.* 36, 781-795.
[44] Paleologos, E.K. (1998). *Analyst* 123, 1005-1009.
[45] Anthemidis, A.N., Zachariadis, G.A., Kougoulis, J.S. & Stratis, J.A. (2002).Talanta 57, 15-22.
[46] Pazos-Capeans, P., Barciela-Alonso, M.C., Bermejo-Barrera, A., Bermejo-Barrera, P., Fisher, A. & Hill, S.J. (2006). *At. Spectrosc.* 27, 107-116.
[47] Yamada, M., Kanai, H. & Suzuki, S. (1985). *Bull. Chem. Soc. Japan* 58, 1137-1142.
[48] Yuan, Y. (1987). *Haiyang Xuebao (Zhongwenban)* 9, 51-55.
[49] Murakami, K., Okamoto, Y. & Kumamaru, T. (1992). *J. Flow Injection Anal.* 9, 195-203.
[50] Wang, Z.H., Wang, Z.P., Zhang, Z.P.; Liu, L.W. & Yan, X.P. (2005). *At. Spectrosc.* 26, 34-39.
[51] Yuan, Y. (1987). *Haiyang Yu Huzhao* 18, 156-161.
[52] Amini, N., Cardwell, T.J., Cattrall, R.W. & Kolev, S. (2005). *Anal. Chim. Acta* 539, 203-207.
[53] Wang, Y., Wang, X., Wang, J. & Guo, Y. (2003). *Lihua Jianyan, Huaxue Fence* 39, 641-642, 645.
[54] Tseng, C.M., Hammerschmidt, C.R. & Fitzgerald, W.F. (2004). *Anal. Chem.* 76, 7131-7136.
[55] Shi, J.B., Liao, C.Y., Wang, Y.W. & Jiang, G.B. (2006). *Guangpuxue Yu Guangpu Fenxi* 26, 336-339.
[56] Yin, Y., Liang, J., Yang, L. & Wang, Q. (2007). *J. Anal. At. Spectrom.* 22, 330-334.
[57] Liang, L., Hu, J., Jiang, G. & Shi, J. (2005). *Fenxi Huaxue* 33, 229-232.
[58] Fostier, A.H., Ferreira, J.R. & Oeterrer de Andrade, M. (1995). *Quimica Nova* 18, 425-430.
[59] Hanna, C.P. & McIntosh, S.A. (1995). *At. Spectros.* 16, 106-114.
[60] Murphy, J., Jones, P. & Hill, S. J. (1996). *Spectrochim. Acta*, 51B, 1867-1873.
[61] Zhou, C.Y., Wong, M.K., Koh, L.L. & Wee, Y.C. (1996). *Anal. Sci.* 12, 471-476.
[62] Bauza de Mirabo, F.M., Thomas, A.Ch., Rubi, E., Forteza, R. & Cerdà, V. (1997). *Anal. Chim. Acta* 355, 203-210.
[63] Magalhaes, C.E.C., Krug, F.J., Fostier, A.H. & Berndt, H. (1997). *J. Anal. At. Spectrom.* 12, 1231-1234.

[64] Martinez-Garcia, M.L., Carlosena, S., Lopez-Mahia, P., Muniategui, S. & Prada, D. (1999). *Analusis* 27, 61-65.
[65] Li, Y., Jiang, Y., Yan, X.P. & Ni, Z.M. (2002). *Environ. Sci.Technol.* 36, 4886-4891.
[66] Collasiol, A., Pozebon, D. & Maia, S.M. (2004). *Anal. Chim. Acta* 518, 157-164.
[67] Navarro, P., Raposo, J.C., Arana, G. & Etxebarria, N. (2006). *Anal. Chim. Acta* 566, 37-44.
[68] Emteborg, H., Bjoerklund, E., Oedman, F., Karlsson, L., Mathiasson, L., Frech, W. & Baxter, D.C. (1996). *Analyst* 121, 19-29.
[69] Woller, A., Garraud, H., Martin, F., Donard, O.F.X. & Fodor, P. (1997). *J. Anal. At. Spectrom.* 12, 53-56.
[70] Colodner, D.C., Boyle, E.A. & Edmond, J.M. (1993). *Anal Chem.* 65, 1419-1425.
[71] de Almeida, M.D., Leandro, K.C., Da Costa, C.V., Santelli, R.E. & de La Guardia, M. (1997). *J. Anal. At. Spectrom.* 12, 1235-1238.
[72] Mesquita, M., Jacintho, A.O., Zagatto, E.A.G. & Antonio, R.F. (1990). *J. Brazilian Chem. Soc.* 1, 28-34.
[73] Gong, L. & Yu, Z. (1995). *Fenxi Shiyanshi* 14, 11-13.
[74] Cheng, X., Qin, X., Liu, F. & Bu, J. (2005). *Haiyang Xuebao (Zhongwenban)* 27, 177-179.
[75] Cheng, X.S., Qin, X.G., Xu, R. & Wang, J.H. (2004). *Haiyang Huanjing Kexue* 23, 72-74.
[76] Yan, X.P., Sperling, M. & Welz, B. (1999). *Anal. Chem.* 71, 4216-4222.
[77] Zachariadis, G.A., Anthemidis, A.N., Bettas, P.G. & Stratis, J.A. (2002). *Talanta* 57, 919-927.
[78] Praveen, R.S., Naidu, G.R.K. & Rao, T.P. (2007). *Anal. Chim. Acta* 600, 205-213.
[79] Levine, N., Kim, K.E., Nitz, L.H., Valdes-Hevia y Temprano, M.C., de la Campa, M.R.F. & Sanz-Medel, A. (1995). *Anal. Chim. Acta* 309, 369-378.
[80] Lasztity, A., Viczian, M., Wang, X. & Barnes, R.M. (1989). *J. Anal. At. Spectrom* 4, 761-766.
[81] Persaud, A.T., Beauchemin, D., Jamieson, H. E. & McLean, R.J.C. (1999). *Canadian J. Chem.* 77, 409-415.
[82] Li, J., Lu, F., Umemura, T. & Tsunoda, K.I. (2000). *Anal. Chim. Acta* 419, 65-72.
[83] Wang, Z.H., Yan, X.P., Wang, Z.P., Zhang, Z.P. & Liu, L.W. (2006). *J.Am. Soc. Mass Spectrom.* 17, 1258-1264.
[84] Halicz, L., Bar-Matthews, M., Ayalon, A. & Kaufman, A. (1997). *At. Spectrosc.* 18, 175-179.
[85] Godoy, M.L.D.P., Godoy, J. M., Kowsmann, R., dos Santos, G.M. & Petinatti da Cruz, R.. (2006). *J. Environ. Radioact.* 88, 109-117.
[86] Epov, V.N., Evans, R.D., Zheng, J., Donard, O.F.X. & Yamada, M. (2007). *J. Anal. At. Spectrom* 22, 1131-1137.
[87] De la Calle Guntinas, M.B., Madrid, Y. & Camara, C. (1992). *Mikrochim. Acta* 109, 149-155.
[88] D'ulivo, A., Lampugnani, L., Faraci, D., Tsalev, D. L. & Zamboni, R. (1998). *Talanta* 45, 801-806.
[89] Bolea, E., Arroyo, D., Laborda, F. & Castillo, J.R. (2006). *Anal. Chim. Acta* 569, 227-233.

[90] Liu, H., Peng, T., Jiang, Z., Xu, Y. & Li, W. (1997). *Wuhan University Journal of Natural Sciences* 2, 348-352.
[91] Ahmed, M.J., Stalikas, C.D., Veltsistas, P.G., Tzouwara-Karayanni, S.M. & Karayannis, M.I (1997). *Analyst* 122, 221-226.
[92] Chan, C.C.Y. & Sadana, R.S. (1992). *Anal. Chim. Acta* 270, 231-238.
[93] Moor, C. & Kobler, J. (2001). *J. Anal. At. Spectrom.* 16, 285-288.
[94] Zhu, X. & Yu, Z. (1996).*Yejin Fenxi* 16, 19-22.
[95] Scriver, C., Mester, Z., Sturgeon, R.E., Peter, L. & Willie, S. (2002). *J. Anal. At. Spectrom* 17, 1511-1515.
[96] Siu, K.W.M., Gardner, G.J. & Berman, S.S. (1989). *Anal. Chem.* 61, 2320-2322.
[97] Liu, Y., Lopez-Avila, V., Alcaraz, M. & Beckert, W.F. (1994). *J. High Resol. Chromatogr.* 17, 527-536.
[98] Moya, M.J., Torralba, R., Palacios, M.A. & Bonilla, M.M. (2001). *Quim. Anal.* 20, 125-130.
[99] Kashiwabara, K., Hobo, T., Kobayashi, E. & Suzuki, S. (1985). *Anal. Chim. Acta* 178, 209-215.
[100] Perez-Ponce, A. & Garrigues, S. (1998). *Analyst* 123, 1817-1821.
[101] McKelvie, I.D., Hart, B.T., Cardwell, T.J. & Cattrall, R.W. (1993). *Talanta* 40, 1981-1993.
[102] Galhardo, C.X. & Masini, J.C. (2000). *Anal. Chim. Acta* 417, 191-200.
[103] Anderson, L.D. & Delaney, M.L. (2000). *Limnol. Oceanogr.* 45, 509-515.
[104] Makishima, A. & Nakamura, E. (2001). *Anal. Chem.* 73, 2547-2553.
[105] Xia, C., He, X. & Shi, H. (2003). *Fenxi Huaxue* 31, 767.

Chapter 6

SEAWEEDS

ABSTRACT

This chapter presents a comprehensive review of flow injection (FI) methodologies proposed for the determination of organic and inorganic analytes (cationic and anionic species) in seaweeds samples. Thus, are described FI methods applied to the determination of intracellular free amino acids, β-dimethylsulfoniopropionate, laminarin, arsenic, germanium, mercury, molybdenum, tin, iodide, phosphate, nitrate, nitrite and silicate. Analytical figures of merit, characteristics, features and interferences are also discussed for each analyte.

INTRODUCTION

The sample preparation for the determination of organic substances in seaweeds samples includes analyte extraction. The desiccated seaweeds are stirred in deionized water or diluted hydrochloric acid (about 2 h). Then, the solution is centrifuged at 5000 rpm for about 5 min and the resulting precipitate is washed twice with diluted hydrochloric acid. The supernatant and washing solution are combined and the resulting solution is kept in freezer at -20 °C until the moment of analysis.

Most of the existing procedures for preparing seaweeds samples for analysis of cationic species involve acid digestion. Thus, powdered seaweed samples are weighed accurately and digested in open vessels at 80°C with HNO_3-H_2SO_4 to near-clearness, addition of H_2O_2, and continuously heated until the fume of H_2SO_4 is produced. Usually, the residual is dissolved and diluted with distilled water. If the acid digestion is performed in closed vessels by using a microwave oven, nitric acid and hydrogen peroxide are the acid/oxidant reagents used. The acid digests are kept at 4 °C until the moment of analysis. For the analysis of seaweeds by using FI systems, microwave digestion is employed for off-line sample digestion. Nevertheless, for mercury determination it was designed and evaluated an automated on-line pressurized sample digestion method based on an electromagnetic induction heating technique. This on-line pressurized digestion system is on-line coupling to cold vapor generation atomic fluorescence spectrometry (Figure 6.1).

For analysis of an anionic species as iodide in seaweeds, samples are dried in an oven and sample preparation is performed by alkaline digestion. These alkaline digestion procedures partially decompose the sample matrix and allow the integrity of the different chemical forms to be retained.

ORGANIC SPECIES

The organic analytes that have been determined in seaweeds by using FI methodologies were the following: intracellular free amino acids, β-dimethylsulfoniopropionate and laminarin. Different features of FI methods for the determination of organic species in seaweeds are illustrated in Table 6.1. In the following paragraphs, some points observed in this table are highlighted due to their interest.

Table 6.1. Features of FI determinations of organic species in sea and estuarine sediments

Analyte	Detection	DL (μM)	Linear range (μM)	SF (s/h)	Recovery (%)	RSD (%)	Ref
Asp	F	0.27	1-25	No data	94	1-22	1
Glu		0.24	1-25		99		
Asn		0.24	1-25		98		
Ser		0.24	1-25		98		
Gln		0.11	1-25		103		
Gly		0.080	1-25		102		
Thr		0.060	0.5-10		95		
Ctr		0.20	0.5-10		98		
Arg		0.41	0.5-10		96		
Ala		0.093	0.5-10		102		
Tyr		0.092	0.5-10		101		
Met		-	0.5-10		-		
Val		-	0.5-10		-		
Phe		0.17	0.5-10		109		
Ile		-	0.5-10		-		
Leu		-	0.5-10		-		
Orn		0.51	1-25		112		
Lys		0.15	1-25		106		
β-DMSNP	FPD	0.1	1-1000	No data	No data	5.2	2
Laminarin	A	99.1	0.99-99.12 mM	12	86.2-88.3	2-2.5	3

A: amperometry; β-DMSNP: β-dimethylsulfoniopropionate; DL: detection limit; F: fluorescence; FPD: flame photometric detection; RSD: relative standard deviation; SF: sampling frequency; SP: spectrophotometry.

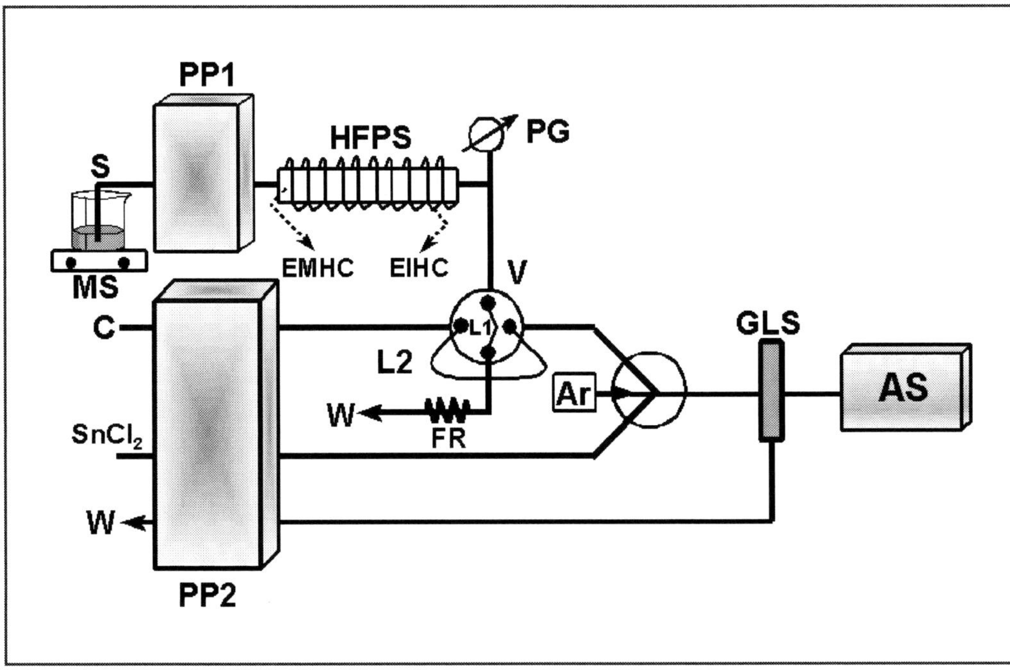

Figure 6.1. Schematic diagram of an continuous digestion manifold coupled to on-line cold vapor generation-atomic spectrometric detection. AS, atomic spectrometer; HFPS, high-frequency induction heating power supply; EIHC, electromagnetic induction heating coil; EMHC, electromagnetic heating column; FR, flow restrictor; GLS, gas-liquid separator; L1 and L2, sample loops; MS, magnetic stirring; PG, pressure gauge; PP1 and PP2, peristaltic pumps; V, 4-way valve.

Intracellular Free Amino Acids

Rigobello-Masini et al. [1] exploited the concept of sequential injection chromatography (SIC) to automate the fluorimetric determination of amino acids after pre-column derivatization with o-phthaldialdehyde in presence of 2-mercaptoethanol using a reverse phase monolithic C18 stationary phase. The method is low-priced and based on five steps of isocratic elutions. The first step employs the mixture methanol: tetrahydrofuran and the other steps use methanol in a phosphate buffer (pH 7.2). At a flow rate of 10 mL/s a 25 mm long-column is able to separate aspartic acid (Asp), glutamic acid (Glu), asparagine (Asn), serine (Ser), glutamine (Gln), glycine (Gly), threonine (Thr), citruline (Ctr), arginine (Arg), alanine (Ala), tyrosine (Tyr), phenylalanine (Phe), ornithine (Orn) and lysine (Lys) with resolution >1.2 as well as methionine (Met) and valine (Val) with resolution of 0.6. Under these conditions isoleucine (Ile) and leucine (Leu) co-eluted. The entire cycle of amino acids derivatization, chromatographic separation and column conditioning at the end of separation lasted 25 min. The method is applied to the determine intracellular free amino acids in the green alga Tetraselmis gracilis during a period of seven days of cultivation.

Table 6.2. Features of FI determinations of cationic species in seaweeds

Analyte	Detection	Separation Technique	DL (µg/L)	SF (s/h)	RSD (%)	Ref
As(V)	Ch		10	No data	No data	4
Inorg. As As	AAS	On-line gas-liquid separation	0.014 µg/g 0.025 µg/g	No data	4 1	5
Inorg. As As	ICP-AES	On-line gas-liquid separation	0.014 µg/g 0.33 µg/g	No data	4 <10	6
Inorg. As As	AF	On-line gas-liquid separation	0.041	No data	1.73	7
Ge	Ch		50	No data	No data	4
Ge	SP		10	30	4.4-7.8	8
Hg	AAS	On-line gas-liquid separation	0.86	No data	9.3	9
Hg	FIAS-400[a] AAS	On-line preconcentration (SCF) On-line gas-liquid separation	0.025 5.23 ng/g	40	11	10
Hg	AF	On-line gas-liquid separation	1.2 ng/g	50	4.2	11
Mo	SP		60	30	4.4-7.8	8
Mo	SP		0.08	60	4.1	12
Sn	SP		40	30	4.4-7.8	8

[a] Perkin-Elmer.

AAS: atomic absorption spectrometry; DL: detection limit; ICP-AES: induced plasma atomic emission spectrometry; AF: atomic fluorescence; Ch: chemiluminescence; RSD: relative standard deviation; SCF: sulfydryl cotton fiber; SF: sampling frequency; SP: spectrophotometry.

β-Dimethylsulfoniopropionate

Howard et al. [2] developed a FIA system for the routine measurement of β-dimethylsulfoniopropionate (DMSP). The analytical system combines an in-line base hydrolysis step with transport of generated dimethylsulfide to the gas phase where it is measured by sulfur-specific flame photometric detection. This combination of chemical, physical and spectroscopic steps results in a system, which is highly selective to DMSP. The system is evaluated for its ability to distinguish DMSP from other dimethylsulfonium compounds. S-methyl methionine is not detected by the system. Only with very closely related compounds, such as the 2-methyl derivative of DMSP, generated false responses.

Laminarin

Miyanishi et al. [3] proposed a novel sensor system equipped with a reactor packed with beads containing immobilized β-1,3-glucanase and glucose oxidase for the amperometric determination of laminarin. The proposed sensor system consisted of a reactor, an oxygen electrode, a flow cell, a pump, a buffer tank, and a recorder. The measurement is performed with a FI system. Furthermore, this sensor can be used not only for the analysis of seaweed and health-enhancing foods but also for monitoring the initial pollution of the marine environment.

Cationic Species

The cationic species that have been determined in seaweeds by using FI methodologies were the following: arsenic, germanium, mercury, molybdenum and tin.

Different features of FI methods for the determination of cationic species in sea and estuarine sediments are illustrated in Table 6.2. In the following paragraphs, some points observed in this table are highlighted due to their interest.

Arsenic

Fujiwara et al. [4] established a FI chemiluminescence method for sensitive determination of arsenate, germanate, phosphate and silicate, after separation by ion chromatography (IC). The post-column detection system involved formation of heteropoly acid in a sulfuric acid medium before the chemiluminescence reaction with luminol in a sodium hydroxide medium. For separation, heteropoly acid formation and the chemiluminescence detection reaction, pH requirements are not compatible. When present as a heteropoly acid complex with molybdenum(VI), arsenic(V) and phosphorus(V) caused chemiluminescence emission from oxidation of luminol with the addition of metavanadate ion to the acid solution of molybdate. Laparra et al. [5] realized a study to examine the bioaccessibility of total (AsT) and inorganic (AsI) arsenic contents and the effects of cooking edible seaweed. An in vitro gastrointestinal digestion (pepsin, pH 2, and pancreatin-bile extraction, pH 7) is applied to obtain the mineral solution fraction of three seaweeds. AsT is determined by dry-ashing FI hydride generation atomic absorption spectrometry. AsI is determined by acid digestion, solvent extraction, and FI hydride generation atomic absorption spectrometry. Farias et al. [6] analyzed total and inorganic As Antarctic macroalgae. Total As is determined by inductively coupled plasma-optical emission spectrometry after microwave-assisted acid digestion. Inorganic As is determined by acid digestion, solvent extraction, FI-hydride generation-atomic absorption spectrometry. Xu et al. [7] studied the digestion method and the acid of carrier solution, concentration of KBH_4 and carrier gas capacity to determine inorganic and total arsenic in seaweed.

Germanium

Fujiwara et al. [4] proposed a FI chemiluminescence (CL) method is for determination of arsenate, germanate, phosphate and silicate, after separation by ion chromatography (IC). The post-column detection system involved formation of heteropoly acid in a H_2SO_4 medium before the chemiluminescence reaction with luminol in an NaOH medium. When present as a heteropoly acid complex with molybdenum(VI), germanium(IV) and silicon(IV) caused chemiluminescence emission from oxidation of luminol. Zou et al. [8] performed the simultaneous determination of Sn, Ge and Mo by FI-charge coupled detector (CCD) diode array detection spectrophotometry with partial least squares (PLS) algorithm. The method is based on the chromogenic reaction of metal ions and salicylfluorone in the presence of cetyltrimethyl ammonium bromide. The overlapping spectra of these complexes are collected by CCD diode array detector and the multi-wavelength absorbance data are processed using partial least squares algorithm. The method is applied to directly determination of Ge, Mo and Sn after digestion with satisfactory results.

Mercury

Zenebon et al. [9] used a mixture of 50% H_2O_2-H_2SO_4 (3 + 1, V/V) for decomposition of marine algae and other foods in open vessels at 80°C. The treatment allows rapid total mercury determination by FI cold vapor atomic absorption spectrometry. Fernandez-Fernandez et al. [10] evaluated the possibilities for the use of an inexpensive sulfydryl cotton fiber (SCF) adsorbent to separate and preconcentrate traces of Hg from acid digests from edible seaweed. After mercury elution, the eluate is mixed with sodium tetrahydroborate and the Hg cold vapor is swept through the atomization cell with an argon flow. Han et al. [11] developed a novel, automated on-line pressurized sample digestion system based on electromagnetic induction heating technique to perform solid sample decomposition in acid medium. The efficiency of acid digestion is increased with 0.36 MPa pressure built up in-line by a 0.5 mm id x 25 m length flow restrictor and up to approx. 135 °C reaction temperature. The system's performance is evaluated for on-line digestion of edible seaweeds and subsequent determination of mercury by cold vapor generation atomic fluorescence spectrometry.

Molybdenum and Tin

Zou et al. [8] performed the simultaneous determination of Sn, Ge and Mo by FI-charge coupled detector (CCD) diode array detection spectrophotometry with partial least squares (PLS) algorithm. Zou et al. [12] presented a sensitive spectrophotometric method for the determination of Mo by FI-CCD (charge coupled detector) diode array detection. The method is based on the chromogenic reaction between Mo ions and salicylfluorone in the presence of cetyltrimethyl ammonium bromide (CTMAB).

Anionic Species

The anionic species that have been determined in seaweeds by using FI methodologies were the following: iodide, iodate, phosphate, nitrate, nitrite and silicate.

Table 6.3. Features of FI determinations of anionic species in seaweeds

Analyte	Detection	DL (μM)	Linear range (μM)	SF (s/h)	RSD (%)	Ref
I^-	A	0.1	0.5-1000	No data	3.2	12
I^-	A	50	10-1000	No data	No data	13
I_2	Ch	0.013	0.03-8	70	2.6	14
IO_3^-	SP	0.6	2-20	84	No data	15
I^-	SP	No data	0-7.8	80	No data	16
I^-	SP	0.6	0-23.6	25	2.2	17
I^-	SP	0.16	0.4-7.9	No data	0.85	18
I_2	ICP-AES	No data	39.4-394.0	40	No data	19
NO_3^-	SP	No data	No data	No data	No data	20
NO_2^-	SP	No data	1-700	No data	No data	20
PO_4^{3-}	SP	No data	6-1000	No data	No data	20
PO_4^{3-}	Ch	0.032	0.032-32.3	No data	No data	4
PO_4^{3-}	SP	1.1	2.1-73.7	75	No data	21
PO_4^{3-}	SP	No data	No data	No data	No data	22
SiO_3^{2-}	Ch	0.35	1.8-355.9	No data	No data	4
SiO_3^{2-}	SP	1.3	65.7-657.0	40	No data	21
SiO_3^{2-}	SP	No data	No data	No data	No data	22

A: amperometry; Ch: chemiluminescence; DL: detection limit; ICP-AES: induced plasma atomic emission spectrometry; RSD: relative standard deviation; SF: sampling frequency; SP: spectrophotometry.

Different features of FI methods for the determination of anionic species in seaweeds are illustrated in Table 6.3. In the following paragraphs, some points observed in this table are highlighted due to their interest.

Iodine/iodide/iodate

Electrochemical Detection

Tian et al. [12] fabricated a vanadium oxide-modified glassy carbon electrode by casting vanadium tri(isoproxide) oxide ($VO(OC_3H_7)_3$) and poly(propylene carbonate) (PPC) onto the glassy carbon electrode surface. The electrochemical properties of iodide at the $VO(OC_3H_7)_3$-PPC film-modified glassy carbon electrode were investigated by cyclic voltammetry. An anodic peak was observed at approximately +0.71 V (vs. a saturated calomel electrode). Based on this, a sensitive and convenient electrochemical method is proposed for the determination of iodide by FI amperometry. Wang et al. [13] investigated and prepared the electrode modified by multilayer films of polyoxometalate (POM) $K_{17}[Ce(P_2Mo_{17}O_{61})_2]$ and polyelectrolyte by means of layer-by-layer assembly. The stable multilayer films are assembled by alternate adsorption of negative charged POM and positive charged polyelectrolyte from their aqueous dispersions. The proposed novel electrode exhibited a good electrocatalytic oxidation on iodide anions in solutions. Based on this, FI amperometry is proposed for the determination of iodide.

Chemiluminescence Detection

Gao et al. [14] described a simple, fast sequential injection system with chemiluminescence detection for the determination of iodine based on iodide ion oxidized by potassium permanganate with formaldehyde as an enhancer.

Spectrophotometric Detection

Chen et al. [15] proposed a rapid spectrophotometric FI method for the determination of iodate based on the red-violet ion association formed between 2-(5-bromo-2-pyridylazo)-5-diethylamino-o-phenol, iodate and thiocyanate. Wang et al. [16] carried out a FI-spectrophotometric method for the determination of iodide based on the color reaction of I_3^- with starch solution. In the FIA manifold, bromine is introduced as an oxidant for species change of iodide and the excess of bromine is eliminated by sulfosalicylic acid solution at room temperature. Lin et al. [17] developed and used a glass gravity separator for a continuous liquid-liquid extraction system for the determination of trace iodide ion by online extraction-spectrophotometry. Xiao et al. [18] described a new, simple and rapid spectrophotometric FI technique for accurate determination of iodide. The method is based on the decrease of absorbance intensity of rhodamine B due to the complexation with $[I_2Br]^-$, which I^- could be selectively oxidized to form I_2 by Ce(IV) in acidic medium where I_2 reacted with Br to form $[I_2Br]^-$. At this moment, some changes of the color of the solution took place. The absorption peak of the complexation increased linearly by addition of I^-, which occurred at 585 nm. Proper amounts of inorganic acid radical ions, metal ions and other halide ions would not interfere the determination.

Inductively Coupled Plasma Atomic Emission Spectrometric Detection

Fan [19] reported a new method for determination of I by using FI on-line precipitation ICP-AES. It is based on the continuous precipitation of silver iodide after silver nitrate reaction with iodide ion. The method is free from the interference of CO_3^{2-}, SO_3^{2-}, $C_2O_4^{2-}$.

Nitrate/nitrite

Masini et al. [20] presented the potential application of sequential injection analysis for on-line automated monitoring of phosphate, nitrate and nitrite in the culture medium of the marine alga Tetraselmis gracilis. Spectrophotometric detection is used in all cases. Nitrite is determined by reaction with sulfanilamide and N-1-naphthyl-ethylenediamine dihydrochloride. Nitrate is determined by the same reaction after reduction in a cadmium column. The sequential injection in-line dilution modes for concentrated samples are described, allowing the analytes in a wide range of concentrations.

Phosphate/silicate

Fujiwara et al. [4] proposed a FI chemiluminescence method for sensitive determination of arsenate, germanate, phosphate and silicate, after separation by ion chromatography (IC). The post-column detection system involved formation of heteropoly acid in a sulfuric acid medium before the chemiluminescence reaction with luminol in an sodium hydroxide medium. Masini et al. [20] presented the potential application of sequential injection analysis for on-line automated monitoring of phosphate, nitrate and nitrite in the culture medium of the marine alga Tetraselmis gracilis. Phosphate is determined by reaction with ammonium molybdate and ascorbic acid. Galhardo et al. [21] presented a spectrophotometric sequential injection (SI) determination of phosphate and silicate in environmental samples and cell cultivation medium using the molybdenum blue reaction. The interference of silicate in the determination of phosphate is eliminated by using a reagent composed of ammonium molybdate in diluted nitric acid, containing oxalic acid to avoid the formation of molybdosilicic acid. The interference of phosphate in the determination of silicate is avoided adding a solution of oxalic acid to the reaction zone where the molybdophosphoric and molybdosilicic acids are previously formed, to destroy the molybdophosphoric acid. To perform this task in the single line SI system, obtaining a total sample and reagent zones penetration, it is used as a combination of sandwiching the sample zone between reagent zones and flow reversal through an auxiliary reaction coil. Narusawa et al. [22] described a matrix effect on the simultaneous determination of Si and P in mussel, tea leaves, sargasso, and rice flour by on-line column FIA spectrometry. Volatile constituents in the materials are easily removed by ashing. The resulting ash is fused and then dissolved in hydrochloric acid. Interfering cations could be removed by cation exchange. The acidic effluent is evaporated to dryness. The residue is fused, taken up in diluted EDTA, and analyzed for Si and P by FIA. For simultaneous determination of these elements, TSK-gel SAX and NaCl-NH_3-EDTA eluent is used.

Conclusion

FIA has been applied a few times for determination of analytes in seaweeds. In most instances FIA methodologies are used simply for sample introduction into the analytical instrument and to perform on-line different chemical reactions. Thus, only was reported an on-line pretreatment for sample digestion, which is based on an electromagnetic induction heating technique. Furthermore, only was proposed a FIA methodology involving an on-line separation technique to separate and preconcentrate traces of mercury from acid digests of seaweeds samples.

References

[1] Rigobello-Masini, M., Penteado, J.C.P., Liria, C.W., Miranda, M.T.M., Masini, J.C. (2008). *Anal. Chim.Acta* 628, 123-132.

[2] Howard, A.G., Freeman, C.E., Russell, D.W., Arbab-zavar, M.H. & Chamberlain, P. (1998). *Anal. Chim.Acta* 377, 95-101.

[3] Miyanishi, N., Inaba, Y., Okuma, H., Imada, C. & Watanabe, E. (2004). *Biosensors & Bioelectronics* 19, 557-562.

[4] Fujiwara, T., Kurahashi, K., Kumamaru, T. & Sakai, H. (1996). *Appl. Organometal. Chem.* 10, 675-681.

[5] Laparra, J.M., Velez, D., Montoro, R., Barbera, R. & Farre, R. (2003). *J. Agric. Food Chem.* 51, 6080-6085.

[6] Farias, S., Smichowski, P., Velez, D., Montoro, R., Curtosi, A. & Vodopivez, C. (2007). *Chemosphere* 69, 1017-1024.
[7] Xu, Q., Liu, J., Zhao, N. & Gong, Y.X. (2008). *Guangpu Shiyanshi* 25, 380-383.
[8] Zou, X., Li, Y., Li, M., Zheng, B. and Yang, J. (2004). *Talanta* 62, 719-725.
[9] Zenebon, O., Sakuma, A.M., Dovidauskas, S., Okada, I. A., De Maio, F.D. & Lichtig, J. (2002). *J. AOAC Int.* 85, 149-152.
[10] Fernandez-Fernandez, A.M., Moreda-Pineiro, A. & Bermejo-Barrera, P. (2007). *J. Anal. At. Spectrom.* 22, 573-577.
[11] Han, S.P., Gan, W.E., Jiang, X.J., Zi, H.J. & Su, Q.D. (2008). *J. Anal. At. Spectrom.* 23, 773-776.
[12] Tian, L., Liu, L., Chen, L., Lu, N. & Xu, H. (2005). *Talanta* 66, 130-135.
[13] Wang, L., Wang, F., Dong, L. & Cai, H.N. (2008). *Yingyong Huaxue* 25, 868-870.
[14] Gao, C.Y., Fan, S.H. & Yao, J.B. (2006). *Fenxi Shiyanshi* 25, 38-41
[15] Chen, X., Zhao, X., Kou, Z. & Hu, Z. (1991). *Mikrochim. Acta* 1, 279-83.
[16] Wang, P. & Shi, S. (1996). *Fenxi Huaxue* 24, 720-723.
[17] Lin, S. & Zheng, Q. (1998). *Yankuang Ceshi* 17, 29-32
[18] Xiao, X., Zhang, X., Chen, G., Gong, Z. & Luo, Y. (2006). *Shipin Kexue* 27, 162-166.
[19] Fan, Z. (1999). *Fenxi Kexue Xuebao* 15, 309-311.
[20] Masini, J.C., Rigobello-Masini, M., Salatino, A. & Aidar, E. (2001). *Latin American Appl. Res.* 31, 463-468.
[21] Galhardo, C.X. & Masini, J.C. (2000). *Anal. Chim. Acta* 417, 191-200.
[22] Narusawa, Y. & Nawa, Y. (1991). *J. Flow Injection Anal.* 8, 21-31.

Chapter 7

MARINE ANIMALS/SEAFOOD

ABSTRACT

This chapter covers the review of the application of flow injection (FI) methodologies for the determination of organic and inorganic analytes (cationic and anionic species) in marine animals samples/seafoods. Thus, are described FI methods applied to the determination of amino acids (histidine, L-lysine and tyrosine), DNA/RNA, formaldehyde, histamine, hypoxanthine, polycyclic aromatic hydrocarbons (PAHs), diarrheic shellfish poisoning (DSP), paralytic shellfish poisoning (PSP), succinate/glutamate, trimethylamine/ total volatile basic nitrogen (TVBN), total lipid hydroperoxides (TLP), total volatile acids (TVA), uric acid, vitamin B_{12}, silver, aluminium, arsenic, boron, calcium, cadmium, cobalt, chromium, copper, iron, gallium, mercury, indium, lithium, manganese, molybdenum, nickel, lead, antimony, selenium, tin, strontium, thallium, vanadium, zinc, nitrate/nitrite, phosphorous/phosphate and silicate. Analytical figures of merit, characteristics, features and interferences are also discussed for each analyte.

INTRODUCTION

Mainly, the marine animals analyzed by FI methodologies are seafood (fish and shellfish): cod (*Gadus morrhua*), monkfish (*Lophius piscatorius*), skate (*Raia clavata*), sardine (*Sardina pilchardus*), mackerel (*Scomber scombrus, L.*), hake (*Merluccius merluccius*), red mullet (*Mullus barbatus*), squid (*Loligo vulgaris*), gilthead bream (*Sparus aurata*), halibut (*Hippoglossus*), sea bream (*Sparus aurata*), salmon (*Salmo onchorhynchus*), tunny (*Thunnus*), bonito (*Katsuwonus*), oyster (*Ostrea virginica*), mussel (*Mytillus edulis*), crab (*Cancer pagurus*), scampi (*Nephrops norvegicus*), lobster (*Homarus*), swordfish (*Xiphias gladius*), clam, prawn, etc.

Preparation of solid samples with an organic matrix as seafood is often the most time-consuming step of the analytical process. In addition, sample pretreatment involves potential drawbacks, such as contamination or losses of some elements.

The sample preparation for the determination of organic substances in seafood samples includes analyte extraction. After removing head, bones, gut and skin, the fish or shellfish extracts are obtained by a) pressing fish muscle cut into small pieces (without skin) placed in a membrane of nylon with a suitable squeezer, b) trituration and homogenization with a blender until a fine paste, c) homogenization with deionized water with microwave heating, d) homogenization by ultrasonication of the solution formed by the dried muscle tissue of fish (powder) and diluted hydrochloric acid. After, takes place immediately a centrifugation step and the supernatant is filtered through a Whatman No.1 filter. The extracts are analyzed immediately and in some cases stored at -40°C until use. Furthermore, in some cases, the analyte is extracted from the supernatant with off-line liquid-liquid extraction.

By using a FIA system is possible to perform on-line sample pretreatment:

For L-lysine determination, an on-line sampling is achieved by a sampling probe with an in-line filter (Figure 7.1). The sampling volume is controlled by the time interval between the off/on action of the aeration. The mass transfer in the sample tank is accelerated by force convection produced by the aeration. The extraction process is initiated by adding dry lysine-enriched feed pellets into a cylindrical sample tank containing a phosphate buffer (0.1 M, pH 7.3).

The low boiling point of trimethylamine is exploited to generate the gaseous amine from an alkalized sample. The simultaneous generation of ammonia and other primary and secondary volatile amines potentially interfering with the determination is avoided by adding formaldehyde to the medium. The gaseous TMA sample is directly introduced into a FIA manifold designed for reaction and detection of the amine (Figure 7.2).

Polycyclic aromatic hydrocarbons are extracted from environmental solid samples (fish samples) by using a dynamic pressurized hot water extraction (Figure 7.3). This extraction procedure involves the following steps: (1) the sample is placed in the extraction cell; (2) the unit is connected to the system and filled with an aqueous solution of sodium dodecyl sulfate by closing the outlet valve in order to pressurize the system; (3) the oven is brought up to the working temperature (225 °C), and the outlet valve is then opened; (4) dynamic extraction is then performed by pumping the extractant through the oven using a restrictor to maintain constant pressure on the system; (5) for kinetics experiments, the extract (cooled in the refrigerant at about 25°C) is collected at preset intervals in two reservoirs, which are alternately exchanged.

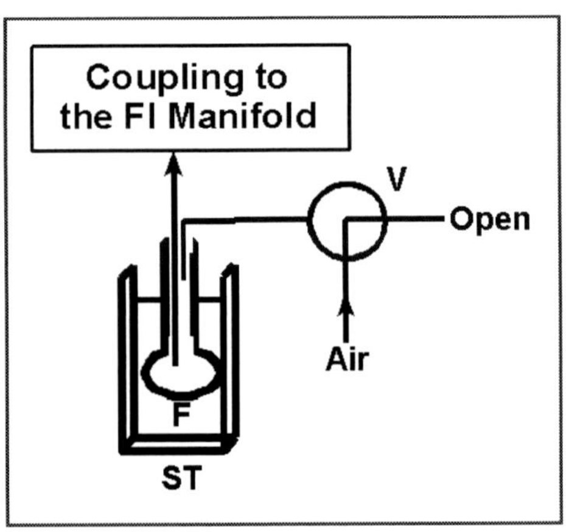

Figure 7.1. Schematic manifold of an on-line sampling proposed for FI L-lysine determination. F: filter; ST: sample tank; V: valve to control the influx of air into the sampling unit.

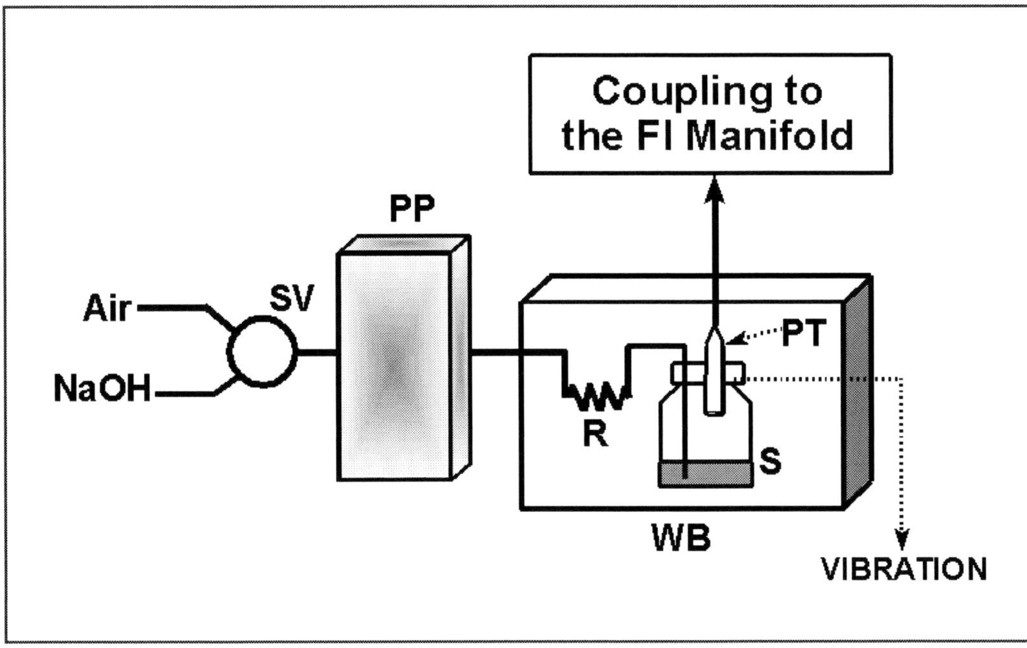

Figure 7.2. On-line gas extraction sampling device developed for FI determination of trimethylamine. PP: peristaltic pump; PT: pipette tip; R: reactor; S: sample vessel; SV, selecting valve; WB: water bath.

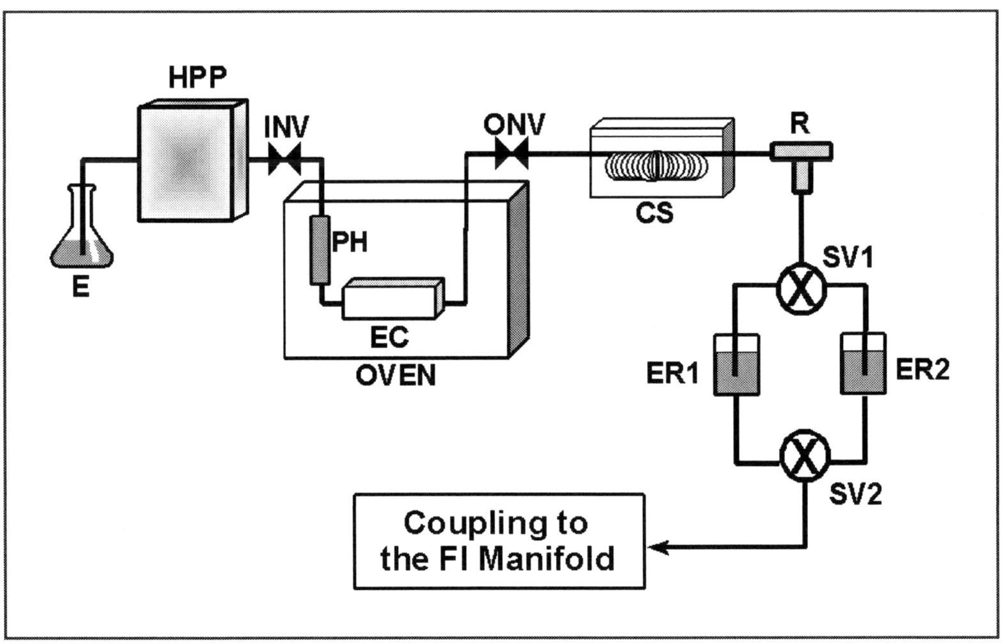

Figure 7.3. Dynamic pressurized hot water extraction system developed for FI determination of polycyclic aromatic hydrocarbons. E: extractant; EC: extraction cell; ER1 and ER2: extract reservoirs; CS: cooler system; HPP: high-pressure pump; INV: inlet needle valve; ONV: outlet needle valve; PH: preheater; R: restrictor; SV1 and SV2: selecting valves.

Usually, FI methods proposed to determine trimethylamine including an on-line gas/diffusion step. Thus, sample extracts are injected into a sodium hydroxide stream. The evolved volatile bases diffuse via a Teflon micro-porous membrane, inserted in a gas diffusion cell, into a phosphate buffer acceptor stream, where the analyte is continuously monitored (Figure 7.4). The gas diffusion cell consisted of two identical methacrylate blocks each with S-shape grooves between which a microporous PTFE membrane is placed. The two blocks are pressed together by eight screws.

After removing head, bones, gut and skin, the fish or shellfish samples are triturated, homogenized, and freeze-dried, and then kept in clean, dry containers. The determination of trace metals in seafood samples can be carried out with sample slurries, where the sample is introduced into the atomizer as an aliquot of a stable suspension. This method allows calibration with standard solutions, to which matrix modifiers may be added. The main problem of the slurry technique is heterogeneity of the samples. By the other hand, conversion of the analyte to the liquid phase is possible by acid digestion with nitric acid and hydrogen peroxide in open vessels carried out in a hot plate or in closed vessels into a stove at 130 °C for about 16 h or by using microwave energy (few minutes), alkaline digestion with potassium hydroxide or tetramethylammonium hydroxide (TMAH), or by ultrasound-assisted acid extraction.

FI digestion procedures involving microwave energy present some advantages over off-line methods, such as reduced digestion times, digestion of complex matrices and solubilization in a closed vessel (decreasing the losses of volatile analytes and atmospheric contamination, and increasing personal security). Due to the advantages offered, on-line microwave digestion has been used for the determination of aluminium, iron and chromium in seafood samples (Figure 7.5). However, these systems present disadvantages, such as matrix interferences because sample matrix is introduced into the detector, and the need to insert a cooling area in the FI system due to the high temperature of the microwave oven. Ultrasound-assisted extraction is another way of converting the analyte to a liquid phase. This technique offers some advantages, such as a wide application field that includes any sample independent of form or size, the use of diluted acids and quantitative extraction of the analyte. This process increases the representative sample mass, and improves the precision and accuracy of slurry sampling. It also avoids total sample matrix introduction into the nebulizer of the spectrometer, which occurs with solid sampling, slurry sampling and digestion procedures [1-3]. Nevertheless, these procedures are laborious and time-consuming because they involve long sonication times and a centrifugation step to separate the liquid phase. Nevertheless, ultrasound assisted extraction can be developed in a continuous mode. For this, a continuous ultrasound-assisted extraction system (CUES) is proposed (Figure 7.6). This on-line system is combined with a FI manifold, and it has been demonstrated to be a rapid, precise and accurate sample pre-treatment procedure for the acid leaching of several trace metals from seafood samples for their determination by FAAS. The main goals obtained with this methodology are a reduction in sample contamination and analyte losses because less manipulation of the sample is required, reduction in sample and reagent amounts and a reduction in the sample preparation time. Compared to off-line ultrasonic-assisted extraction methods, the sonication time is reduced by a factor of 6-12 and the centrifugation step to separate the liquid phase is avoided. These two last advantages significantly increase the sample throughput. This procedure combines the benefits of ultrasound-assisted extraction with those obtained for FI systems.

On-line matrix removal and analyte preconcentration has been the most active area in the field of FI analysis, especially for atomic absorption spectrometry determinations because this analytical technique is insensitive (FAAS) or nonselective due to pronounced matrix interference (ETAAS). FI methods involving a separation technique offer higher sample throughput and much better precision and accuracy compared to their manual counterparts. The use of on-line solid phase extraction (SPE) preconcentration techniques coupled with FAAS (FI-FAAS) have been shown to be a promising alternative to ETAAS thus, became increasingly important for trace metal determinations. These techniques are based predominantly on incorporation of a minicolumn, packed with a chelating resin or a sorbent material to adsorb analyte complexes. Usually, as the sample is in acid medium because comes from an acid digestion or an acid extraction is required that the quantitative retention of the analyte(s) on the micro(mini)column takes place at a wide range of sample acidity. In any case, the pH is raised first off-line with ammonia, and then can be on-line adjusted to the required value with a buffer solution to proper retention of analyte(s) on the micro(mini)column. Typical configurations for on-line preconcentration using a FI manifold are shown in the Figure 5.2. (Chapter 5).

The coupling between continuous ultrasound-assisted extraction, preconcentration (using a chelating resin) and flame atomic absorption spectrometry using a FI manifold as interface for coupling these three analytical steps, allowing automation of the whole analytical process. This manifold can be seen in Figure 7.7. Seafood samples (particle size less than 30 μm are directly weighed into a glass minicolumn. Then, the minicolumn is connected to the continuous manifold. First, the extraction circuit is loaded with the acid leaching solution (diluted nitric acid). Once the extraction circuit is closed by means of SV1, the leaching solution circulates through the minicolumn subjected to ultrasound energy during a precise time. The direction of the flow is changed each 30 s in order to avoid sample accumulation at the end of the minicolumn. Then, the switching valve (SV2) is switched to its opposite position and the acid extract is homogenized in the mixing coil. After this, the acid extract channel converges with a buffer solution stream in order to obtain an adequate pH for analyte retention. The mixture is homogenized in a second mixing coil and then, is passed through the preconcentration minicolumn. Then, the analyte is eluted by injection of a small volume of diluted hydrochloric acid into a water carrier stream, which swept it to the detector where is continuously monitored. This procedure combines the benefits of ultrasound-assisted extraction with the high sensitivity provided by the preconcentration step and the inherent advantages of FI systems. The system presents the advantage of simplicity and avoids the use of expensive and sophisticated instruments. High sample throughput, good accuracy and precision, low detection limit, easy of use and automation, freedom from interferences, safety conditions (concentrated acids and carcinogenic nitrous vapors were avoided) make this methodology very suitable for trace metals determination in solid samples as seafoods.

For analysis of anionic species, seafood samples all sample pretreatments proposed (acid digestion and dry ashing) are performed off-line.

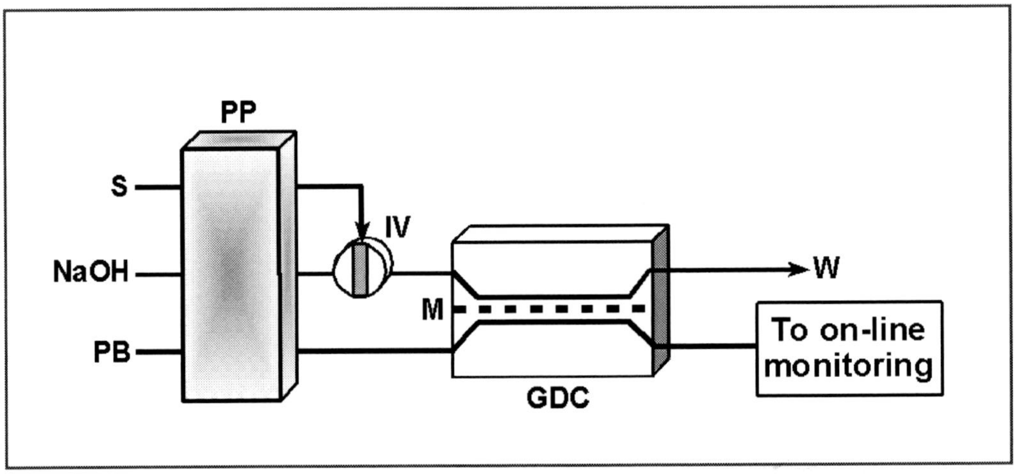

Figure 7.4. Schematic representation of FIA manifold including an on-line gas/diffusion step proposed for determination of trimethylamine. GDC: gas diffusion cell; IV: injection valve; M: membrane; PB: phosphate buffer; PP: peristaltic pump; S: sample; W: waste.

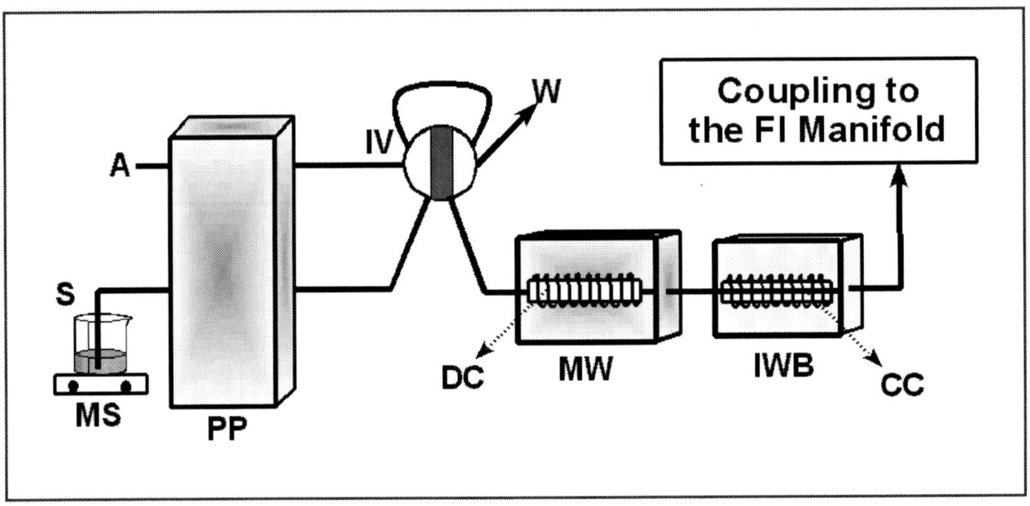

Figure 7.5. Diagram of a flow system for the on-line wet digestion of slurry samples assisted by microwave energy. A: acid solution; CC: cooling coil; DC: digestion coil; IV: injection valve; IWB: ice-water bath; MS: magnetic stirrer; MW: focused microwave oven; PP: peristaltic pump; S: sample slurry; W: waste.

ORGANIC SPECIES

The organic analytes that have been determined in seafood by using FI methodologies were the following: amino acids (histidine, L-lysine and tyrosine), DNA/RNA, formaldehyde, histamine, hypoxanthine, polycyclic aromatic hydrocarbons (PAHs), diarrheic shellfish poisoning (DSP), paralytic shellfish poisoning (PSP), succinate/glutamate, trimethylamine/ total volatile basic nitrogen (TVBN), total lipid hydroperoxides (TLP), total volatile acids (TVA), uric acid and vitamin B_{12}. Different features of FI methods for the determination of organic species in seafood are illustrated in Table 7.1. In the following paragraphs, some points observed in this table are highlighted due to their interest.

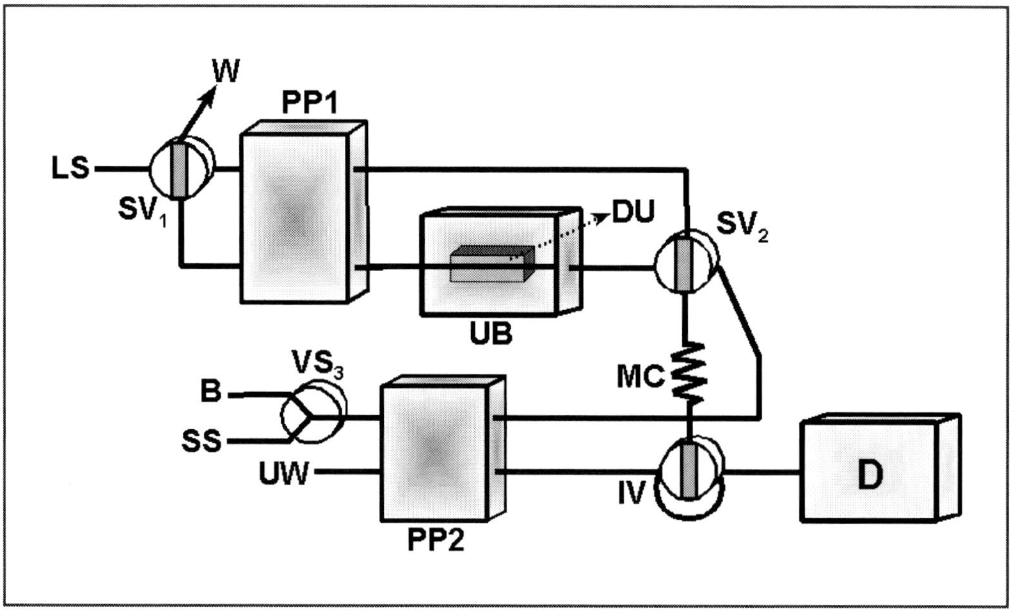

Figure 7.6. Schematic diagram of the continuous ultrasound assisted extraction system (CUES). B: blank solution; D: detector; DU: digestion unit; IV: injection valve; LS: leaching solution; MC: mixing coil; PP1 and PP2: peristaltic pumps; SS: standard solution; SV1, SV2 and SV3: selecting valves; UB: ultrasonic bath; UW: ultrapure water; W: waste.

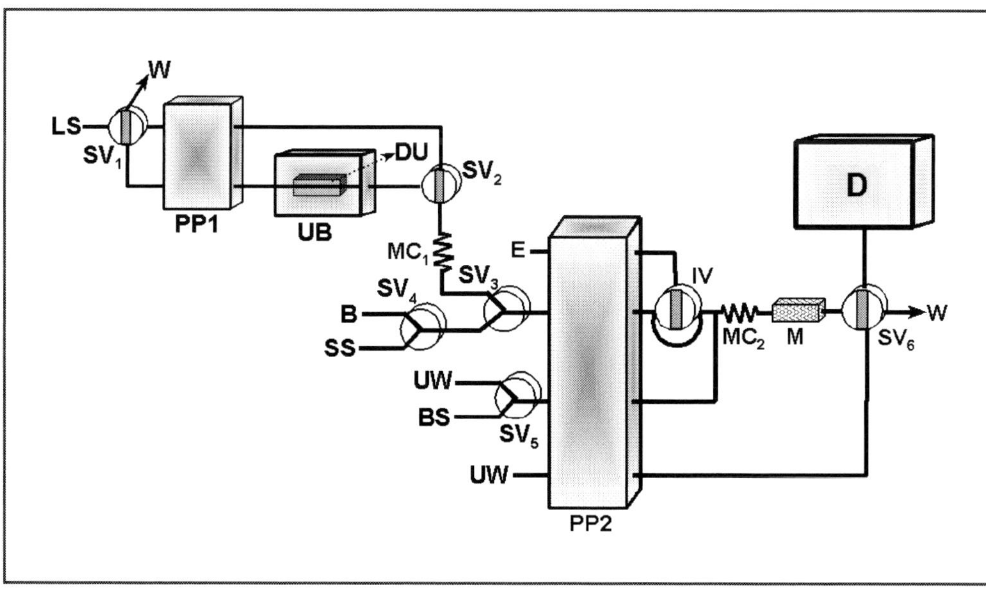

Figure 7.7. Flow injection manifold for the whole on-line procedure (continuous acid extraction system-preconcentration device and detector). B: blank solution; BS: buffer solution; D: detector; DU: digestion unit; E: eluent; IV, injection valve; LS: leaching solution; M: minicolumn containing a chelating resin; MC1 and MC2: mixing coils; PP1 and PP2, peristaltic pumps; SS: standard solution; SV1- SV6: selecting valves; UB: ultrasonic bath; UW: ultrapure water; W, waste.

Table 7.1. Features of FI determinations of organic species in marine animals (seafood)

Analyte	Detection	DL	Linear range	SF (s/h)	Recovery (%)	RSD (%)	Ref
Histidine	SP	No data	1-5 mM	No data	No data	No data	4
Histidine	Ch	0.01 mM	0.05-5 mM	10	97-99	0.93	5
L-lysine	B	No data	No data	12	No data	1.45	6
L-tyrosine	B	0.01	0.1-1 mM	No data	No data	1.9	7
DNA RNA	F	0.01 μg 0.10 μg	No data	No data	97.9-101.5 98-99	No data	8
FA	SP	2.5 mg/L	No data	10	No data	4	9
Histamine	Ch	No data	0.1-50 μM	30	No data	1.25	4
Histamine	F	6 μg/g	Up to 4.5 mM	60	89-105	2-5	10
Histamine Putrescine Cadaverine	ER	0.5 μM 0.07 μM -	Up to 60 μM Up to 400 μM -	20-30	103-110 93-95 -	1 1 -	11
Histamine	F	0.8	Up to 340 μM	60	97-118	<3	12
Histamine	A	5 μM	Up to 0.6 mM	No data	No data	1.1	13
Histamine	A	0.6 μM	2-100 μM	No data	97-99	2.5-3.6	14
Histamine	SP	3 μM	10-100 μM	20	No data	No data	15
Hx	A	No data	No data	No data	No data	No data	16
Hx	A	No data	No data	15	No data	No data	17
Hx	Ch	No data	0.2-200 μM	No data	No data	No data	18
Hx	Ch	No data	No data	No data	No data	No data	19
Hx	SP	4.4 μM	Up to 110 μM	30	92-99	<5	20
Hx Inosine IMP	A	1 μM	Up to 50 μM	100	No data	2-3	21
Hx	A	0.1 μM	Up to 20 μM	12	No data	2-3	22
Hx	SP	No data	2.5-1000 μM	30	No data	<1	23
Hx Inosine IMP	B	No data	Up to 4 mM	2	No data	No data	24
Hx	A	No data	No data	22	95.9-102	0.6-1.2	25
PAHs: Pyr B[a]ant B[ghi]per B[a]pyr B[k]flu	F	0.022 μg/mL	No data	No data	97.6-101.3 99.8-101.2 99.3-104.6 98.6-101.7 99.2-100.7	1.9-2.8 2.1-2.4 2.4-4.5 3.3-5.2 0.9-3.2	26
PAHs	F	0.022 μg/mL	No data	No data	98.3-102.1	4.3-5.8	27
Okadaic acid	Ch	0.1 μg/L	0.2-200 μg/g	3	No data	11.7	28
Saxitioxins	LC/MS CE/MS	No data	No data	No data	No data	No data	29
Succinate	F	No data	5-200 μM	No data	No data	1.01	30
Glutamate	F SP	0.4 μM 0.7 μM	Up to 50 μM	No data	99-105.3	4.3-7.3 8.4-10.9	31
TMA	SP	No data	No data	No data	No data	No data	32
TMA	SP	1 mg/L 4 mg/Kg	3-250 mg/L	No data	93.6-96.1	1.2	33
TMA	SP	0.20 μg/mL	0.2-10 mg/mL	10	No data	2.2	34
TMA	SP	6 μM	0-200 μM	30	99.6-100.5	1.15	35

TMAO	IO	1.35 nM	No data	No data	No data	3	36
TMA	SP	No data	No data	No data	No data	No data	37
TMA TVB	SP	No data	No data	No data	No data	No data	38
TMA TVB	SP	No data	0.7-42 μg/mL 3-56 μg/mL	No data	97.7 96.1	1.4-6 0.8-5.6	39
TMA	SP	No data	No data	No data	No data	No data	40
TMA	P	0.05 μg/mL	1-10 μg/mL	20±2	99.2-100	1.2-1.6	41
TMA	B	No data	No data	No data	No data	4.39	42
TMA	P	0.02 mg/L	1-20 mg/L	40	100-103.6	1.2-2.2	43
TLH	F	No data	0.35-77.8 meq/Kg	No data	87.2-102	4.3-11.2	44
TVA	SP	No data	No data	No data	No data	No data	45
Urea	A	0.2 μM	1-50 μM	No data	No data	5	21
Urea	B	No data	No data	No data	No data	No data	46
Vitamin B_{12}	Ch	0.05 ng/L	0.2-1200 ng/L	120	99.3-101.4	5	47

A: amperometry; B: biosensor; B[a]ant: benzo[a]anthracene; B[ghi]per: benzo-[ghi]perylene; B[a]pyr: benzo[a]pyrene; B[k]flu: benzo-[k]fluoranthene; CE/MS: capillary electrophoresis/mass spectrometry; Ch: chemiluminescence; DL: detection limit; ER: enzyme reactor; F: fluorescence; FA: formaldehyde; Hx: hypoxanthine; IMP: inosine monophosphate; IO: ion chromatography; LC/MS: liquid chromatography/mass spectrometry; OE: oxygen electrode; P: pontetiometry; PAHs: polycyclic aromatic hydrocarbons; Pyr: pyrene; RSD: relative standard deviation; SF: sampling frequency; SP: spectrophotometry; TLH: total lipid hydroperoxides; TMA: trimethylamine; TMAO: trimethylamine-N-oxide; TVA: total volatile acids; TVB: total volatile bases

Amino Acids

Histidine

Yamamura et al. [4] developed a simple, selective and rapid method to determine the quantity of L-histidine and histamine in fish by utilizing two enzymes, L-histidine oxidase and histamine oxidase. The quantity of L-histidine was determined by colorimetric analysis with 4-aminoantipyrine, N-ethyl-N-2(2-hydroxy-3-sulfopropyl)-3,5-dimethoxyaniline, peroxidase and L-histidine oxidase. Kiba et al. [5] described a chemiluminometric FIA system for the quantitation of L-histidine. Histidine oxidase from Brevibacillus borstelensis KAIT-B-022 is immobilized on tresylated poly(vinyl alcohol) beads and packed into a stainless-steel column. The hydrogen peroxide produced is detected chemiluminometrically by a flow-through sensor containing immobilized peroxidase. The system is applied to the determination of L-histidine in fish meat, such as salmon, tunny, bonito, and mackerel.

L-lysine

Chen et al. [6] proposed an automated FIA system, equipped with an immobilized L-lysine-a-oxidase reactor and having a hydrogen peroxide electrode in the manifold for biosensing L-lysine. A substantial improvement in the selectivity for lysine is achieved by lowering the system temperature to 10°C. The sensor response is linearly dependent on the L-lysine concentration. The leaching of lysine from L-lysine-enriched fish feed into an aqueous buffer is automatically monitored by the aid of an in-line filter.

L-tyrosine

Gao et al. [7] immobilized L-Tyrosinase to determine L-tyrosine in proteolytic oyster extracts for monitoring the proteolysis degree. A FIA system is developed for the purpose by measuring O_2 consumption in the tyrosinase reaction. The system consists of a micro-tube pump, air-damper, injection valve, sample loop, O_2 electrode, reactor potentiostat, and recorder. The carrier solution is 0.1M phosphate buffer (pH 7.2).

DNA/RNA

Caldarone et al. [8] described an automated two-dye FIA system to quantitate DNA and RNA in crude extracts of tissues. The method uses the fluorochrome dyes ethidium bromide and Hoechst 33258. DNA concentration is determined directly from its fluorescence in Hoechst dye. RNA is estimated from fluorescence in ethidium bromide after subtraction of the fluorescence due to DNA. The method presents a high degree of sensitivity and elimination of the homogenization step make it ideal for the analysis of large number of samples (larval fish and small organisms).

Formaldehyde

Bechmann [9] reported a FIA system for determination of formaldehyde in frozen fish products. The system provides a rapid and selective determination of formaldehyde in aqueous fish extracts by the combination of a deproteinization procedure and a stopped-flow enzymic approach in a FIA system. The FIA system is furnished with a gel-filtration chromatographic column for on-line removal of the proteins from the extract before the enzymic analysis is performed. Compared with the standard methods for determination of formaldehyde in fish products this method is much faster and less affected by interferences.

Histamine and Biogenic Amines

Hungerford et al. [10] described a FIA method for the determination of histamine. Control of reaction timing allows exploitation of a transient, chemical-kinetic increase in selectivity that occurs when o-phthalaldehyde reacts with histamine. The molar fluorescence ratio (selectivity) of histamine/histidine reaches a maximum value of 800 in 32 s, precluding the need for separation of histamine from histidine, spermidine, and other potential interferences. On-line dilution prevents matrix effects and affords a linear response up to approximately 4.45 mM histamine, or 500 mg of histamine free base/100 g. Carsol et al. [11] developed enzyme reactors for the determination of biogenic amines using diamine oxidase (DAO) from porcine kidney and from lentil and putrescine oxidase (PUO) from microorganism (*Micrococcus roseus*). Determination is based on the electrochemical oxidation of enzymically produced hydrogen peroxide at a platinum electrode poised at 600 mV vs. Ag/AgCl. The enzymes are immobilized on controlled pore glass beads activated by glutaraldehyde in a small reactor and included in a FIA assembly. The reactor using DAO from porcine kidney as the biochemical component responds mainly to histamine, and it can be used for the evaluation of fish spoilage. The PUO reactor shows a significant response only to putrescine. The reactor using DAO from lentil is sensitive to several amines and it could be useful to evaluate a total value. Yamamura et al. [4] proposed a simple, selective and rapid method to determine the quantity of L-histidine and histamine in fish by utilizing two enzymes, L-histidine oxidase and histamine oxidase. The quantity of histamine is determined by using a chemiluminescent flow sensor with histamine oxidase and peroxidase. Histamine oxidase and peroxidase are co-immobilized covalently on tresylated hydrophilic vinyl polymer beads and packed into transparent PTFE, which is used as a flow cell. Hungerford et al. [12] used an automated, kinetically-enhanced FI method for histamine to screen thousands of sub-samples of various fish and seafood products. Selectivity (expressed as molar fluorescence ratios) for histamine vs. histidine in the sample matrix is a function of reaction kinetics. Fine-tuning this effect is sufficient to remove the need for ion-exchange sample conditioning. This fine-tuning of reaction kinetics is easily accomplished via control of flow rates, reaction temperature, etc. Takagi et al. [13] developed a FIA system for the determination of histamine using histamine dehydrogenase (HmDH)-based electrode. Histamine dehydrogenase is immobilized in an Os-derivatized redox polymer, poly(1-vinylimidazole) complexed with Os(4,4'-dimethylbipyridine)$_2$Cl$_2$ (PVI-dmeOs), film on a glassy carbon electrode. As expected from the characteristics of this enzyme in a solution, this electrode exhibits high selectivity to histamine and is not sensitive to other primary amines including common biogenic amines as putrescine, cadaverine and tyramine. Carralero et al. [14] prepared and used gold nanocrystal-modified glassy carbon electrodes (nAu-GCE) for the determination of histamine by FI and HPLC using pulsed amperometric detection (PAD) as the detection mode. A catalytic enhancement of the histamine voltammetric response is observed at the nAu-GCE when compared with that obtained at a conventional Au disk electrode, as a consequence of the microdispersion of gold nanocrystals on the glassy carbon substrate. PAD using a very simple potential waveform consisting of an anodic potential (+700 mV for 500 ms) and a cathodic potential (-300 mV for 30 ms), is used to avoid the electrode surface fouling when histamine is detected under flowing conditions. FI amperometric responses showed much higher I_p values and signal-to-noise ratios at the nAu-GCE than at a conventional gold disk electrode. HPLC-PAD at the nAu-GCE is used for the determination of histamine in the presence of other biogenic amines and indole. Watanabe et al. [15] determined histamine in fish sauce with a single-line FI system including an immobilized histamine oxidase/peroxidase reactor. Hydrogen peroxide produced by histamine oxidase reaction is monitored photometrically by using N-ethyl-N-(2-hydroxy-3-sulfopropyl)-3-methoxyaniline and 4-aminoantipyrine as color reagents with peroxidase at 530 nm.

Hypoxanthine

Moody et al. [16] described FIA systems incorporating amperometric bienzyme electrodes for the determination of glucose and hypoxanthine. The sensor for the determination of xanthine and hypoxanthine is based on xanthine oxidase-peroxidase. The enzyme is immobilized on nylon mesh and held over a platinum electrode. The hydrogen peroxide product of the enzymic catalysis is monitored amperometrically in a 3-electrode

Stelte cell, adapted for FIA, after its peroxidase-catalyzed reaction with hexacyanoferrate(II). The xanthine oxidase-peroxidase bienzyme electrode is used to determine hypoxanthine in fish meat as an indicator of deterioration during storage. Yao et al. [17] incorporated at fixed positions in a flow system a 5'-adenylic acid deaminase reactor, alkaline phosphatase reactor, and nucleoside phosphorylase-xanthine oxidase coimmobilized reactor. This FI system is based on the splitting of the flow after sample injection and subsequent confluence before reaching the peroxidase electrode. Because each channel has a different residence time, two peaks are obtained. The first peak corresponded to the total of hypoxanthine and inosine, and the second to the total of hypoxanthine, inosine, inosine-5'-monophosphate, and adenosine-5'-monophosphate. The index of fish freshness, K, is established by: $K = [(s_2/s_1)(i_1/i_2)] \times 100$, where s_1 and s_2 represent the sensitivity (nA/mM) and i_1 and i_2 represent the peak current (nA) of the first and second peaks, respectively. Yagiuda et al. [18] developed for quality control in the food industry a low-cost chemiluminescence detector with a photodiode for FIA of hypoxanthine in fish and meat. In this system, hypoxanthine is oxidized enzymically to produce hydrogen peroxide during passage through a xanthine oxidase immobilized reactor. The chemiluminescence caused by mixing hydrogen peroxide with chemiluminescence reagent is detected by the flow-through photodiode cell that was designed to detect efficiently weak chemiluminescence. Hayashi et al. [19] carried out direct measurements of hypoxanthine in fish flesh using sardine (*Sardinops melanosticta*) and a grunt (*Paraprisripoma trilineatum*) fish twice with the same sample, and the average of ten successive measurements (from 120 s to 480 s after pressing the probe against fish tissue) is taken as the hypoxanthine concentration by direct measurement. Balladin et al. [20] analyzed fish muscle extracts (*Scomberomorus brasiliensis*; carite) to determine their hypoxanthine content using a FI system incorporating an immobilized xanthine oxidase bioreactor. The xanthine oxidase is immobilized under mild conditions to a 2-fluoro-1-methylpyridinium Fractogel support. The uric acid produced from the oxidation of hypoxanthine by the immobilized xanthine oxidase at pH 7.0 and 35°C is monitored at 290 nm. Carsol et al. [21] immobilized xanthine oxidase either directly on the surface of the electrode or in a reactor with aminopropylsilane controlled pore glass beads in a FIA assembly. This reactor has high reproducibility and lifetime. The uric acid produced from the oxidation of hypoxanthine, inosine and inosine monophosphate by the immobilized xanthine oxidase is detected with carbon-based screen-printed electrodes. Carsol et al. [22] developed and applied a continuous system for the determination of fish freshness with double enzyme reactors to the determination of the freshness indicator (K): $K = 100(HxR + Hx)/(IMP + HxR + Hx)$, where IMP, HxR and Hx are inosine monophosphate, inosine and hypoxanthine, respectively. The system is assembled with a three electrode screen-printed element (graphite as working electrode, silver as counter and silver, silver chloride as reference electrode) placed in a flow cell, a sample injection valve and two enzyme reactors. The determination of the total amount of HxR and Hx is realized by flowing the sample through two reactors in series: one reactor is packed with nucleoside phosphorylase and the other with xanthine oxidase immobilized on aminopropyl glass. Similarly, the other term of the equation is evaluated by flowing through the two reactors the sample treated by alkaline phosphatase for 5-10 min at 45°C. Sarker et al. [23] determined hypoxanthine in fish by a FIA system using immobilized xanthine oxidase and catalase reagents, with spectrophotometric detection. Park et al. [24] developed a biosensor system for simultaneous determination of hypoxanthine, inosine and inosine monophosphate to measure fish freshness indexes such as K_i and H. The biosensor, which is composed of three enzyme reactors packed with enzyme-bonded chitosan porous beads (Chitopearl) is operated in a FIA mode with a serial circuit system. The concentrations of the three compounds are determined simultaneously based on the sensor response from each enzyme reactor. Nanjyo et al. [25] proposed a novel FI system for the rapid measurement of the fish freshness indexes K_1 and K_2: K_1=[([HxR]+[Hx])x100/([IMP]+[HxR]+[Hx])] and K_2=[[Hx] x 100/([HxR]+[Hx])]. For the estimation of index K_1, 5'-nucleotidase immobilized reactor and nucleoside phosphorylase (NP)/xanthine oxidase (XO) coimmobilized reactor are incorporated in series in the FI line made up by a 16-way switching valve with two sample loops. For the estimation of index K_2, NP and XO immobilized reactors are also incorporated in the similar flow-line. Two sample portions passed through the flow-line with different residence times so that two peaks were obtained. The first and second peaks obtained in the K_1-determining system corresponded to the total of HxR and Hx and the total of Hx, HxR and IMP, respectively. Similarly, the first and second peaks obtained in the K_2-determining system corresponded to Hx and the total of Hx and HxR, respectively. Therefore, the indexes K_1 and K_2 can be estimated by K_1 and $K_2(\%)=(i_1/i_2) \times f \times 100$, where i_1 and i_2 present the peak current of the first and second peaks, respectively, and f the ratio of the peak currents of the first and second peaks for a Hx standard solution.

Polycyclic Aromatic Hydrocarbons

Morales-Muñoz et al. [26] performed a comparison of the feasibility of the three operational modes of pressurized hot solvent extraction (static, where a fixed extractant volume is used; dynamic, where the extractant continually flows through the sample; and static-dynamic mode, which consists of a combination of the two previous modes) for the extraction of polycyclic aromatic hydrocarbons (PAHs) from environmental solid samples (such as soil, sediment, trout, and sardine). In all cases, an aqueous solution of sodium dodecyl sulfate is used as the leaching agent. The use of a FI manifold between the extractor and a molecular fluorescence detector allowed real-time on-line fluorescence monitoring of the PAHs extracted from the samples, thus working as a screening system and providing qualitative and semiquantitative information on the target analytes extracted from both natural and spiked samples. The on-line monitoring option allowed the extraction kinetics to be monitored and the end of the leaching step to be determined independently of the sample matrix, thereby reducing extraction times. Efficiencies close to 100% have been provided by the three modes, which differ in the extraction time required for total removal of the target compounds. The time needed for the dynamic mode is shorter than that for the static mode. However, the establishment of a static extraction step prior to dynamic extraction is the key to shortening the time required for complete extraction. Morales-Muñoz et al. [27] establish that the coupling of a static pressurized liquid extractor to a FI manifold allowed real-time on-line fluorescence monitoring of the polycyclic aromatic hydrocarbons extraction from environmental solid samples, which can be used for either screening or semiquantitative purposes. Sodium dodecyl sulfate is added to the water for favoring the extraction of the low-polar analytes. The results demonstrated that fluorometric monitoring of static pressurized liquid extraction constitutes an approach as efficient as conventional Soxhlet for the extraction of PAHs from solid samples but with the following positive features: (a) drastic reduction of the extraction time as the extraction kinetics can be monitored and thus the end of the leaching step determined independently of the sample matrix; (b) use of water as extractant thus given place to an environmentally friendly method; and (c) coupling of static extraction to subsequent dynamic steps.

Diarrheic Shellfish Poisoning: Okadaic Acid

Marquette et al. [28] developed a chemiluminescent immunosensor integrated in a FIA system for the detection of the "diarrheic shellfish poisoning" (DSP) toxin okadaic acid (OA). Anti-OA monoclonal antibodies are labeled with horseradish peroxidase for their use in a competitive assay, in which the free antigen of the sample competes with immobilized OA. Based on commercially available polyethersulfone membranes, this bioanalytical system exhibits low non-specific binding of antibodies in the presence of mussel homogenate. The immunosensor is used in a semi-automated analysis procedure in which the free OA containing sample is injected in the flow system concomitantly with the labeled antibodies.

Paralytic Shellfish Poisoning: Saxitioxins

Pleasance et al. [29] used ion-spray mass spectrometry to monitor the purification of saxitoxin, the parent compound in the family of toxins responsible for "paralytic shellfish poisoning" (PSP), from a strain of the dinoflagellate *Alexandrium excavatum.* Quantitative results obtained by FIA are compared to those obtained by high-performance liquid chromatography with post-column oxidation and fluorescence detection. The coupling of liquid chromatography and capillary electrophoresis with ion-spray mass spectrometry is described for the separation of mixtures of PSP toxins and the highly potent pufferfish toxin tetrodotoxin. Thus, these studies demonstrated that the application of these methodologies provide structural information, and the ability to distinguish isomeric PSP toxins.

Succinate/glutamate

Tsukatani et al. [30] proposed a method for the quantification of succinate by FIA using an immobilized-enzyme reactor and a fluorescence detection. Succinate is quantified using a co-immobilized isocitrate lyase (ICL) and isocitrate dehydrogenase (ICDH) reactor. Succinate is converted to isocitrate by ICL in the presence of glyoxylate, and then the produced isocitrate is oxidized with $NADP^+$ by ICDH. The NADPH produced by the ICL-ICDH reactor is monitored fluorometrically at 455 nm (excitation at 340 nm).

Khampha et al. [31] developed a FIA system for specific determination of L-glutamate in food samples based on a bi-enzymic amplification system. The content of L-glutamate in the sample is amplified by cycling between L-glutamate dehydrogenase (GlDH) and a novel enzyme, D-phenylglycine aminotransferase (D-PhgAT). In this system, GlDH converts L-glutamate to 2-oxoglutarate with concomitant reduction of NAD^+ to NADH. D-PhgAT transfers an amino group from D-4-hydroxyphenylglycine to 2-oxoglutarate, thus recycling L-glutamate. Accumulation of NADH in the course of the enzymic recycling is monitored both by fluorescence and UV absorbance and used for quantification of L-glutamate.

Trimethylamine/total Volatile Basic Nitrogen

Nakashima et al. [32] examined the color reaction of a crowned 2,4-dinitrophenylazophenol (HCDPA)-Ba(II) complex with 28 different amines in methanol. Absorbances of the resulting chromogens are measured and it is found that a linear relationship exists between the apparent molar absorptivities (a) of the complexes and the pKa values of the corresponding amines, and that amines having a large pK_a value give a large a value. Then, they developed a FI spectrophotometric method for the determination of trimethylamine (TMA) with the HCDPA-Ba complex as chromogenic reagent. The proposed method is simple and sensitive (0.6-3.0 mg per 50 mL for TMA), and can be applied to the determination of total volatile amines as the content of TMA in fish. León et al. [33] used FIA to automate the AOAC method 971.14 for the determination of trimethylamine (TMA) in fish by reaction with nonaqueous picric acid following removal of major interfering compounds (dimethylamine and ammonia) by blocking with formaldehyde in an alkaline medium, and organic solvent extraction. The proposed analytical unit allows automated addition of interfering blocking reagents, extraction of TMA into an organic solvent (with the aid of a PTFE hydrophobic membrane phase separator) and colorimetric determination of the analyte at 410 nm. The results provided by the proposed automated method were consistent with those obtained by using the original AOAC method, as were temporal changes in the amounts of TMA in fish, which is thus a suitable index for fish freshness. Zhi et al. [34] proposed a new method for the direct determination of trimethylamine in fish samples. The method uses a gas extraction sampling device where samples are treated to release gaseous trimethylamine species; the gaseous analyte is then driven to the loop of an injection valve located in a straightforward single-channel manifold, where the flow direction is repeatedly reversed (no debubbler is needed), and the derivatizing reaction in the carrier solution is monitored in-situ with a spectrophotometer. A pH indicator (Bromothymol Blue) is used to detect the amine by a gas-liquid (aqueous) dissolution reaction. Sadok et al. [35] reported a direct method for the determination of TMA in aqueous solutions and in fish muscle extracts, which involves the extraction of a small quantity of tissue in perchloric acid and the subsequent analysis of the extract by an inexpensive FI/gas-diffusion system. Interferences from other volatile amines present in the extract are suppressed by the use of formaldehyde. Hatton et al. [36] presented a new, safe and sensitive method for the determination of trimethylamine-N-oxide (TMAO) in aqueous and biological media, where TMAO is enzymically reduced to TMA and subsequently quantified using FI-Gas diffusion-Ion chromatography. Ruiz-Capillas et al. [37] studied the modifications to a FIA method necessary to make it suitable for the determination of nitrogen of trimethylamine (TMA-N) and nitrogen of total volatile bases (TVB-N). These modifications involve the concentrations of sodium hydroxide and formaldehyde for releasing the volatile bases from the acidic fish extracts and sequestering the non-TMA volatile bases, respectively. Ruiz-Capillas et al. [38] used a combination of FIA, gas diffusion cell and a laboratory built photometer for monitoring trimethylamine (TMA) and total volatile bases (TVB) in fish sauce. Pons-Sanchez-Cascado et al. [39] performed a collaborative project among different European countries tries to validate a new methodology applying FI/gas diffusion techniques (FIGD), to automate the determination of TMA-N and the TVB-N for different fish species. The proposed FIGD methods are simple, rapid and can be applied to the on-line monitoring control of fish factories. Ruiz-Capillas et al. [40] compared the official methods to determine TVB-N, TMA-N and TMAO-N with the FIA technique using three different fish species over their storage in melting ice for a period of 16 days. The results of this study are that there is a good correlation between determinations made using official and FIA methods with no apparent significant differences. FIA thus appears to be a suitable method for TMAO-N, TMA-N and TVB-N in fish extracts and the use of this system is much faster and simpler than the official methods for determination of these parameters. Adhoum et al. [41] developed a simple FI gas/diffusion method for the determination of trimethylamine in seafood with potentiometric detection using tungsten oxide electrode. The method is based on the diffusion of TMA through a PTFE membrane from a sodium hydroxide donor stream to a phosphate buffer acceptor stream. The TMA in the acceptor stream passes through an electrochemical flow cell containing a tungsten oxide wire and a Ag/AgCl electrode, where TMA is sensitively detected. Common interferences from those species that can diffuse through the membrane are removed by the addition of formaldehyde to the seafood extract. The method is simple, feasible with satisfactory accuracy and precision. Mitsubayashi et al. [42] constructed a trimethylamine biosensor by immobilizing flavin-containing monooxygenase type-3 (FMO3), as one of drug metabolizing enzymes in human liver, onto a sensitive area of a dissolved oxygen electrode. The sensor output increased with the holding time of fish samples at 25°C because of their decomposition. Dhaouadi et al. [43] described a potentiometric FI method, using a gas-diffusion cell, for the determination of total volatile basic nitrogen (TVB-N) in seafood products. The method is based on the change of the potential of a tungsten oxide electrode when volatile basic compounds, liberated from the fish extracted sample, diffuses via a permeable membrane into a phosphate buffer acceptor stream and locally shifts the pH.

Total Lipid Hydroperoxides

Sohn et al. [44] developed a FIA system coupled with a fluorescence detection system using diphenyl-1-pyrenylphosphine (DPPP) as a highly sensitive and reproducible quantitative method of total lipid hydroperoxide analysis. Fluorescence analysis of DPPP oxide generated by the reaction of lipid hydroperoxides with DPPP enabled a quantitative determination of the total amount of lipid hydroperoxides. Use of 1-myristoyl-2-(12-((7-nitro-2-1,3-benzoxadiazol-4-yl)amino) dodecanoyl)-sn-glycero-3-phosphocholine as the internal standard improved the sensitivity and reproducibility of the analysis.

Total Volatile Acids

Hollingworth et al. [45] described a FIA method, using a gas diffusion cell at elevated temperature and spectrometric detection at 560 nm, for the determination of total volatile acids in fish and applied to halibut.

Uric Acid

Carsol et al. [21] used carbon-based screen-printed electrodes for uric acid detection. Schwedt et al. [46] developed an efficient procedure for routine determination of urea in food by integration of immobilized urease (epoxide polymer carrier) into a common ammonium FIA with gas diffusion and simplified sample preparation (extraction with perchloric acid). The enzyme reactor used is stable for approximately one week when stored at 6°C.

Vitamin B_{12}

Song et al. [47] proposed a sensitive chemiluminescence method, based on the enhancive effect of cobalt(II) on the chemiluminescent reaction between luminol and dissolved oxygen in a FI system, for determination of Vitamin B_{12}.

Cationic Species

The cationic species that have been determined in seafood by using FI methodologies were the following: silver, aluminium, arsenic, boron, calcium, cadmium, cobalt, chromium, copper, iron, gallium, mercury, indium, lithium, manganese, molybdenum, nickel, lead, antimony, selenium, tin, strontium, thallium, vanadium and zinc.

Different features of FI methods for the determination of cationic species in seafood are illustrated in Table 7.2. In the following paragraphs, some points observed in this table are highlighted due to their interest.

Silver

Stefanka et al. [48] developed a FI method using hydraulic high-pressure nebulization as a sample introduction system, coupled to inductively coupled plasma time-of-flight mass spectrometer (ICP-TOFMS) for rapid and simultaneous determination of 19 elements.

Table 7.2. Features of FI determinations of cationic species in marine animals (seafood)

Analyte	Detection	Sample preparation	DL (µg/L)	SF (s/h)	RSD (%)	Ref
Ag	ICP-MS	Off-line pressure-cooker digestion	0.098	No data	1.3	48
Al	ETAAS	On-line microwave digestion	10	3	4.3	49
Al	ICP-MS	Off-line microwave digestion	1.46[a]	No data	4.2	50
As	ICP-MS	Off-line pressure-cooker digestion	0.064	No data	2.5	48
As	ICP-MS	Off-line microwave digestion	0.89[a]	No data	1.20	50
As	AF	Off-line digestion (open vessels) On-line gas-liquid separation	0.009	No data	No data	51
As	AAS	Off-line dry ashing Off-line high-pressure ashing Off-line microwave digestion On-line gas-liquid separation	No data	No data	No data	52
As	AAS	Off-line digestion (KDS) On-line gas-liquid separation	0.0135[b]	No data	2.1	53

Table 7.2. Continued

Analyte	Detection	Sample preparation	DL (μg/L)	SF (s/h)	RSD (%)	Ref
As	AAS	Off-line ultrasonic extraction On-line gas-liquid separation	0.19-0.5[b]	No data	No data	54
As	AAS	Off-line digestion (open vessels) On-line gas-liquid separation	0.14	No data	4.1	55
As	AAS	Off-line microwave digestion On-line gas-liquid separation	15000	No data	1.84	56
As	AAS	Off-line microwave digestion On-line gas-liquid separation	0.068[a]	No data	No data	57
As	ICP-MS	Off-line microwave digestion On-line minicolumn (alumina) separation	9 0.9[b]	No data	No data	58
B	ICP-MS	Off-line microwave digestion	1.97[a]	No data	No data	50
Ba	ICP-MS	Off-line pressure-cooker digestion	3.7	No data	5.4	48
Ba	ICP-MS	Off-line microwave digestion	0.07[a]	No data	1.6	50
Ca	FAAS	On-line ultrasonic extraction	44.4[b]	40	0.9	59
Cd	ICP-MS	Off-line pressure-cooker digestion	0.029	No data	1.3	48
Cd	ICP-MS	Off-line microwave digestion	0.16[a]	No data	2.4	50
Cd^{2+}	FAAS	Off-line digestion (open vessels)	20	No data	2.2	60
Cd	FAAS	Off-line digestion (open vessels) On-line minicolumn (AG50W-X8 resin) separation	$7x10^{-4}$	80	<2	61
Cd	FAAS	Off-line microwave digestion On-line minicolumn (PAPhA) separation	0.56	No data	1.4-6.6	62
Cd	FAAS	Off-line microwave digestion On-line minicolumn (PAPhA) separation	0.56	No data	1.4	63
Cd	FAAS	Off-line digestion (open vessels) On-line minicolumn adsorption: analyte complexes with Phen on RP-C18	0.5	90	1.4	64
Cd	FAAS	Off-line digestion (open vessels) On-line coprecipitation	2.5	60	3.5	65
Cd	FAAS	On-line minicolumn adsorption: analyte complexes with APDC on PT-C18	0.45	No data	2.5	66
Cd-MT Cd	FAAS	Off-line extraction Off-line digestion (open vessels) On-line minicolumn (C_{60} fullerene) separation	0.1	No data	4 3	67
Cd	FAAS	Off-line microwave digestion On-line minicolumn (HQ azo-immobilized on CPG) separation	0.4-0.7	41-63	5	68

Analyte	Detection	Sample preparation	DL (μg/L)	SF (s/h)	RSD (%)	Ref
Cd	FAAS	Off-line microwave digestion Off-line microwave digestion On-line minicolumn (chitosan with HQ) separation	0.1	No data	1.5	69
Cd	FAAS	Off-line microwave digestion On-line minicolumn (ion-imprinted thiol-functionalized SG sorbent) separation	0.07	No data	0.9	70
Cd	AAS	Off-line microwave digestion On-line gas-liquid separation	0.0018	12	5	71
Cd	FAAS	Off-line microwave digestion On-line minicolumn adsorption: analyte complexes with ADDP on sugar cane bagasse	0.697	25	0.96	72
Cd	FAAS	On-line ultrasonic extraction On-line minicolumn (Chelite P resin) separation	0.011[b]	16	2-3	73
Cd	ICP-MS	No data	No data	No data	No data	74
Cd	ICP-MS	Off-line microwave digestion	No data	No data	0.2	75
Cd	ICP-MS	Off-line microwave digestion On-line minicolumn (Toyopearl AFChelate650M resin)separation	0.0055	No data	1.2	76
Co	ICP-MS	Off-line pressure-cooker digestion	0.128	No data	3.2	48
Co	ICP-MS	Off-line microwave digestion	0.15[a]	No data	1.9	50
Co	FAAS	Off-line digestion (open vessels) On-line minicolumn adsorption: analyte complexes with Phen on RP-C18	6	90	2.2	64
Co	Ch	Off-line ultrasonic extraction	4×10^{-6}	120	3	77
Co	FAAS	On-line ultrasonic extraction On-line minicolumn (Serdolit Che and Chelite P resins) separation	0.11[b]	16	1.9	78
Co	ICP-MS	Off-line microwave digestion On-line minicolumn (Toyopearl AFChelate 650M resin)separation	0.0001	No data	3.3	76
Cr	ICP-MS	Off-line microwave digestion	1.36[a]	No data	5	50
Cr	FAAS	On-line ultrasonic extraction On-line minicolumn (Serdolit Che and Chelite P resins) separation	0.09[b]	13	2.7	78
Cr	FAAS	On-line ultrasonic extraction	0.12[b]	11	1.7	79
Cr(III)	FAAS	Off-line microwave digestion On-line minicolumn (HQ azo-immobilized on CPG) separation	2.2-2.5	72	1.2-2.8	80

Table 7.2. Continued

Analyte	Detection	Sample preparation	DL (μg/L)	SF (s/h)	RSD (%)	Ref
Cr	ICP-MS	Off-line microwave digestion On-line minicolumn (alumina) separation	6 0.6[b]	No data	No data	58
Cr	ICP-MS	No data	No data	No data	No data	74
Cu	ICP-MS	Off-line pressure-cooker digestion	1.101	No data	6.5	48
Cu	SV	Off-line digestion (open vessels)	No data	No data	3.9	81
Cu	FAAS	Off-line digestion (open vessels) On-line minicolumn adsorption: analyte complexes with Phen on RP-C18	0.3	90	3	64
Cu	FAAS	Off-line digestion (open vessels) On-line minicolumn adsorption: analyte complexes with 1N2N on RP-C18	2.0	90	1.7	82
Cu	FAAS	Off-line digestion (open vessels) On-line coprecipitation	0.5	60	3.0	65
Cu	FAAS	On-line ultrasonic extraction	0.06[b]	11	2.7	83
Cu	FAAS	Off-line microwave digestion On-line minicolumn (chitosan with HQ) separation	0.4	No data	0.7	69
Cu	FAAS	On-line ultrasonic extraction	0.3[b]	46	1.6	84
Cu	FAAS	Off-line microwave digestion On-line minicolumn separation (acrylic acid-grafted PTFE fibers)	0.20	55	1.2	85
Cu	ICP-MS	Off-line microwave digestion On-line minicolumn (Toyopearl AFChelate 650M resin) separation	0.003	No data	1.4	76
Fe	F	No data	No data	No data	<8	86
Fe	SP	On-line microwave digestion	7.5	3-5	3.7	87
Fe	FAAS	Off-line digestion (open vessels) On-line coprecipitation	2.5	60	2.8	65
Fe	FAAS	On-line ultrasonic extraction	0.6[b]	18	0.3	84
Fe	FAAS	On-line ultrasonic extraction	0.7[b]	11	0.5	88
Fe	ICP-MS	Off-line microwave digestion On-line minicolumn (Toyopearl AF Chelate650M resin) separation	0.011	No data	1.1	76
Ga	ICP-MS	Off-line pressure-cooker digestion	0.024	No data	1.3	48
In	ICP-MS	Off-line pressure-cooker digestion	0.011	No data	1.1	48
Li	ICP-MS	Off-line pressure-cooker digestion	0.023	No data	1.8	48

Analyte	Detection	Sample preparation	DL (μg/L)	SF (s/h)	RSD (%)	Ref
Org. Hg Inorg. Hg	AF	No data	2 0.02	17	No data	89
Hg	AF	Slurry technique	0.002		3.9	90
Hg MeHg	AF	On-line slurry technique On-line gas-liquid separation	7[a]	No data	6.8	91
MeHg	AF	Off-line extraction with TMAH On-line gas-liquid separation	0.01[a]	No data	<5	92
Hg	AF	On-line microwave digestion On-line gas-liquid separation	0.06	No data	No data	93
Org. Hg Inorg. Hg	AF	Off-line ultrasonic extraction Off-line microwave digestion On-line sorption separation in a Knotted reactor On-line gas-liquid separation	0.002 0.0036	20 30	2.8 2.2	94
MeHg	AF	No data	No data	No data	No data	95
Hg	AAS	On-line gas-liquid separation	No data	No data	No data	96
Hg	AAS	Off-line microwave digestion On-line gas-liquid separation	0.34	30	0.95	97
Hg	AAS	Off-line microwave digestion On-line gas-liquid separation	7.7[a]	No data	6.7	98
Hg	AAS	Off-line digestion with TMAH	0.1	100[c]	1.3	99
Hg	AAS	Off-line microwave digestion On-line gas-liquid separation	No data	No data	No data	100
MeHg EtHg PhHg Inorg. Hg	AAS	On-line solid phase extraction HPLC On-line gas-liquid separation	0.009 0.006 0.010 0.005	2.3	3.6 5.5 10.4 7.6	101
MeHg Inorg. Hg	AAS	Off-line ultrasonic extraction On-line gas-liquid separation	11[a] 5[a]	No data	5-10	102
Hg Inorg. Hg	AAS	Off-line ultrasonic extraction On-line gas-liquid separation	0.11 0.12	No data	2.2-2.8 8.8-9	103
Hg	AAS	On-line slurry technique On-line gas-liquid separation	3[a]	No data	4-5	104
Hg Inorg. Hg	AAS	Off-line digestion with TMAH On-line gas-liquid separation	0.1 0.2	100	<2	105
MeHg Inorg. Hg	AAS	Off-line ultrasonic extraction Off-line microwave digestion On-line gas-liquid separation	0.25 0.17	No data	1.16 0.9-1.4	106
Hg	AAS	Off-line microwave digestion On-line gas-liquid separation	0.4	No data	1.5	107
MeHg Inorg. Hg	AAS	Off-line ultrasonic extraction Off-line microwave digestion HPLC On-line gas-liquid separation	$3.4x10^{-5}$ $6.8x10^{-5}$	12	2.9 5.7	108

Table 7.2 Continued

Analyte	Detection	Sample preparation	DL (µg/L)	SF (s/h)	RSD (%)	Ref
Hg	AAS	Off-line digestion (open vessels) On-line gas-liquid separation	0.46	45	0.9	109
Hg	AAS	Off-line microwave digestion On-line gas-liquid separation	No data	No data	2.4-14	110
Hg	AAS	Off-line digestion (open vessels) On-line gas-liquid separation	No data	No data	No data	111
Hg	ETAAS	Off-line digestion (open vessels) On-line sorption separation in a Knotted reactor	0.0062	22	1.1	112
MeHg	ETAAS	Off-line alkaline digestion On-line minicolumn sorption	0.0068	30	2.3	113
Hg Inorg. Hg MeHg	AAS	On-line slurry technique On-line gas-liquid separation	4.04[a] 0.977[a] 11[a]	No data	1.1-2.5 0.3-1.8 0.7	114
Hg	AAS	Off-line solubilization with formic acid On-line gas-liquid separation	0.001[b,c] -0.01[b,d]	No data	2.7	115
Hg	AAS	Off-line microwave digestion On-line gas-liquid separation	0.005	11	1.4	116
Hg	AAS	Off-line digestion (open vessels) On-line gas-liquid separation	0.65 4.8[a]	60	6	117
Hg Inorg. Hg Org. Hg	AAS	Off-line microwave digestion On-line minicolumn preconcentration On-line gas-liquid separation	0.01	20	3.5	118
Org. Hg	ICP-MS	Off-line extraction with organic solvents	2.2-4	No data	7	119
Hg	ICP-MS	Off-line microwave digestion	5.1	No data	9.5	120
MeHg EtHg PhHg Inorg. Hg	UV	On-line minicolumn sorption HPLC	10-25[a]	No data	2-3	121
Mn	ICP-MS	Off-line pressure-cooker digestion	1.184	No data	9.6	48
Mn(II)	FAAS	Off-line microwave digestion On-line minicolumn (HQ azo-immobilized on CPG) separation	0.9-1.1	60	1.4	80
Mn	FAAS	On-line ultrasonic extraction	0.4[b]	60	0.9	122
Mo	ICP-MS	Off-line pressure-cooker digestion	0.558	No data	3.9	48
Mo	ICP-MS	Off-line microwave digestion	0.44[a]	No data	3.3	50
Ni	ICP-MS	Off-line microwave digestion	0.59[a]	No data	3	50
Ni	ICP-MS	No data	No data	No data	No data	74

Analyte	Detection	Sample preparation	DL (μg/L)	SF (s/h)	RSD (%)	Ref
Ni	ICP-MS	Off-line microwave digestion On-line minicolumn separation (Toyopearl AFChelate 650M resin)	0.011	No data	0.8	76
Ni	FAAS	Off-line microwave digestion On-line minicolumn separation (acrylic acid-grafted PTFE fibers)	0.25	55	1.6	85
Ni	FAAS	On-line ultrasonic extraction On-line minicolumn (Serdolit Che resin) separation	0.12[b]	13-28	1.9-3.6	123
Pb	ICP-MS	Off-line pressure-cooker digestion	0.023	No data	1.8	48
Pb	ICP-MS	Off-line microwave digestion	0.898[a]	No data	3.7	50
Pb	SV	Off-line digestion (open vessels)	No data	No data	1.4	81
Pb	AF	On-line gas-liquid separation	0.31	No data	<5.6	128
Pb	FAAS	Off-line digestion (open vessels)	90	No data	1.9	60
Pb	FAAS	Off-line digestion (open vessels) On-line coprecipitation	2.7	60	2.0	65
Pb	FAAS	On-line minicolumn adsorption: analyte complexes with APDC on PT-C18	6.9	No data	2.8	66
Pb	ETAAS	Off-line digestion (open vessels) On-line sorption separation in a Knotted reactor	0.0048	31	2.1	124
Pb	FAAS	Off-line microwave digestion On-line minicolumn (macrocycle immobilized on silica gel) separation	5	63	1.9	125
Pb	FAAS	Off-line microwave digestion On-line minicolumn (poly(aminophosphonic acid)) separation	0.25[b]	45	2.3	126
Pb	FAAS	Off-line digestion (closed vessels) On-line minicolumn (Amberlite XAD-2 modified by BTAC) separation	3.7	No data	2.3-4.4	127
Pb	ICP-MS	No data	No data	No data	No data	74
Pb	ICP-MS	Off-line microwave digestion On-line minicolumn (Toyopearl AFChelate 650M resin)separation	0.001	No data	0.1	76
Sb	ICP-MS	Off-line pressure-cooker digestion	0.026	No data	1.8	48
Sb	AAS	Off-line microwave digestion On-line gas-liquid separation	0.15[a]	No data	No data	57

Table 7.2. Continued

Analyte	Detection	Sample preparation	DL (μg/L)	SF (s/h)	RSD (%)	Ref
Sb	AAS	Off-line digestion (open vessels) On-line gas-liquid separation	No data	No data	No data	129
Se	ICP-MS	Off-line pressure-cooker digestion	3.401	No data	0.5	48
Se	AF	Off-line digestion (open vessels) On-line gas-liquid separation	0.0087	No data	No data	51
Se	AAS	Off-line microwave digestion On-line gas-liquid separation	0.06[a]	No data	No data	57
Se	FIAS200[a] AAS	Off-line digestion (open vessels) On-line gas-liquid separation	No data	No data	No data	130
Se	AAS	Off-line digestion (open vessels) On-line gas-liquid separation	0.3 75[a]	No data	2.6	131
Se	AAS	Off-line microwave digestion On-line gas-liquid separation	10	30	<1.5	132
Se	AAS	Off-line microwave digestion On-line gas-liquid separation	160[a]	No data	7.5	133
Se	ICP-MS	Off-line microwave digestion On-line minicolumn (alumina) separation	4 6.5[b]	No data	No data	58
Sn	ICP-MS	Off-line pressure-cooker digestion	0.067	No data	2	48
Sn	ICP-MS	Off-line microwave digestion	0.39[a]	No data	No data	50
Sn	AAS	Off-line digestion (open vessels) On-line gas-liquid separation	0.08	72	2.5	134
Sr	ICP-MS	Off-line pressure-cooker digestion	0.023	No data	1	48
Sr	ICP-MS	Off-line microwave digestion	0.08[a]	No data	1.0	50
Tl	ICP-MS	Off-line pressure-cooker digestion	0.011	No data	1.3	48
V	ICP-MS	Off-line pressure-cooker digestion	0.126	No data	3.4	48
V	ICP-MS	Off-line microwave digestion	0.11[a]	No data	2.0	50
V	ICP-MS	Off-line microwave digestion On-line minicolumn (alumina) separation	0.10 0.12[b]	No data	No data	58
V	SP	Off-line ashing On-line minicolumn (anion exchanger packed into a flow-through cell) separation	14-24	10-15	1.2-2.2	135
V(V)	F	Off-line ashing On-line minicolumn (anion exchanger loaded with Alizarin Red S packed into a flow-through cell) separation	0.45	No data	4.22	136
Zn	ICP-MS	Off-line pressure-cooker digestion	0.799	No data	3.9	48

Analyte	Detection	Sample preparation	DL (μg/L)	SF (s/h)	RSD (%)	Ref
Zn	ICP-MS	Off-line microwave digestion On-line minicolumn (Toyopearl AFChelate 650M resin) separation	12	No data	1.8	76
Zn	F	No data	5	No data	1.8	137

[a] ng/g
[b] μg/g
[c] with amalgamation.
[d] without amalgamation.
[e] Perkin-Elmer.

1N2N: 1-nitroso-2-naphthol; AAS: atomic absorption spectrometry; ADDP: ammonium diethyldithiophosphate; AF: atomic fluorescence; APDC: ammonium pyrrolidine dithiocarbamate; BTAC: 2-(2-benzothiazolylazo)-2-p-cresol; Cd-MT: metallothionein-bound cadmium; Ch: chemiluminescence; CPG: controlled-pore glass; DL: detection limit; ETAAS: electrothermal atomic absorption spectrometry; EtHg: ethylmercury; F: fluorescence; FAAS: flame atomic absorption spectrometry; GG: silica gel; HQ: 8-hydroxyquinoline; HPLC: high performance liquid chromatography; ICP-MS: inductively coupled plasma-mass spectrometry; KDS: Kjeldahl digestion system; MeHg: methyl mercury; PAPhA: poly(aminophosphonic acid) resin; Phen: 1,10-Phenanthroline; PhHg: phenylmercury; RSD: relative standard deviation; SF: sampling frequency; SP: spectrophotometry; SV: stripping voltammetry; UV: ultraviolet.

Aluminium

Arruda et al. [49] constructed a FI system incorporating a microwave oven for the digestion of food samples for the subsequent determination of aluminum by electrothermal atomic absorption spectrometry. A volume of sample slurry is simultaneously injected with diluted nitric cid and transferred into a PTFE reactor (coiled around an Erlenmeyer flask filled with water) located inside a microwave oven. The digested sample and flushing solution are collected in an autosampler cup. The sample is diluted to 20-fold within the FI manifold. Harrington et al. [50] developed a protocol, based on off-line microwave digestion, followed by FI coupled to inductively coupled mass spectrometry (FI-ICP-MS) for the simultaneous determination of up to 17 metals present in food digests. The method, which involves a combined nitric acid-hydrogen peroxide microwave digestion step.

Arsenic

Atomic Fluorescence Detection

Wei et al. [51] used the hydride vapor generator-coupled atomic fluorescence spectroscopic system to replace the atomic absorption spectroscopic detector, which was used in the ordinary hydride generation technique. For better sensitivity, membrane dryers were used that use a hygroscopic, ion-exchange membrane in a continuous drying process between hydride generator (separator) and atomic fluorescence detector to selectively remove water vapor from mixed hydride gas streams.

Atomic Absorption Spectrometric Detection

Damkroger et al. [52] tested high-pressure digestion and a closed-vessel microwave heated system, both employing a mixture of nitric acid and hydrogen peroxide as digesting agent for decomposing seafood samples to determine arsenic by use of FI-hydride generation-atomic absorption spectrometry. They proved that the microwave system is insufficient to mineralize arsenic in marine samples, high-pressure ashing at 300°C results in no quantitative recovery percentages. However, a dry ashing procedure is given as a reference digestion, yielding a complete recovery of As. Oygard et al. [53] developed a simple, rapid, and reliable method for determination of inorganic arsenic in biological samples such as fish fillet. Inorganic arsenic is distilled from the sample as $AsCl_3$ with hydrochloric acid. The separated inorganic arsenic is determined by FI hydride-generation atomic absorption spectrometry after prereduction with KI and HCl. Capelo et al. [54] developed a new sample pretreatment method based on ultrasonic extraction in hydrochloric acid medium and subsequent oxidation of the extracts by sonozone (i.e., sonolysis-ozonolysis) for determination of reactive arsenic toward sodium tetrahydroborate [mainly As(III) + As(V)] by FI-hydride generation atomic absorption spectrometry. A sonozone process at room temperature is optimized to break the bond of arsenic to proteins and macromolecules constituents, who is an essential requirement for effective reduction by L-cysteine prior to arsine generation. Low arsenic recoveries were observed for fish samples, as a result of the nondegradability of arsenobetaine by sonozone. Sakamoto et al. [55] described a method for the determination of total arsenic by hydride generation-atomic absorption spectrophotometry using a mixed acid as a pretreatment. Hydride generation is done by the FI method. These authors investigated in detail the temperature and time of decomposition using inorganic, organic arsenic and environmental standard samples, pretreated with nitric-perchloric-sulfuric mixed acid. By using a mixed acid as a pretreatment agent at 220°C, the decomposition time could be shortened and the blank value of arsenic from the reagents used is reduced. The mixed acid of nitric-perchloric-sulfuric is also found to be effective as a pretreatment agent for organic arsenic compounds in which a dimethylated compound, sodium cacodylate or biological samples, is known to be one of the indecomposables. This approach is proved to be satisfactory as a pretreatment for the quantitative analysis of trace amounts of total arsenic. Ringmann et al. [56] described a microwave assisted wet digestion method for organoarsenic compounds and subsequent determination of total arsenic by FI hydride generation electrothermal atomic absorption spectrometry (FI-HG-ETAAS). Sodium persulfate, sodium fluoride and nitric acid serve as digestion reagents, which allow a quantitative transformation of organoarsenic compounds to hydride forming species in a commercial microwave sample preparation system. If medium pressure vessels is used, seafood samples usually require either the use of high pressure vessels or a second digestion step. Korenovska [57] proposed a FI hydride generation atomic absorption spectrometric (FI-HG-AAS) method for the determination of arsenic in foods consumed in Slovakia.

Inductively Coupled Plasma Mass Spectrometric Detection

Ebdon et al. [58] described a method involving the retention of the analytes as anions on activated alumina (acidic form) in a microcolumn using an on-line FI system, with simultaneous matrix removal and detection by inductively coupled plasma mass spectrometry.

Boron

Harrington et al. [50] developed a protocol, based on microwave digestion, and followed by FI coupled to ICP-MS for the simultaneous determination of up to 17 metals present in food digests.

Barium

Stefanka et al. [48] described a FI method using hydraulic high-pressure nebulization as a sample introduction system, coupled to ICP-TOFMS for rapid and simultaneous determination of 19 elements. Harrington et al. [50] developed a protocol, based on microwave digestion, and followed by FI coupled to ICP-MS for the simultaneous determination of up to 17 metals present in food digests.

Calcium

Moreno-Cid et al. [59] extracted calcium from solid seafood samples by a simple and rapid continuous ultrasound-assisted extraction system. This system is connected to a FI manifold, which allows the on-line flame atomic absorption spectrometric determination of calcium. The on-line manifold for calcium determination is the simplest possible, because a small volume of acid extract is injected into an ultrapure water carrier stream. The acid extract is diluted on-line with lanthanum, which also acts as masking agent in order to avoid chemical interferences.

Cadmium

Atomic Absorption Spectrometric Detection

Becerra et al. [60] proposed a simple and rapid method is for the determination of lead and cadmium in foodstuffs. The samples are digested in hydrogen peroxide and nitric acid, and then lead and cadmium are determined in a FI atomic absorption system. Saraviva-Miranda et al. [61] designed a FI system comprised of four mini-columns packed with cation exchange resin (AG50W-X8 resin) to implement an on-line preconcentration procedure for cadmium determination by flame atomic absorption spectrometry. Enriquez-Dominguez et al. [62] developed a FI preconcentration system with a chelating resin to determine trace and ultratrace amounts of cadmium in mussels by flame atomic absorption spectrometry. The metal is preconcentrated on a microcolumn packed with poly(aminophosphonic acid) resin and eluted with diluted hydrochloric acid into the nebulizer-burner system of an atomic absorption spectrometer. Yebra et al. [63] reported a simple and rapid acid sample digestion method by microwave heating in high-pressure Teflon bombs for the determination of cadmium in mussels by flame atomic absorption spectrometry coupled on-line with a FI preconcentration system. Ali et al. [64] tested 1,10-Phenanthroline (phen) as a complexing agent for online preconcentration of Cu, Cd and Co, on RP-C18 material in a microcolumn with a FI-flame at. absorption spectrometric system. The on-line formed metal-phen complexes can be adsorbed on the C18 sorbent material. Liu et al. [65] used a FI on-line coprecipitation system with diethyldithiocarbamate (DDTC) Ni (II) as a carrier coupled to flame atomic absorption spectrometry for the determination of trace copper, lead, cadmium and iron in environmental and biological samples. Metal ions are on-line coprecipitated with DDTC-Ni (II) in nitric acid medium, and the precipitate is collected in a knotted reactor. Then, the precipitate is dissolved by iso-buthylmethylketone (IBMK), and the concentrated zone is transported directly into the nebulizer-burner system of a flame atomic absorption spectrometer. Wang et al. [66] determined lead and cadmium by flame atomic absorption spectrometry with on-line PT-C18 precolumn-FI separation and enrichment technique. In this system, ammonium pyrrolidinedithiocarbamate, and Na diethyldithiocarbamate are used as a chelating agent, and methanol is used as an eluent. Muñoz et al. [67] assessed for the first time the adsorptive potential of C60 fullerene for the preconcentration of metallothionein-bound cadmium (Cd-MT) traces. Thus, they developed a simple FI system for the on-line preconcentration of Cd-MT. The compounds are directly retained on the sorbent column and subsequently eluted, by hydrolysis of the complexes with diluted nitric acid, and the cadmium is directly transferred to a flame atomic absorption spectrometer. Total cadmium (inorganic and Cd-MT) is also determined by preparing the sample in acid medium and complexing total cadmium with Na-DDC. The neutral chelates of cadmium, retained also on C60, are eluted with IBMK and transferred to the atomic absorption instrument. By the difference between both determinations, inorganic and organometallic forms of cadmium could be distinguished. Bruhn et al. [68] developed a simple low-cost FI-FAAS methodology for determination of cadmium traces in food samples using preconcentration by on-line extraction in a chelating resin. In the case of food samples, a more complex matrix medium, the pH of acid digested solutions (pH 1) is raised first to pH 3-5 off-line with ammonia, and then is on-line adjusted to pH 8.0 with ammonium acetate buffer to proper retention of cadmium as Cd-8-hydroxyquinoline complex, minimizing interferences. Cadmium is preconcentrated in a minicolumn filled with 8-hydroxyquinoline azo-immobilized on controlled-pore glass. Martins et al. [69] describes the functionalization of biopolymer chitosan, using the complexing agent 8-hydroxyquinoline (oxine) by reaction of diazotization. The chelating resin is characterized by degree of deacetylation, IR, Raman spectroscopy. The efficiency of the chelating resin and accuracy of the proposed method is evaluated by the metal ion recovery technique in the analysis of a certified sample of oyster tissue. The metal ions Cd(II) and Cu(II) in the samples were previously enriched in a mini-column and FI flame atomic absorption spectrometry is used to determine the concentrations of the analytes. The chelating resin exhibited high selectivity for Cd(II) at pH 7 and for Cu(II) at pH 10. The eluent concentration is 0.5 mol/L nitric acid. Fang et al. [70] synthesized a new ion-imprinted thiol-functionalized silica gel sorbent by a surface imprinting technique in combination with a sol-gel process for selective online, solid-phase extraction of Cd(II). The Cd(II)-imprinted thiol-functionalized silica sorbent is characterized by FTIR, the static adsorption-desorption experiment, and the dynamic adsorption-desorption method. The maximum static adsorption capacity of the ion-imprinted functionalized sorbent is 284 mmol/g. The imprinted functionalized silica gel sorbent offered a fast kinetics for the adsorption and desorption of Cd(II). The prepared ion-imprinted functionalized sorbent is promising for on-line, solid-phase extraction coupled with flame atomic absorption spectrometry for the determination of trace cadmium in environmental and biological samples. All competitive ions studied did not interfere with the determination of Cd(II). Korkmaz et al. [71] designed a quartz trap for on-line preconcentration of Cd species. The cold vapor generation technique is used for the generation of cadmium species. The trapping medium is formed by external heating of the inlet arm of a quartz T-tube. The generated analyte species are trapped on a quartz surface heated to the collection temperature, 350°C, and the collected species are revolatilized when the trap is heated further to revolatilization temperature, 1000°C, and hydrogen gas is introduced in the trapping medium. Borges et al. [72] describes the use of sugar cane bagasse as solid phase extractor for cadmium determination after complexation of the analyte with ammonium diethyldithiophosphate (ADDP) and sorption of the Cd-DDP complexes on the solid support. The concomitants are separated using a FIA system coupled to flame atomic absorption spectrometry for analyte detection. Yebra-Biurrun et al. [73] coupled a continuous ultrasound-assisted extraction system with preconcentration and flame atomic absorption spectrometry for the determination of cadmium and lead in mussel samples. A minicolumn containing a chelating resin (Chelite P, with aminomethylphosphoric acid groups) is proved as an excellent material for the quantitative preconcentration of cadmium and lead prior to their flame atomic absorption detection. A FI manifold is used as interface for coupling the three analytical steps, which allowed the automation of the whole analytical process.

Inductively Coupled Plasma Mass Spectrometric Detection

Harrington et al. [74] described a method using FIA for the on-line dilution and spiking with internal standard for the analysis of Cr, Ni, Cd, and Pb in food digests. The FIA method improved the reporting limits and accuracy for the analysis of a proficiency testing standard and the reporting limits and accuracy for the analysis of an oyster tissue certified reference material, compared to the conventional inductively coupled plasma optical emission spectroscopy (ICP-MS) method. The developed method is also quicker than the conventional method, as no off-line sample dilution is needed. Valles-Mota et al. [75] characterized for first time the experimental parameters governing the instrumental precision and accuracy for isotope ratio measurements of cadmium in ICP-MS, including sampling time, mass bias, detector dead-time and spectroscopic interferences. Two alternative flow approaches for the determination of ultratrace concentrations of cadmium by isotope dilution (ID) were explored and compared with the more conventional ID methodology: first, on-line mixing of the sample solution with the spike solution just before the ICP-MS nebulizer using a peristaltic pump and, second, the generation of volatile cadmium species using sodium tetraethylborate by merging zones FI-ICP-MS. Thus, the three approaches were successfully applied to the determination of ultratrace levels of cadmium in biological and environmental certified reference materials. The online ID method proved to be the most convenient for the determination of cadmium in such samples because it is fast, provides similar results to those of conventional ID and requires less sample preparation. Willie et al. [76] presented a method for the determination of trace elements in seawater and fish otoliths by inductively coupled plasma mass spectrometry (ICP-MS) with FI on-line separation and preconcentration by using a minicolumn packed with Toyopearl AF-Chelate-650M iminodiacetate resin.

Cobalt

Chemiluminescence Detection

Song et al. [77] determined femtogram-level cobalt based on its significantly catalyzed effect on luminol-dissolved oxygen chemiluminescence reaction in the flow system.

Atomic Absorption Spectrometric Detection

Ali et al. [64] used 1,10-Phenanthroline (phen) as a complexing agent for on-line preconcentration of Cu, Cd and Co, on RP-C18 material in a microcolumn with a FI-flame atomic absorption spectrometric system. Yebra-Biurrun et al. [78] proposed a rapid and sensitive method for the determination of chromium and cobalt in seafood samples by flame atomic absorption spectrometry combined with dynamic ultrasound-assisted acid extraction and an on-line minicolumn preconcentration using a chelating resin (Serdolit Che or Chelite P). A FI manifold is used as interface for coupling all analytical steps, which allowed the automation of the whole analytical process.

Inductively Coupled Plasma Mass Spectrometric Detection

Willie et al. [76] presented a method for the determination of trace elements by inductively coupled plasma mass spectrometry with FI on-line separation and pre-concentration.

Chromium

Atomic Absorption Spectrometric Detection

Yebra-Biurrun et al. [78] developed a rapid and sensitive method for the determination of Cr and Co in seafood samples by flame atomic absorption spectrometry combined with a dynamic ultrasound-assisted acid extraction and an on-line minicolumn preconcentration. Yebra et al. [79] extracted total chromium from solid mussel samples by an on-line methodology using a simple, rapid, and continuous ultrasound-assisted extraction system (CUES). CUES is connected to a FI manifold, which permits the on-line determination of chromium. Bruhn et al. [80] coupled a FI system with a preconcentration minicolumn based on a chelating resin to a flame atomic absorption spectrometer. The focus of the work of these investigators was the study of interference effects and the analytical applicability of the azo-immobilized 8-hydroxyquinoline on controlled-pore glass for the determination of chromium and manganese in mussel. All studied concomitants affected the retention of Cr(III). These effects are probably related to the formation of hydroxo-complexes at the optimum pH range 9.0-10.

Inductively Coupled Plasma Mass Spectrometric Detection

Ebdon et al. [58] developed a method involving the retention of the analytes as anions on activated alumina (acidic form) in a microcolumn using an on-line FI system, with simultaneous matrix removal and ICP-MS detection. Harrington et al. [74] described a method using FIA for the on-line dilution and spiking with internal standard for the analysis of Cr, Ni, Cd, and Pb in food digests by ICP-MS.

Copper

Electrochemical Detection

Izquierdo et al. [81] proposed an automatic-continuous method for the simultaneous determination of copper and lead based on FIA and stripping voltammetry.

Atomic Absorption Spectrometric Detection

Ali et al. [64] used 1,10-Phenanthroline (phen) as a complexing agent for on-line preconcentration of Cu, Cd and Co, on RP-C18 material in a microcolumn with a FI-flame atomic absorption spectrometric system. The on-line formed metal-phen complexes are adsorbed on the C18 sorbent material. Ali et al. [82] tested the suitability of 1-nitroso-2-naphthol as a complexing agent for on-line preconcentration of copper using RP-C18 material in a microcolumn with FI coupled with flame atomic absorption spectrometry. The adsorbed complexes in the microcolumn are eluted with ethanol. Liu et al. [65] described a FI on-line coprecipitation system with diethyldithiocarbamate (DDTC) Ni (II) as a carrier coupled to flame atomic absorption spectrometry for the determination of trace copper, lead, cadmium and iron in environmental and biological samples. Moreno-Cid et al. [83] carried out the on-line extraction of copper from solid mussel samples by a simple and rapid continuous ultrasound-assisted extraction system (CUES). The CUES is connected to a FI manifold, which permits the on-line flame atomic absorption spectrometric determination of copper. Martins et al. [69] described the functionalization of biopolymer chitosan, using the complexing agent 8-hydroxyquinoline (oxine) by reaction of diazotization. The metal ions Cd(II) and Cu(II) in the samples are previously enriched in a mini-column and FI flame atomic absorption spectrometry determination of concentrations of the analytes. Yebra et al. [84] performed the on-line extraction of copper and iron from solid seafood samples by a robust, fast and simple continuous ultrasound-assisted extraction system. Wang et al. [85] applied a newly prepared acrylic acid-grafted PTFE fiber sorbent for FI on-line microcolumn preconcentration-flame atomic absorption spectrometric determination of trace copper and nickel in environmental and biological samples. On-line preconcentration of trace analytes is achieved on the microcolumn packed with acrylic acid-grafted PTFE fibers, and the retained analytes are eluted with diluted hydrochloric acid for on-line flame atomic absorption spectrometric determination.

Inductively Coupled Plasma Mass Spectrometric Detection

Willie et al. [76] presented a method is for the determination of trace elements in seawater and fish otoliths by inductively coupled plasma mass spectrometry with FI on-line separation and preconcentration.

Iron

Fluorescence Detection

Luan et al. [86] described an analytical system based on continuous and automatic fluorescence spectroscopy set up by conjunction of FI with charge coupled device array detector and laser induced fluorescence spectrometer. Thus, a new catalytic fluorimetric method is developed for determination of Fe(III) based on the catalysis of Fe(III) to oxidize rhodamine B with hydrogen peroxide in hydrochloric acid medium.

Spectrophotometric Detection

Pereira-Fo et al. [87] developed a method for the spectrophotometric determination of iron in seafood slurry samples based on microwave assisted digestion. The stabilized slurry is introduced in a flow system and transported by an air carrier stream to a digestion coil positioned inside the microwave oven. After the digestion step the flushing solution is collected in a calibrated flask. A sample aliquot is introduced in a FI system and the iron is determined at 512 nm with 1,10-phenanthroline.

Atomic Absorption Spectrometric Detection

Liu et al. [65] applied a FI on-line coprecipitation system with diethyldithiocarbamate (DDTC) Ni (II) as a carrier and coupled to flame atomic absorption spectrometry for the determination of trace copper, lead, cadmium and iron in environmental and biological samples. Metal ions are on-line coprecipitated with DDTC-Ni (II) and the precipitate is collected in a knotted reactor. The precipitate is then dissolved by IBMK. Yebra et al. [84, 88] proposed a continuous ultrasound-assisted extraction system for on-line extraction of copper and iron from solid seafood samples. This system is connected to a FI manifold, which allows on-line flame atomic absorption spectrometric determination of copper and iron.

Inductively Coupled Plasma Mass Spectrometric Detection

Willie et al. [76] presented a method is for the determination of trace elements in seawater and fish otoliths by inductively coupled plasma mass spectrometry with FI on-line separation and preconcentration.

Gallium/Indium/Lithium/Molybdenum/Strontium/Thallium

Stefanka et al. [48] developed a FI method using hydraulic high-pressure nebulization as a sample introduction system, coupled to ICP-TOFMS for simultaneous determination of 19 elements including gallium, indium, molybdenum strontium and thallium. Harrington et al. [50] developed a protocol, based on microwave digestion, and followed by FI coupled to ICP-MS for the simultaneous determination of up to 17 metals present in food digests including molybdenum and strontium.

Mercury

Atomic Fluorescence Detection

Edwards et al. [89] coupled FIA involving online oxidation of organomercury species to Hg^{2+}, followed by reduction to Hg^0 with acidified Sn(II) chloride to an atomic fluorescence spectrophotometer for determination of total organomercury in environmental samples. Cheng et al. [90] proposed a simple and rapid method for the determination of trace mercury in marine biological samples by FI atomic fluorescence spectrometry using a slurry technique. Cava-Montesinos et al. [91] developed a fully mechanized procedure for the speciation of mercury in fish samples by using cold vapor atomic fluorescence spectrometry. For this, sample slurries in an acid mixture (formed by a solution containing diluted hydrochloric acid, nitric acid, sulfuric acid and hydrogen peroxide) in the presence of a surfactant and with traces of potassium dichromate, are injected into a flow system, sonicated and merged with an oxidant mixture of bromide/bromate heated at 50°C in a water bath. Sonicated sample slurries were also measured, in the absence of bromide/bromate, in order to obtain a second series of data, which could be employed to establish the concentrations of free Hg(II). Tseng et al. [92] improved the precision and bias of monomethylmercury (MMHg) determinations in environmental samples by directly coupling and automating the numerous steps involved with analysis of this toxic Hg species. The authors developed a simple and robust mercury speciation analyzer for measurement of MMHg in environmental matrixes. This on-line hyphenated system couples the main anal. steps, including sample introduction, aqueous-phase ethylation, Tenax preconcentration, and gas chromatographic separation, to cold vapor atomic fluorescence detection and data acquisition. Liang et al. [93] proposed the method of slurry sampling FI-microwave digestion with cold vapor generation-atomic fluorescence detection for the determination of mercury in biological and environmental samples. Thus, samples are dispersed in aqua regia and agitated magnetically to obtain well-proportioned and stable sample solutions. Wu et al. [94] described a novel nonchromatographic approach for direct speciation of mercury, based on the selective retention inorganic mercury and methylmercury on the inner wall of a knotted reactor by using ammonium di-ethyldithiophosphate (DDP) and dithizone as complexing agents, respectively, for FI on-line sorption preconcentration coupled with chemical vapor generation nondispersive atomic fluorescence spectrometry. With the sample pH kept at 2.0, the preconcentration of inorganic mercury on the inner walls of the knotted reactor is carried out based on the exclusive retention of Hg-DDP complex in the presence of methylmercury via on-line merging the sample solution with ammonium di-ethyldithiophosphate solution. Selective preconcentration of methylmercury is achieved with dithizone instead of ammonium di-ethyl dithiophosphate. Diluted hydrochloric acid is introduced into the FI system to elute the retained Hg species. After, the eluate merges with a solution of potassium borohydride for atomic fluorescence spectrometry detection. Carvalho et al. [95] evaluated the exposure to methylmercury (MeHg). Thus, fish samples are collected in the dock of Sesimbra and total mercury is determined by FI cold vapor atomic fluorescence spectroscopy.

Atomic Absorption Spectrometric Detection

Luo et al. [96] realized the hydride-graphite furnace determination of mercury by using fluidity injector binding with graphite furnace and successful by selected the determination conditions. Bauza de Mirabo et al. [97] reported a SI analytical system for the determination of mercury by cold vapor atomic absorption spectrometry. Both the sample and the reagent are sequentially aspirated using a Crison automatic Compact Titrator and impelled into a gas-liquid separation cell. Once there, a nitrogen flow sweeps the reduced mercury into a measuring cell of an atomic absorption spectrometer. This system allows the detection of mercury in addition to data acquisition and treatment in an automatic way. Aduna de Paz et al. [98] studied a FIA system involving cold vapor atomic absorption spectrometry preceded by a wet digestion in a microwave oven, as a method for measuring mercury in fish. Tao et al. [99] developed a simple, rapid and reliable method for the determination of total mercury in biological samples. Samples are solubilized by tetramethylammonium hydroxide (TMAH). The organically bound mercury is cleaved and converted to inorganic mercury by on-line addition of potassium permanganate. The decomposed mercury together with inorganic mercury originally present in samples is determined by FI cold vapor atomic absorption spectrometry after reduction to elemental mercury vapor using sodium borohydride. Chessa et al. [100] described a method to determine mercury in fish using the FIAAS technique. Mineralization, the most critical step, is achieved with nitric acid and microwave heating. Yin et al. [101] proposed a fully automated system for the direct determination of methylmercury (MeHg), ethylmercury (EtHg), phenylmercury (PhHg), and inorganic mercury (Hg(II)) at the ng/L level. It is based on solid phase extraction preconcentration incorporated in a FI system, high performance liquid chromatography separation, reduction combined with thermolysis and determination by cold vapor atomic absorption spectrometry. For preconcentration a microcolumn of bonded silica with octadecyl functional groups (C18 reversed phase material) is used as a sorbent for the mercury complexes formed on-line with ammonium pyrrolidine dithiocarbamate. Retained mercury species are eluted with a mixture of methanol-acetonitrile-water and subjected to separation on an octadecyl silane column before determination by cold vapor atomic absorption spectrometry. The sensitivity of organo-mercury determination could be improved by sodium borohydride as a reductant combined with a thermolysis step. Rio-Segade et al. [102] developed a simple and rapid ultrasound-assisted extraction method with hydrochloric acid for mercury speciation in fish tissues. Centrifuged extracts are directly injected into a FI-cold vapor atomic absorption spectrometry system. First, methylmercury is separately determined using sodium tetrahydroborate as reducing agent after selective extraction with diluted hydrochloric acid. Second, inorganic mercury is determined by selective reduction with stannous chloride in diluted hydrochloric acid extracts containing both mercury species. Total mercury could not be determined in the sonicated acid extracts using sodium tetrahydroborate as reducing agent because the methylmercury and inorganic mercury sensitivities are different. Burguera et al. [103] used an on-line time-based injection system in conjunction with cold vapor generation atomic absorption spectrometry and microwave-aided oxidation with potassium persulfate for the determination of the different mercury species. Inorganic mercury is determined after reduction with sodium borohydride, while total mercury is determined after an oxidation step with persulfate prior to the reduction step to elemental mercury with the same reducing agent. The difference between total and inorganic mercury determines the organomercury content in samples. Rio-Segade et al. [104] developed a method for the determination of total mercury in solid biological and environmental samples by slurry sampling combined with FI-cold vapor-atomic absorption spectrometry. The slurries are prepared in diluted nitric acid. Triton X-100 is used as dispersing agent, and slurries are subjected to ultrasonic pretreatment before being injected into the FI manifold. Tao et al. [105] presented a rapid and simple method for the quantification of inorganic and total mercury in biological tissues using FI cold vapor generation atomic absorption spectrometry. Samples are solubilized using tetramethylammonium hydroxide. The inorganic mercury is released by the on-line addition of L-cysteine and then reduced to metallic Hg by stannous chloride. Rio-Segade et al. [106] proposed a FI cold vapor atomic absorption spectrometry method to determine inorganic mercury and total mercury in mussel samples obtained from the Galicia coasts. The mussel samples are digested in a microwave oven using an nitric acid/hydrogen peroxide mixture and then total mercury is determined using sodium borohydride as reducing agent. Inorganic mercury is determined in a separate subsample, following ultrasonic extraction in hydrochloric acid medium, by selective reduction using stannous chloride in acid medium as reducing agent. Capelo et al. [107] optimized a FI manifold for the determination of mercury in seafood after microwave-assisted digestion (by using a mixture of nitric acid and hydrogen peroxide) by cold vapor atomic absorption spectrometry. Qvarnstrom et al. [108] further developed a previously described system for determination of low concentrations of mercury species in environmental samples using FI-HPLC cold vapor atomic absorption spectrometry with respect to time of analysis, long term signal stability, memory effects, detection limits, and environmental friendliness. Methyl and inorganic mercury are determined without pre-treatment in digested biological certified reference materials. Doering et al. [109] presented a SI method for the determination of mercury via cold vapor atomic absorption spectrophotometry. The method differs from FI cold vapor methods for the determination of mercury because of the simplicity of the system required for the method: one pump, one valve, a gas-liquid separator, and an atomic absorption spectrophotometer equipped with a quartz cell. Julshamn et al. [110] reported the study of ten laboratories, which participated in an interlaboratory method-performance (collaborative) study of a method for the determination of mercury in foods of marine origin by FI-cold vapor atomic absorption spectrometry after wet digestion using a microwave oven technique. Zenebon et al. [111] used a mixture of hydrogen peroxide-sulfuric acid (3 + 1, V/V) for decomposition of food in open vessels at 80°C. The treatment allowed rapid total mercury determination by FI cold vapor atomic absorption spectrometry. Li et al. [112] developed a novel methodology for the determination of trace mercury in environmental and foods samples by on-line coupling of FI displacement sorption preconcentration in a knotted reactor (KR) to electrothermal atomic absorption spectrometry (ETAAS). The developed methodology involves the on-line formation of copper pyrrolidine dithiocarbamate (Cu-PDC), presorption of the resulting Cu-PDC onto the inner walls of the KR, and selective retention of the analyte Hg(II) onto the inner walls of the KR through on-line displacement reaction between Hg(II) and the presorbed Cu-PDC. The retained analyte is subsequently eluted by ethanol and on-line is detected by ETAAS. Yan et al. [113] established a novel nonchromatographic speciation technique for ultratrace methylmercury in biological materials by FI microcolumn displacement sorption preconcentration and separation coupled on-line with ETAAS. In the developed technique, Cu(II) was first on-line complexed with diethyldithiocarbamate (DDTC), and the resultant Cu-DDTC is presorbed onto a microcolumn packed with the sorbent from a cigarette filter. Selective preconcentration of methylmercury (MeHg) in the presence of Hg(II), ethylmercury (EtHg), and phenylmercury (PhHg) is achieved at pH 6.8 through loading the sample solution onto the microcolumn due to a displacement reaction between MeHg and the presorbed Cu-DDTC. The retained MeHg is subsequently eluted with ethanol and on-line determined by ETAAS. Rio Segade et al. [114] used two simple FI systems for mercury speciation analysis in slurried fish tissue samples by cold vapor atomic absorption spectrometry. The solid samples are first suspended in hydrochloric acid containing Triton X-100 as dispersing agent, and then the slurries are injected in an acid carrier stream, which is either sequentially mixed with sulfuric acid, potassium persulfate oxidizing agent and stannous chloride reducing agent streams or merged with sodium borohydride reducing agent stream. Mercury speciation analysis is carried out as function of the oxidation coil temperature or sodium borohydride concentration, respectively. The FI system without on-line oxidation resulted in being simpler, faster and more sensitive for inorganic Hg and total Hg determination. Furthermore, the last system permitted the separate determination of Me-Hg and inorganic Hg by selecting an adequate hydrochloric acid concentration in the suspension medium and the suitable sodium borohydride reducing agent. Kan et al. [115] determined total mercury in biological samples by FI cold vapor atomic absorption spectrometry following tissue solubilization with formic acid. A mixture of potassium bromide and potassium bromate is used to decompose organomercury compounds prior to their reduction with sodium borohydride. Leal et al. [116] proposed a new software-controlled time-based multisyringe FI system for mercury determination by cold-vapor atomic absorption spectrometry. Precise known volumes of sample, reducing agent (stannous chloride in diluted hydrochloric acid) and carrier (diluted hydrochloric acid) are dispensed into a gas-liquid separation cell with a multisyringe burette coupled with one 3-way solenoid valve. An argon flow delivers the reduced mercury to the spectrometer. Silva et al. [117] developed a flow system for the determination of total mercury concentrations in fish samples by cold vapor atomic absorption spectrophotometry, based on the multicommuted FIA approach. The system uses independently controlled solenoid valves for the introduction of reagents and samples. When not injected, solutions are recirculating to the reservoir bottles, in this way reducing the waste produced by the analytical system and also the sample consumption. Vereda Alonso et al. [118] studied an on-line inorganic and organomercury species separation, preconcentration and determination system coupled to cold vapor atomic absorption spectrometry. The inorganic mercury species are retained on a column packed with a chelating resin aminopropyl-controlled pore glass functionalized with [1,5-bis (2 pyridyl)-3-sulfophenyl methylene thiocarbonohydrazyde] placed in the injection valve of a simple flow manifold. Methylmercury is not directly determined. Previous oxidation of the organomercurial species permitted the determination of total mercury. The separation of mercury species is obtained by the selective retention of inorganic mercury on the chelating resin. The difference between total and inorganic mercury determined the organomercury content in the sample.

Inductively Coupled Plasma Mass Spectrometric Detection

Beauchemin et al. [119] used inductively coupled plasma mass spectrometry for the determination of organomercury in two marine biological standard reference materials for trace metals (dogfish muscle tissue DORM-1 and lobster hepatopancreas TORT-1). Organomercury is extracted as the chloride from the material with toluene and back extracted into an aqueous medium of cysteine acetate. Since the final extracts contained more than 4% sodium, isotope dilution and FIA are used to respectively counter the effect of concomitant elements and avoid clogging the interface. Harrington et al. [120] demonstrated that using a combination of FI sample introduction and a sulfur-containing compound (2-mercaptoethanol) in the carrier solution, it is possible to decrease the memory effect of mercury in the sample introduction system to that for the internal standard (rhodium) produced in its determination by inductively coupled plasma mass spectrometry.

High-performance Liquid Chromatography

Dong et al. [121] developed a simple and cost-effective method for speciation analysis of trace mercury in seafood by on-line coupling FI microcolumn displacement sorption preconcentration to HPLC with UV detection. The methodology involves the presorption of the Cu-PDC (pyrrolidine dithiocarbamate) chelate onto a microcolumn packed with a cigarette filter sorbent, simultaneous preconcentration of Hg(II), methylmercury (MeHg), ethylmercury (EtHg), and phenylmercury (PhHg) onto the microcolumn through a displacement reaction with the presorbed Cu-PDC, and their subsequent elution from the microcolumn for on-line HPLC separation.

Manganese

Bruhn et al. [80] coupled a FI system with a preconcentration minicolumn based on a chelating resin (azo-immobilized 8-hydroxyquinoline on controlled-pore glass) to a flame atomic absorption spectrometer. Yebra et al. [122] carried out the on-line extraction of manganese from solid seafood samples by a simple continuous ultrasound-assisted extraction system (CUES). This system is connected to an on-line manifold, which permits the FI-FAAS detection of manganese.

Nickel

Atomic Absorption Spectrometric Detection

Wang et al. [85] used a newly prepared acrylic acid-grafted PTFE fiber sorbent for FI on-line microcolumn preconcentration-flame atomic absorption spectrometric determination of trace copper and nickel in environmental and biological samples. Yebra et al. [123] described a sensitive and low cost FI method that combines acid extraction, preconcentration and flame atomic absorption spectrometric determination of nickel in food samples at mg/g levels. The dynamic acid extraction step is carried out by using a continuous ultrasound-assisted extraction system. The acid extract is preconcentrated. on-line on a minicolumn packed with a chelating resin (Serdolit Che, with iminodiacetic groups) and nickel is eluted with diluted hydrochloric acid, being continuously monitored by flame atomic absorption spectrometry.

Inductively Coupled Plasma Mass Spectrometric Detection

Harrington et al. [74] developed a method using FIA for the on-line dilution and spiking with internal standard for the analysis of Cr, Ni, Cd, and Pb in food digests by ICP-MS. Willie et al. [76] presented a method for the determination of trace elements in fish otoliths by ICP-MS with FI on-line separation and preconcentration.

Lead

Electrochemical Detection

Izquierdo et al. [81] proposed an automatic-continuous method for the simultaneous determination of copper and lead based on FIA and stripping voltammetry.

Atomic Fluorescence Detection

Cheng et al. [128] proposed a method for the determination of lead in marine biological samples by the hydride generation-atomic fluorescence spectrometry. Lead hydride is generated from the merging of an acidified sample solution with potassium tetrahydroborate reducing solution containing potassium ferricyanide.

Atomic Absorption Spectrometric Detection

Becerra et al. [60] developed a simple and rapid method for the determination of lead and cadmium in foodstuffs. The samples are digested in hydrogen peroxide and nitric acid, and then lead and cadmium are determined in a FI atomic absorption system. Liu et al. [65] described a FI on-line coprecipitation system with diethyldithiocarbamate (DDTC) Ni (II) as a carrier and coupled to flame atomic absorption spectrometry for the determination of trace copper, lead, cadmium and iron in environmental and biological samples. Wang et al. [66] determined lead and cadmium by flame atomic absorption spectrometry with on-line PT-C18 precolumn-FI separation/preconcentration technique using ammonium pyrrolidine-dithiocarbamate, and Na diethyldithiocarbamate as a chelating agent, and methanol as eluent. Yan et al. [124] developed a selective and robust FI on-line sorption separation and preconcentration procedure for the ETAAS determination of (ultra)trace amounts of lead in biological and environmental samples. With the use of diethyldithiophosphate as complexing agent and citric acid as masking agent, the analyte complex is selectively formed and adsorbed onto the inner walls of a PTFE knotted reactor with removal of the high-content salts in saline water and most of the other metals in digested biological and environmental samples. The collected analyte complex is eluted quantitatively with ethanol and all the eluate was directly introduced into the graphite tube without a L'vov platform by an air flow. Neither a pre-heating step nor precise timing is needed during elution and eluate introduction. Yan et al. [125] reported a simple and highly selective FI online preconcentration and separation-flame atomic absorption spectrometric method for routine analysis of trace amounts of lead in biological and environmental samples. The selective preconcentration of lead is achieved in a wide range of sample acidity on a microcolumn packed with a macrocycle immobilized on silica gel. The lead retained on the column is effectively eluted with a solution of EDTA. Yebra et al. [126] incorporated a minicolumn packed with poly(aminophosphonic acid) chelating resin in an on-line preconcentration system with flame atomic absorption spectrometric detection to determine ultratrace amounts of lead in mussel samples. Ferreira et al. [127] proposed the use of Amberlite XAD-2 modified by 2-(2-benzothiazolylazo)-2-p-cresol (BTAC) as a sorbent in an on-line preconcentration system for lead determination. The procedure is based on the sorption of lead(II) ions onto a minicolumn packed with modified Amberlite XAD-2, followed by elution with diluted hydrochloric acid and detection by flame atomic absorption spectrometry. The on-line flow system uses four three-way solenoid valves and is electronically controlled in a time-based mode.

Inductively Coupled Plasma Mass Spectrometric Detection

Harrington et al. [74]developed a method using FIA for the on-line dilution and spiking with internal standard for the analysis of Cr, Ni, Cd, and Pb in food digests. Willie et al. [76] presented a method for the determination of trace elements in seawater and fish otoliths by (ICP-MS) with FI on-line separation/preconcentration by using Toyopearl AF-Chelate-650M iminodiacetate resin.

Antimony

Atomic Absorption Spectrometric Detection

Korenovska [57] developed a FI hydride generation at. absorption spectrometric method for the determination of antimony after microwave digestion of samples. Krachler et al. [129] described an analytical procedure based on an open vessel acid digestion of freeze-dried biological samples and subsequent antimony quantification in the digests by FI hydride generation atomic absorption spectrometry.

Selenium

Atomic Fluorescence Detection

Wei et al. [51] applied the hydride vapor generator-coupled atomic fluorescence spectroscopic system to replace the atomic absorption spectroscopic detector, which is used in the ordinary hydride generation technique for selenium determination. The methodology proposed is carried out by using FIA. For better sensitivity, membrane dryers are used that use a hygroscopic, ion-exchange membrane in a continuous drying process between hydride generator (separator) and atomic fluorescence detector to selectively remove water vapor from mixed hydride gas streams.

Atomic Absorption Spectrometric Detection

Korenovska [57] developed a FI hydride generation atomic absorption method for the determination of selenium after microwave digestion of samples. Guo et al. [130] determined hydride-forming elements with the FIAS-200 flow injection-mercury/hydride system. Chan et al. [131] applied FIA to sample introduction in conjunction with automated hydride generation and AAS for the determination of arsenic and selenium in environmental samples. Machado et al. [132] presented a system for electrochemical hydride generation using FI and atomic absorption spectrometry to determine selenium in biological materials. The electrolytic cell is constructed by assembling two reservoirs, one for the sample and the other for the electrolytic solution separated by a Nafion membrane. Each compartment has a Pt electrode. The atomization system used a T quartz tube in an air-liquid petroleum gas flame. Lavilla, et al. [133] evaluated microwave-assisted wet digestion procedures of seafood based on nitric acid or the mixture nitric acid/hydrogen peroxide and further thermal reduction of the Se(VI) formed to Se(IV). These procedures were as follows: (I) without H_2O_2 and without heating to dryness; (II) without H_2O_2 and with heating to dryness; (III) with H_2O_2 and without heating to dryness; (IV) with H_2O_2 and with heating to dryness. In general, low recoveries of Se are obtained for several marine species (e.g., crustaceans and cephalopods), which may be ascribed to the presence of Se forms mainly associated with nonpolar proteins and lipids. Post-digestion UV irradiation proved very efficient since not only complete organoselenium decomposition is achieved but also the final step required for prereduction of Se(VI) into Se(IV) could be avoided. With the microwave-assisted wet digestion/UV procedure, the use of strong oxidizing agents (persulphate, etc.) or acids (e.g. perchloric acid) which are typically applied prior to selenium determination by hydride generation techniques is overcome, and as a result, sample pre-treatment is significantly simplified.

Inductively Coupled Plasma Mass Spectrometric Detection

Ebdon et al. [58]described a method involving the retention of the analytes as anions on activated alumina (acidic form) in a microcolumn using an on-line FI system, with simultaneous matrix removal.

Tin

Fang et al. [134] described a hydride generation atomic absorption spectrometric method with FI, aimed at developing a practical routine assay for the determination of tin in food digests. In order to modify the sample matrix and to achieve optimized and reproducible conditions for the hydride generation reaction, the analyte is initially converted into its chlorostannate complex, thereby allowing it to be separated and preconcentrated on-line on an incorporated micro-column packed with a strongly basic anion-exchanger and subsequently to be eluted by diluted nitric acid.

Vanadium

Fluorescence Detection

Ruedas Rama et al. [136] determined V(V) by a simple bead injection spectroscopy-FIA system with spectrofluorometric detection using a commercially available flow. A volume of a homogeneous bead suspension of an anionic resin (Sephadex QAE A-25) previously loaded with the fluorogenic reagent 1,2-dihydroxyanthraquinone-3-sulfonic acid (Alizarin Red S) is injected to fill the flow cell. Next, V(V) is injected into the carrier and reacts with the immobilized reagent on the active solid support placed in the flow cell to form a fluorescent chelate in the absence of surfactant agents. The complex is so strongly retained on the beads that the regeneration of the solid support becomes extraordinarily difficult, so needing the renovation of the sensing surface, which is achieved by bead injection. At the end of the analysis, beads are automatically discarded from the flow cell and transported out of the system by reversing the flow. The measurement of fluorescence is continuously performed at an excitation wavelength of 521 nm and an emission wavelength of 617 nm.

Spectrophotometric Detection

Ayora-Cañada et al. [135] developed a flow analysis method based on direct absorptiometric measurement in solid phase of the violet complex formed between V(V) and 5-bromosalicylhydroxamic acid. The measurement is continuously performed at 555 nm, while the colored species is being concentrated on-line on the beads of an anion exchanger packed into a flow-through cell.

Inductively Coupled Plasma Mass Spectrometric Detection

Ebdon et al. [58]described a method involving the retention of the analytes as anions on activated alumina (acidic form) in a microcolumn using an on-line FI system, with simultaneous matrix removal.

Zinc

Fluorescence Detection

Gañan Gutierrez et al. [137] presented a selective room-temperature FI spectrofluorometric method for the determination of Zn(II), based on the use of salicylaldehyde thiocarbohydrazone in the presence of Triton X-100 and sodium acetate-acetic acid buffer.

Table 7.3. Features of FI determinations of anionic species in marine animals (seafood)

Analyte	Detection	DL	Linear range	SF (s/h)	RSD (%)	Ref
NO_2^-	F	51 μg/L	0.23-4.6 μg/mL	No data	No data	138
NO_2^-	A	0.25 μM	Up to 30 μM	10	0.42	139
NO_3^- NO_2^-	SP	0.025 μg/mL 0.01 μg/mL	0.06-1.6 μg/mL 0.05-1.15 μg/mL	25 ± 2	<2	140
P	SP	No data	No data	No data	No data	141
SiO_3^{2-} PO_4^{3-}	SP	No data	No data	No data	No data	142
SiO_3^{2-} PO_4^{3-}	SP	No data	0.1-10 mg/mL 2.0-30 mg/mL	50	<1.8 <1.3	143
SiO_3^{2-} PO_4^{3-}	SP	No data	0-12 mg/L 1.0-24 mg/L	No data	<1.8 <1.3	144

A: amperometry; DL: detection limit; F: fluorescence; RSD: relative standard deviation; SF: sampling frequency; SP: spectrophotometry.

Inductively Coupled Plasma Mass Spectrometric Detection

Willie et al. [76]reported a method for the determination of trace elements in seawater and fish otoliths by ICP-MS with FI on-line separation and preconcentration by using a minicolumn packed with Toyopearl AF-Chelate-650M iminodiacetate resin.

ANIONIC SPECIES

The anionic species that have been determined in seafood by using FI methodologies were the following: nitrate/nitrite, phosphorous/phosphate and silicate.

Different features of FI methods for the determination of anionic species in seafood are illustrated in Table 7.3. In the following paragraphs, some points observed in this table are highlighted due to their interest.

Nitrate/nitrite

Wada et al. [138] developed a simple and rapid FIA methodology with fluorescence detection for the determination of nitrite in meat and fish products. 1,2-diamino-4,5-dimethoxybenzene (DDB) is used as the fluorogenic reagent, and the carrier solution is DDB in diluted sulfuric acid. The DDB derivative is monitored at 402 nm (excitation at 311 nm). Kobayashi et al. [139] reduced nitrite in acidic potassium iodide solution to produce nitrogen oxide, which is measured amperometrically by a modified gold electrode covered with a PTFE-membrane as the permeability barrier for interfering substances in the sample. Monser et al. [140] proposed a direct spectrophotometric FI method for the simultaneous determination of nitrite and nitrate. The method is based on the oxidation of a phosphomolybdenum blue complex by the addition of nitrite and the decrease in absorbance of the blue complex is monitored at 820 nm. The injected sample is split into two segments. One of the streams is directly reacted with the above reagent and detected as nitrite. The other stream is passed through a copperized cadmium reductor column where reduction of nitrate to nitrite occurs, and the sample is then mixed with the reagent and passed through the cell of the spectrophotometer to be detected as nitrite plus nitrate.

Phosphorus/phosphate/Silicate

Munaf et al. [141] reported a FIA method with spectrophotometric detection for the determination of phosphorus in biological samples. Potassium peroxodisulfate is used as an oxidizing reagent with aid of a platinum wire as the catalyst, and organophosphorous compounds are converted to orthophosphate, which reacted with ammonium molybdate in the presence of ascorbic acid to form the molybdenum blue complex. The absorbance of the complex is detected at 880 nm. Narusawa et al. [142] described a matrix effect on the simultaneous determination of silica and phosphorous in mussel, tea leaves, sargasso, and rice flour by on-line column FIA spectrometry. Volatile constituents in the materials are easily removed by ashing. The resulting ash is fused with a lithium carbonate-boric acid mixture and then dissolved in hydrochloric acid. In order to remove interfering cations, an aliquot of the solution is filtered with a cation exchange column. The acidic effluent is evaporated to dryness and the residue is fused with a small amount of sodium carbonate, taken up in diluted EDTA solution, and analyzed by FIA. For simultaneous determination of these elements, TSK-gel SAX and NaCl-NH_3-EDTA as eluent are used. For cross-checking, ICP-AES is used for determination of silica and phosphorous. The agreement of the results between FIA and AES for is satisfactory for silica, but for phosphorous, it is unsatisfactory except for mussel. This discrepancy in phosphorous determination is interpreted as being due to matrix problems in the FIA measurements. Li [143] proposed a FIA method involving a continuous ion-exchange pretreatment and separation for the simultaneous determination of silicate and phosphate in biological samples. To remove interfering cations, the sample solution passes through the mini cation-exchange column connected at the inlet and outlet of one of the loading loops of a two-channel valve, and it again passes through another loading loop of the valve for sampling. After the valve is switched to the inject position, the defined sample is injected and passed through the separation column, and silicate and phosphate ions in the sample zone are separated. At the same time, the mini cation-exchange column is switched to the regeneration position and regenerated by a regenerant. By this manner, the sample pretreatment and determination can be made continuously. Li et al. [144] described the simultaneous determination of silicate and phosphate by using a FI spectrophotometric system based on an on-line ion-exchange pretreatment to remove interfering cations and a separation column to separate silicate and phosphate ions in the sample zone.

CONCLUSION

It is proven that FIA is a powerful strategy for the determination of analytes in seafood samples because this methodology allows accurate and selective determinations. Furthermore, on-line sample pretreatments for determination of organic analytes, on-line microwave digestion, continuous ultrasound assisted extraction and on-line separation procedures for determination of trace metals, solve the drawbacks of batch/manual preliminary operations, and have been shown to be very effective to increase sampling frequency and throughput as well as ease of automation. In addition, have been proposed flow systems incorporating continuous acid extraction, sample preconcentration and analyte detection, which have allowed the total automation of the determination of trace metals in solid samples as seafood. However, flow systems for sample pretreatment in the determination of anionic species have still not been proposed. It is hoped that in the future are developed new FI methodologies including continuous sample pretreatments for these determinations to allow the total automation of the determination of these analytes in seafood samples.

REFERENCES

[1] Luque-Garcia, J.L. & Luque de Castro, M.D (2003). *Trends Anal. Chem.* 22, 41-47.
[2] Priego-Capote, F. & Luque de Castro, M. D. (2004). *Trends Anal. Chem.* 23, 644-653.
[3] Priego-Capote, F. & Luque de Castro, M. D. (2007). *Analytical applications of ultrasound.* Amsterdam: Elsevier.
[4] Yamamura, A., Sekiguchi, Y., Koyama, T., Matsumoto, K. & Kiba, N. (2001). *Chem. Sensors* 17, 430-432.
[5] Kiba, N., Koga, A., Tachibana, M., Tani, K., Koizumi, H., Koyama, T., Yamamura, A., Matsumoto, K., Okuda, T. & Yokotsuka, K. (2006). *Anal. Sci.* 22, 95-98.
[6] Chen, R.L.C., Lee, M. H. & Matsumoto, K. (1996). *Anal. Sci.* 12, 87-90.
[7] Gao, L.Y., Matsumoto, K. & Osajima, Y. (1992). *Nippon Shokuhin Kogyo Gakkaishi* 39, 620-624.
[8] Caldarone, E.M. & Buckley, L.J. (1991). *Anal. Biochem.* 199, 137-141.
[9] Bechmann, I.E. (1996). *Anal. Chim. Acta* 320, 155-164.
[10] Hungerford, J.M., Walker, K.D., Wekell, M.M., LaRose, J.E. & Throm, H.R. (1990). *Anal. Chem.* 62, 1971-1976.
[11] Carsol, M.A. & Mascini, M. (1999). *Talanta* 50, 141-148.
[12] Hungerford, J.M., Hollingworth, T.A. & Wekell, M.M. (2001). *Anal.Chim. Acta* 438, 123-129.
[13] Takagi, K. & Shikata, S. (2004). *Anal.Chim. Acta* 505, 189-193.

[14] Carralero, V., Gonzalez-Cortes, A., Yanez-Sedeno, P. & Pingarron, J. M. (2005). *Electroanal.* 17, 289-297.

[15] Watanabe, S., Matsuo, K., Suzuki, Y., Tachibana, M., Tani, K., Koizumi, H., Matsumoto, K. & Kiba, N. (2007). *Bunseki Kagaku* 56, 1033-1036.

[16] Moody, G.J., Sanghera, G.S. & Thomas, J.D.R. (1987). *Analyst* 112, 65-70

[17] Yao, T. & Matsumoto, Y. (1989). *Electroanalysis* 1, 173-176.

[18] Yagiuda, K., Hemmi, A., Sakamura, K., Ito, S., Asano, Y., Hayashi, K. & Karube, I. (1995). *Kogyo Butsuri Kagaku* 63, 1148-1153.

[19] Hayashi, K., Okugawa, T., Kozuka, Y., Sasaki, S., Ikebukuro, K. & Karube, I. (1996). *J. Food Sci.* 61, 736-745.

[20] Balladin, D.A., Narinesingh, D., Stoute, V.A. & Ngo, T.T. (1997). *Appl. Biochem. Biotechnol.* 62, 317-328.

[21] Carsol, M.A., Volpe, G. & Mascini, M. (1997). *Talanta* 44, 2151-2159.

[22] Carsol, M.A. & Mascini, M. (1998). *Talanta* 47, 335-342.

[23] Sarker, A.K., Ukeda, H., Kawana, D. & Sawamura, M. (1999). *Anal. Sci.* 15, 1141-1144.

[24] Park, I.S. & Kim, N. (1999). *Anal.Chim. Acta* 394, 201-210.

[25] Nanjyo, Y. & Yao, T. (2002). *Anal.Chim. Acta* 470, 175-183.

[26] Morales-Muñoz, S., Luque-Garcia, J.L. & Luque de Castro, M.D. (2002). *Anal. Chem.* 74, 4213-4219.

[27] Morales-Muñoz, S., Luque-Garcia, J.L. & Luque de Castro, M.D. (2002). *J. Chromatogr.* 978A, 49-57.

[28] Marquette, C.A., Coulet, P.R. & Blum, L.J. (1999). *Anal. Chim.Acta* 398, 173-182.

[29] Pleasance, S., Ayer, S.W., Laycock, M.V. & Thibault, P. (1992). *Rapid Commum.. Mass Spectrom.* 6, 14-24.

[30] Tsukatani, T. & Matsumoto, K. (2000). *Anal. Chim.Acta* 416, 197-203.

[31] Khampha, W., Yakovleva, J., Isarangkul, D., Wiyakrutta, S., Meevootisom, V. & Emneus, J. (2004). *Anal. Chim.Acta* 518, 127-135.

[32] Nakashima, K., Muraki, K., Nakatsuji, S., Akiyama, S., Kaneda, T. & Misumi, S. (1989). *Analyst* 114, 501-504.

[33] Leon, A., Chica, A., Garcia-Raurich, J. & Centrich, F. (1994). *Quim. Anal.* 13, 78-81.

[34] Zhi, Z.L., Rios, A. & Valcarcel, M. (1995). *Anal. Chem.* 67, 871-877.

[35] Sadok, S., Uglow, R.F. & Haswell, S.J. (1996). *Anal.Chim. Acta* 321, 69-74.

[36] Hatton, A.D. & Gibb, S.W. (1999). *Anal. Chem.* 71, 4886-4891.

[37] Ruiz-Capillas, C. & Horner, W.F.A. (1999). *J. Sci. Food Agric.* 79, 1982-1986.

[38] Ruiz-Capillas, C., Gillyon, C.M. & Horner, W.F.A. (2000). *European Food Res.Technol.* 210, 434-436.

[39] Pons-Sanchez-Cascado, S., Izquierdo-Pulido, M., Marine-Font, A., Veciana-Nogues, M. T. & Vidal-Carou, M.C. (2000). *Quim. Anal.* 19, 165-170.

[40] Ruiz-Capillas, C., Gillyon, C.M. & Horner, W.F.A. (2001). *J. Food Biochem.* 25, 541-553.

[41] Adhoum, N., Monser, L., Sadok, S., El-Abed, A., Greenway, G.M. & Uglow, R.F. (2003). *Anal.Chim. Acta* 478, 53-58.

[42] Mitsubayashi, K., Kubotera, Y., Yano, K., Hashimoto, Y., Kon, T., Nakakura, S., Nishi, Y. & Endo, H. (2004). *Sensors and Actuators, B: Chemical* B103, 463-467.

[43] Dhaouadi, A., Monser, L., Sadok, S. & Adhoum, N. (2007). *Food Chem.* 103, 1049-1053.
[44] Sohn, J.H., Taki, Y., Ushio, H. & Ohshima, T. (2005). *Lipids* 40, 203-209.
[45] Hollingworth, T.A., Hungerford, J.M., Barnett, J.D. & Wekell, M.M. (1994). *J. Food Protec.* 57, 505-508.
[46] Schwedt, G., Shi, R. & Stein, K. (1994). *Deutsche Lebensmittel-Rundschau* 90, 178-181.
[47] Song, Z. & Hou, S. (2003). *Anal.Chim. Acta* 488, 71-79.
[48] Stefanka, Z., Abranko, L., Dernovics, M. & Fodor, P. (2004). *Talanta* 63, 705-712.
[49] Arruda M.A., Gallego, M. & Valcarcel, M. (1995). *J. Anal. At. Spectrom.* 10, 501-504.
[50] Harrington, C.F., Elahi, S., Ponnampalavanar, P. & D'Silva, T.M. (1999). *At. Spectrosc.* 20, 174-179.
[51] Wei, L., Gupta, P., Hernandez, R. & Farhat, F. (1999). *Microchem. J.* 62, 83-98.
[52] Damkroger, G., Grote, M. & Janssen, E. (1997). *Fresenius' J. Anal. Chem.* 357, 817-821.
[53] Oygard, J. K., Lundebye, A.K. & Julshamn, K. (1999). *J. AOAC Int.* 82, 1217-1223.
[54] Capelo, J.L., Lavilla, I. & Bendicho, C. (2001). *Anal. Chem.* 73, 3732-3736.
[55] Sakamoto, H., Susa, Y., Ishiyama, H., Tomiyasu, T. & Anazawa, K. (2001). *Anal. Sci.* 17, 1067-1071.
[56] Ringmann, S., Boch, K., Marquardt, W., Schuster, M., Schlemmer, G. & Kainrath, P. (2002). *Anal. Chim.Acta* 452, 207-215.
[57] Korenovska, M. (2006). *J. Food Nutrition Res.* 45, 84-88.
[58] Ebdon, L., Fisher, A.S. & Worsfold, P.J. (1994). *J. Anal. At.Spectrom.* 9, 611-614.
[59] Moreno-Cid, A. & Yebra, M.C. (2004). *Anal. Bioanal. Chem.* 379, 77-82.
[60] Becerra, G., Burguera, J.L. & Burguera, M. (1987). *Quim. Anal.* 6, 52-59.
[61] Saraviva Miranda, C.E., Freire dos Reis, B. & Krug, F.J. (1995). *J. Brazilian Chem. Soc.* 6, 387-392.
[62] Enriquez-Dominguez, M.F., Yebra-Biurrun, M.C.; Bermejo, M.P. (1998). *Analyst* 123, 105-108.
[63] Yebra, M.C. & Enriquez, M.F. (1998). *Analusis* 26, 261-263.
[64] Ali, A., Yin, X., Shen, H., Ye, Y. & Gu, X. (1999). *Anal. Chim. Acta* 392, 283-289.
[65] Liu, J., Chen, H., Mao, X. & Jin, X. (2000). *Int. J. Environ. Anal. Chem.* 76, 267-282.
[66] Wang, A., Zhang, H. & Liu, L. (2001). *Fenxi Huaxue* 29, 1284-1287.
[67] Muñoz, J., Baena, J.R., Gallego, M. & Valcarcel, M. (2002). *J. Anal. Atomic Spectrom.* 17, 716-720.
[68] Bruhn, C.G., Campos, V.H., Diaz, V.P., Cid, H.J. & Nobrega, J.A. (2002). *Bol. Soc. Chilena de Quimica* 47, 123-135.
[69] Martins, A.O., da Silva, E. L., Carasek, E., Goncalves, N.S., Laranjeira, M.C.M. & de Favere, V.T. (2004). *Anal. Chim. Acta* 521, 157-162.
[70] Fang, G.Z., Tan, J. & Yan, X.P. (2005). *Anal. Chem.* 77, 1734-1739.
[71] Korkmaz, D., Demir, C., Aydin, F. & Ataman, O.Y. (2005). *J. Anal. At. Spectrom.* 20, 46-52.
[72] Borges, E.C.L., Paiva de Oliveira, A., de Moraes, M. & Gomes Neto, J.A. (2006). *At. Spectrosc.* 27, 139-145.
[73] Yebra-Biurrun, M.C., Cancela-Pérez, S. & Moreno-Cid-Barinaga, A. (2005). *Anal. Chim. Acta* 533, 51-56.

[74] Harrington, C.F., Fairman, B.E. & Catterick, T. (1997). *Spectroscopy Europe* 9, 10-16.
[75] Valles Mota, J. P., Fernandez de la Campa, M.R., Garcia Alonso, J.I. & Sanz-Medel, A. (1999). *J. Anal. At. Spectrom.* 14, 113-120.
[76] Willie, S.N., Lam, J.W.H., Yang, L. & Tao, G. (2001). *Anal. Chim. Acta* 447, 143-152.
[77] Song, Z., Yue, Q. & Wang, C. (2006). *Food Chem.* 94, 457-463.
[78] Yebra-Biurrun, M.C. & Cancela-Perez, S. (2007). *Anal. Sci.* 23, 993-996.
[79] Yebra, M.C., Moreno-Cid, A., Cespon, R. M. & Cancela, S.(2003). *At. Spectrosc.* 24, 31-36.
[80] Bruhn, C.G., Pino, F.E., Campos, V.H. & Nobrega, J.A. (2002). *Anal. Bioanal. Chem.* 374, 131-140.
[81] Izquierdo, A., Luque de Castro, M. D. & Valcarcel, M. (1993). *J. Autom. Chem.* 15, 121-125.
[82] Ali, A., Ye, Y., Xu, G. & Yin, X. (1999). *Fresenius' J. Anal.Chem.* 365, 642-646
[83] Moreno-Cid, A. & Yebra, M.C. (2002). *Spectrochim. Acta* 57B, 967-974.
[84]Yebra, M.C., Moreno-Cid, A. & Cancela, S. (2005). *Int. J. Environ. Anal. Chem.* 85, 315-323.
[85] Wang, Z.H., Wang, Z.P., Zhang, Z.P., Liu, L.W. & Yan, X.P. (2005). *At. Spectrosc.* 26, 34-39.
[86] Luan, C.L. & Li, Y.Q. (2003). *Fenxi Shiyanshi* 22, 74-76.
[87] Pereira Fo, E.R., Rohwedder, J.J.R. & Arruda, M.A.Z. (1998). *Analyst* 123, 1023-1028.
[88] Yebra, M.C. & Moreno-Cid, A. (2002). *J. Anal. At. Spectrom.* 17, 1425-1428.
[89] Edwards, S.C., Macleod, C.L., Corns, W.T., Williams, T.P. & Lester, J.N. (1996). *Int. J. Environ. Anal. Chem.* 63, 187-193.
[90] Cheng, X., Qin, X., Chen, D. & Xu, R. (2003). *Fenxi Ceshi Xuebao* 22, 66-68.
[91] Cava-Montesinos, P., Dominguez-Vidal, A., Cervera, M.L., Pastor, A. & de la Guardia, M. (2004). *J. Anal. At.Spectrom.* 19, 1386-1390.
[92] Tseng, C.M., Hammerschmidt, C.R. & Fitzgerald, W.F. (2004). *Anal. Chem.* 76, 7131-7136.
[93] Liang, L., Hu, J., Jiang, G. & Shi, J. (2005). *Fenxi Huaxue* 33, 229-232.
[94] Wu, H., Jin, Y., Han, W., Miao, Q. & Bi, S. (2006). *Spectrochim. Acta* 61B, 831-840.
[95] Carvalho, C.M. L., Matos, A.I.N.M., Mateus, M.L., Santos, A.P.M. & Batoreu, M.C.C. (2008). *J. Toxicol. Environ. Health* 71A, 1279-1288.
[96] Luo, X., Xiong, W. & Li, J. (1994). *Guangpu Shiyanshi* 11, 19-22.
[97] Bauza de Mirabo, F.M., Thomas, A.C., Rubi, E., Forteza, R. & Cerda, V. (1997). *Anal. Chim. Acta* 355, 203-210.
[98] Aduna de Paz, L., Alegria, A., Barbera, R., Farre, R. & Lagarda, M.J. (1996). *Food Chem.* 58, 169-172.
[99] Tao, G. & Willie, S.N. (1998). *Analyst* 123, 1215-1218.
[100] Chessa, G., Calaresu, G., Orru, A., Ledda, G., Lisca, S., Piras, P. & Sechi, F. (1998). *Bollettino dei Chimici Igienisti, Parte Scientifica* 49, 295-301.
[101] Yin, X.F., Frech, W., Hoffmann, E., Luedke, C. & Skole, J. (1998). *Fresenius' J. Anal.Chem.* 361, 761-766.
[102] Rio-Segade, S. & Bendicho, C. (1999). *J. Anal. At.Spectrom.* 14, 263-268.
[103] Burguera, J.L., Burguera, M., Rondon, C., Carrero, P., Petit de Peña, Y., Quintana, I. A., Salager, J.L. & Anton de Salager, R. (1999). *Analyst* 124, 593-599.
[104] Rio-Segade, S. & Bendicho, C. (1999). *J. Anal. At.Spectrom.* 14, 1907-1912.

[105] Tao, G.,Willie, S.N. & Sturgeon, R.E. (1999). *J. Anal. At.Spectrom.* 14, 1929-1931.
[106] Rio-Segade, S. & Bendicho, C. (1999). *Ecotoxicol. Environ. Safety* 42, 245-252.
[107] Capelo, J.L., Lavilla, I. & Bendicho, C. (2000). *At. Spectrosc.* 21, 229-233.
[108] Qvarnstrom, J., Tu, Q., Frech, W. & Ludke, C. (2000). *Analyst* 125, 1193-1197.
[109] Doering, W.E., James, R.R. & Echols, R.T. (2000). *Fresenius' J. Anal. Chem.* 368, 475-479.
[110] Julshamn, K., Brenna, J., Asman, B., Ekman, A., El-Ghauoui, A., Gjerstad, K.O., Gronningen, D., Hammer, S.H., Jensen, A.M., Liukkonen-Lilja, H., Mikkola, J., Ruth, T., Sedal, L.O., Solli, B., Venalainen, E.R. & Wollertsen, N. (2002). *J. AOAC Int.* 85, 626-631.
[111] Zenebon, O., Sakuma, A.M., Dovidauskas, S., Okada, I.A., De Maio, F.D. & Lichtig, J. (2002). *J. AOAC Int.* 85, 149-152.
[112] Li, Y., Jiang, Y., Yan, X.P. & Ni, Z.M. (2002). *Environ. Sci. Technol.* 36, 4886-4891.
[113] Yan, X.P., Li, Y. & Jiang, Y. (2003). *Anal. Chem.* 75, 2251-2255.
[114] Rio Segade, S. & Tyson, J.F. (2003). *J. Anal. At.Spectrom.* 18, 268-273.
[115] Kan, M., Willie, S.N., Scriver, C. & Sturgeon, R.E. (2006). *Talanta* 68, 1259-1263.
[116] Leal, L.O., Elsholz, O., Forteza, R. & Cerda, V. (2006). *Anal. Chim.Acta* 573+574, 399-405.
[117] Silva, M.F., Toth, I.V. & Rangel, A.O.S.S. (2006). *Anal. Sci.* 22, 861-864.
[118] Vereda Alonso, E., Siles Cordero, M.T., Garcia de Torres, A., Canada Rudner, P. & Cano Pavon, J.M. (2008). *Talanta* 77, 53-59.
[119] Beauchemin, D., Siu, K.W.M. & Berman, S.S. (1988). *Anal. Chem.* 60, 2587-2590.
[120] Harrington, C.F., Merson, S.A. & D' Silva, T.M. (2004). *Anal. Chim. Acta* 505, 247-254.
[121] Dong, L.M., Yan, X.P., Li, Y., Jiang, Y., Wang, S.W. & Jiang, D.Q. (2004). *J. Chromatogr.* 1036A, 119-125.
[122] Yebra, M.C. & Moreno-Cid, A. (2003). *Anal. Chim. Acta* 477, 149-155.
[123] Yebra, M.C., Cancela, S. & Cespón, R.M. (2008). *Food Chem.* 108, 774-778.
[124] Yan X.P. & Adams F. (1997). *J. Anal. At. Spectrom.* 12, 459-464.
[125] Yan, X.P., Sperling, M. & Welz, B. (1999). *Anal. Chem.* 71, 4216-4222.
[126] Yebra, M.C., Enriquez, M.F., Garcia, A. & Moreno-Cid, A. (2001). *Fresenius' J. Anal. Chem.* 370, 64-68.
[127] Ferreira, S.L.C., Lemos, V.A., Santelli, R.E., Ganzarolli, E. & Curtius, A.J. (2001). *Microchem. J.* 68, 41-46.
[128] Cheng, X.S., Qin, X.G. & Xu, R. (2004). *Guangpu Shiyanshi* 21, 955-958.
[129] Krachler, M., Burow, M. & Emaons, H. (1999). *J. Environ. Monitor.* 1, 477-481.
[130] Guo, T., Erler, W., Schulze, H. & McIntosh, S. (1990). *At. Spectrosc.* 11, 24-28.
[131] Chan C.C.Y. & Sadana R.S. (1992). *Anal. Chim.Acta* 270, 231-238.
[132] Machado, L.F.R., Jacintho, A. O. & Gine, M.F. (2000). *Quimica Nova* 23, 30-33.
[133] Lavilla, I., Gonzalez-Costas, J.M. & Bendicho, C. (2007). *Anal. Chim.Acta* 591, 225-230.
[134] Fang, Z., Sun, L., Hansen, E.H., Olesen, J.E. & Henriksen, L.M. (1992). *Talanta* 39, 383-390.
[135] Ayora-Cañada, M.J., Molina-Diaz, A. & Pascual-Reguera, M.I. (2000). *Int. J. Environ. Anal. Chem.* 76, 319-330.

[136] Ruedas Rama, M.J., Ruiz Medina, A. & Molina Diaz, A. (2005). *Talanta* 66, 1333-1339.
[137] Gañan Gutierrez, N., Sanchez Rojas, F. & Cano Pavon, J.M. (1996). *Fresenius' J. Anal. Chem.* 355, 88-91.
[138] Wada, M., Kurose, M., Nakamura, A., Kuroda, N., Tanigawa, M. & Nakashima, K. (2000). *Nippon Kasei Gakkaishi* 51, 115-120.
[139] Kobayashi, H. & Hikuma, M. (2000). *Anal. Lett.* 33, 1013-1023.
[140] Monser, L., Sadok, S., Greenway, G.M., Shah, I. & Uglow, R.F. (2002). *Talanta* 57, 511-518.
[141] Munaf, E., Hu, W. & Haraguchi, H. (1993). *Bull. Chem. Soc. Japan* 66, 966-968.
[142] Narusawa, Y. & Nawa, Y. (1991). *J. Flow Injection Anal.* 8, 21-31.
[143] Li, Y.S. (1997). *Lab. Rob. Autom.* 9, 55-67.
[144] Li, Y.S. & Naruzawa, Y. (2001). *Fenxi Shiyanshi* 20, 53-57.

INDEX

A

D

E

F

G

H

I

J

K

L

M

N

O

P

S

T